BORDERS OF BIODIVERSITY

FLOWS, MIGRATIONS, AND EXCHANGES

MART STEWART | HARRIET RITVO | editors

The Flows, Migrations, and Exchanges series publishes new works of environmental history that explore the cross-border movements of organisms and materials that have shaped the modern world, as well as the varied human attempts to understand, regulate, and manage these movements.

A complete list of books published in Flows, Migrations, and Exchanges is available at https://uncpress.org/series/flows-migrations-exchanges.

BORDERS OF BIODIVERSITY

HOW GRAY WOLVES, MONARCH BUTTERFLIES, AND GIANT SEQUOIAS TRANSFORMED LARGE LANDSCAPE CONSERVATION

Will Wright

THE UNIVERSITY OF
NORTH CAROLINA PRESS
CHAPEL HILL

© 2026 Will Wright
All rights reserved

Designed by April Leidig
Set in Garamond, Muller, and The Sans
by Copperline Book Services

Manufactured in the United States of America

Material in chapters 3 and 4 was previously published in Will Wright, "Monarch Butterfly Conservation (Mexico)," in *Oxford Research Encyclopedia of Latin American History*, edited by Stephen Webre (Oxford University Press, 2022), https://doi.org/10.1093/acrefore/9780199366439.013.1084.

Cover art: *Migrations in Motion* by Dan Majka,
The Nature Conservancy. Used by permission.

Library of Congress Cataloging-in-Publication Data
Names: Wright, Will (Environmental historian) author
Title: Borders of biodiversity : how gray wolves, monarch butterflies, and giant sequoias transformed large landscape conservation / Will Wright.
Other titles: Flows, migrations, and exchanges
Description: Chapel Hill : The University of North Carolina Press, [2026] | Series: Flows, migrations, and exchanges | Includes bibliographical references and index.
Identifiers: LCCN 2025047298 | ISBN 9781469694061 cloth alk. paper | ISBN 9781469694078 paper alk. paper | ISBN 9781469694085 epub | ISBN 9781469694092 pdf
Subjects: LCSH: Biodiversity conservation—North America—International cooperation | Corridors (Ecology)—North America—History | Transfrontier conservation areas—North America | Biodiversity—Environmental aspects | Ecosystem management—International cooperation | Gray wolf—Conservation—Case studies | Monarch butterfly—Conservation—Case studies | Giant sequoia—Conservation—Case studies | BISAC: SCIENCE / Environmental Science (see also Chemistry / Environmental) | SCIENCE / Life Sciences / Biological Diversity
Classification: LCC QH77.N56 W75 2026
LC record available at https://lccn.loc.gov/2025047298

For safety concerns under the European Union's General Product Safety Regulation (EU GPSR), please contact gpsr@mare-nostrum.co.uk or write to the University of North Carolina Press and Mare Nostrum Group B.V., Doelen 72, 4831 GR Breda, The Netherlands.

For Carly

CONTENTS

ix   List of Illustrations

1   **INTRODUCTION**
    Birds Without Borders

Part I | Gray Wolves
(Canis lupus)

21   **CHAPTER ONE**
    Uneven Eradication

55   **CHAPTER TWO**
    (De)stabilizing Technologies

Part II | Monarch Butterflies
(Danaus plexippus)

95   **CHAPTER THREE**
    Tracking Flight

131   **CHAPTER FOUR**
    Stepping Stones

Part III | Giant Sequoias
(Sequoiadendron giganteum)

171   **CHAPTER FIVE**
    Creative Destruction

207   **CHAPTER SIX**
    Baptized by Fire

243   **CONCLUSION**
    Barriers of Biodiversity

251   Acknowledgments
255   Notes
293   Bibliography
325   Index

ILLUSTRATIONS

- 18 Gray wolves and the Yellowstone to Yukon bioregion
- 22 Al Close and dead wolf hanging off his horse
- 26 Wolves eating a poisoned bison carcass
- 28 Portrait of John George Brown
- 39 Alberta cowhands with dead wolf
- 45 Stoney Nakoda wolfer David Bearspaw and family
- 56 Pluie the Wolf's journey
- 69 Biologist Diane Boyd and a collared wolf
- 70 Map of the Northern Rockies Wolf Recovery Plan
- 80 Early map of the Y2Y landscape
- 82 Red Earth Wildlife Overpass
- 83 Trail camera images of wildlife using overpasses
- 92 Monarch butterfly migration and overwintering sites
- 109 Norah and Fred Urquhart
- 111 Map of tagged monarch butterfly flights
- 119 *Hacendado* descendant Jesús Ávila
- 119 CEPANAF forest rangers
- 126 Lincoln Brower and other conservationists
- 137 Pro-Monarca president Carlos Gottfried
- 137 Mexican women and reforestation of oyamel trees
- 140 CEPANAF ranger Genaro Reyna
- 150 Ecologist Karen Oberhauser and milkweed monitoring team
- 152 Biologist Chip Taylor and elementary students
- 162 Map of Interstate 35, or "Monarch Highway"

168	Giant sequoia groves and Indigenous treaty territories
175	Map of Indigenous lands according to 1851–52 treaties
177	Kaweah colonists and Karl Marx Tree
182	Sequoia stump and loggers with crosscut saw
191	1893 map of Sequoia National Park
193	Park Superintendent Walter Fry
202	A. E. Douglass on V-cut stump
213	Margaret Baty (Mono) with acorn harvest
219	Bulldozers in Sequoia National Park
225	NPS fire crew member Ed Nelson with drip torch
234	Researcher Chris Baisan with a chainsaw
234	Sequoia slice with dated fire scars

BORDERS OF
BIODIVERSITY

INTRODUCTION

Birds Without Borders

I heard the whistled notes of a songbird, but a thicket of willows obstructed my view of flying feathers. I was supposed to be inspecting the woody shrub in front of me as Ben Bobowski held back layers of verdant growth to reveal older stems that elk had chewed almost to the ground. We stood on the banks of the Big Thompson River, part of a postglacial meadow called Moraine Park, with stands of aspen and willow fenced off to prevent this type of overbrowsing. Bobowski, a conservation biologist who served as the chief of resource stewardship for Rocky Mountain National Park from 2007 to 2015, explained to me that park managers had erected this ungulate barrier to meet the goals of their 2007 Elk and Vegetation Management Plan. In park parlance, these so-called exclosures allowed riparian flora to regenerate so that beavers would reclaim the place and dam the river. Spillovers not only created wetlands that operate as a natural filtration device, improving water quality for Colorado's Front Range communities, but they also functioned as breeding habitat for many kinds of birds.[1]

"There's the yellow warbler we were listening to," Bobowski remarked, pointing behind us at a research team who had caught a bird in avian netting. We dashed clumsily across the waterlogged terrain to observe what was going on. A scientist with the Rocky Mountain Bird Conservancy held the bright, egg yolk–colored creature while looping a geolocator, a tracking device the size of a pen cap, around its wings. This research tool was employed to decipher the yellow warbler's exact flight route southward. Bobowski shared that of the 282 bird species identified at Rocky, bird banding studies have shown that 150 of them migrate annually from the US-based park to Agua y Paz Biosphere Reserve in Costa Rica. Their flyway demonstrated intimate biological connections between nations.[2]

A "sister park" relationship was born at the nexus of science and society. When the first band was recovered in 2004, community members in Estes Park, Colorado, the gateway town at Rocky's eastern entrance, partnered with Monteverde, Costa Rica, through the Sister Cities International network. Citizen diplomacy made sense given the two towns' similarities: Both had population sizes of about six thousand, both were located on the Continental Divide, and both relied on tourism near protected areas. As a park liaison to the Estes Park Sister Cities chapter, Bobowski started traveling with the civic group to Costa Rica in 2010 to bird-watch and "build out our relationship with shared species."[3]

He learned that biology united what nationality separated. Johnny Pérez Artavia, natural history coordinator at the Santa Elena Cloud Forest Reserve, spoke with Bobowski about their mutual responsibilities to migratory birds: "You don't protect the habitat there, we don't protect the forest here, then we both are in trouble."[4] Bobowski and Santa Elena director Yaxine María Arias Núñez developed a series of personnel exchanges over the next few years, which they called the Naturalmente Juntos–Naturally Together project. In 2015, Rocky Mountain National Park and the conservation areas making up Agua y Paz Biosphere Reserve entered a formal sistering arrangement.[5]

Its meaning for transnational conservation was powerful. Yellow warblers and other neotropical bird species spend eight months in Central America, then migrate northward every May across North America for breeding and return every August for overwintering. Some of them follow the Continental Divide from Monteverde to Estes Park, crossing at least five international borders in the process. What happens when climate disruption causes suitable habitat within one country to move or vanish? Scientists have documented how more frequent El Niño cycles in Costa Rica and intensifying drought periods in the southwestern United States impact the timing and flight patterns of neotropical migrants. While individual bird species adapted to a warming world by moving, there was an overall decline in avian richness for places where hydration, food, and cover became scarcer along migratory routes.[6] "The success of this project, we can't do [it] at Rocky alone," Bobowski commented. "We must work across borders, across bureaucracies."[7] This couldn't be a unique experience, I remembered thinking to myself. How and why did people of different nationalities and their respective governments cooperate, or not, when conserving transboundary ecologies?

My goal became to recover a usable past with three focal species on the North American continent. Gray wolves, monarch butterflies, and giant sequoias were emblematic of the ways in which living things transcended nation-states and disrupted their borders. These border-crossing species came to stand for ecologies that connected human communities, even when international lines administered by Canada, Mexico, and the United States tried to separate them. In a world obsessed with debating borders and the issues they represent—trade, immigration, sovereignty—a wider discussion about conserving transborder species during an era of climate dislocation and mass extinction has been absent. Tracing the origins of large landscape conservation in the past is significant for identifying alternative pathways to a more just, livable future.

The overall arc of this history is that state-centered conservation followed the logic of nation-building around the turn of the twentieth century, creating borders and enclosing space to protect unique lands. Like bird banding, new scientific practices rendered visible geographies for animals, insects, and plants that did not fit the size or shape of those parks and protected areas. Rocky was merely a small piece of habitat embedded in a larger puzzle of avian flight. Civil society, such as Sister Cities groups, rallied behind alternative ways of organizing space by building relationships outside of their home countries for biological well-being. I call this political outlook in the making "ecological internationalism."[8] Once recognized, its strategy becomes obvious: Forge solidarity across borders or face extinction of species.

"Finding commonality was suddenly the name of the game," Bobowski stated, when I asked him to reflect on bridging national divides for bird conservation. Despite differences in culture, language, and history, US Americans and Costa Ricans bonded over shared wildlife. Bobowski and I left the riverbank after the yellow warbler was released and flew away.

From Island Biogeography to Half Earth

I was taking part in a field workshop called Parks as Portals to Learning in August 2015 when Bobowski hosted me and other college participants. We were tasked with telling a history that explained to visitors why park managers had exclosures in Moraine Park, a federally designated wilderness area where, according to the 1964 Wilderness Act, "the earth and its community of life are untrammeled by man."[9] For tourists who enjoyed watching elk herds congregate there every fall, the management decision to keep elk away with fences seemed like a conflicting objective. Rehabilitating the park's aspen-willow-beaver complex—a partnership on which migratory birds relied for habitat—was not acceptable to visitors when it came at the expense of deterring the piercing bugles of bull elk. We looked to the past to clarify the present, assessing the dynamic relationship between species and space.

The best available evidence suggests that Moraine Park had been a wetland for thousands of years, basically since the end of the last Ice Age when beavers dammed the Rocky Mountain headwaters. Beaver populations, estimated at sixty to one hundred million across North America before colonization, were relatively stable until the creation of a European market for peltry during the eighteenth century triggered a trophic cascade. British, French, Spanish, and US trappers, as well as Indigenous allies, harvested fur-bearing animals on colonial peripheries—from beavers and bears to minks and martens—delivering thousands of pelts every year to trading centers in London, Montreal, Santa Fe, and St. Louis. Animal skins were made into hats, robes, and other apparel worn by bourgeois European men and women. In the late 1700s, Nuevomexicano merchants bartered with Nuche (Ute) peoples who lived in the high Rockies of what is today Colorado, exchanging goods like knives for beaver or deer hides. Commercial integration almost eliminated beaver until a change in fashion halted the fur trade—silk hats became all the rage by the early nineteenth century. Similar stories, except of total collapse instead of near extinction, could be told about the passenger pigeon, Stellar's sea cow, or Atlantic gray whale.[10]

The precipitous decline of beaver was reversed momentarily when parks and protected areas became an antidote to extinction. Under the world system of imperialism, conservation of internal biophysical resources—plants and animals, water and soils, timber and minerals—was paramount to an external geopolitical struggle. The first national parks arrived at the end of the late nineteenth century when settler frontiers had very nearly decimated wild lands: not only the example of Yellowstone (1872) in the United States but also Royal (1879) in Australia, Banff (1885) in Canada, and Tongariro (1887) in New Zealand. For the final case, Māori chief Te Heuheu Tunkino IV gifted three volcanic peaks to the British government for a national park so that Pākehā (white New Zealanders) would not settle on the sacred mountains. A comparable concern for protecting the high country around Longs Peak won the day when on September 4, 1915, Rocky Mountain National Park was created in Colorado. Despite tension in the founding legislation between public recreation and nature preservation, beaver colonies rebounded during the following decades due to a hunting ban inside the federal reserve.[11]

The guiding assumption for settler societies was that drawing a "fence of laws" around a place would preserve all biophysical resources therein. Because nation-building and park-making developed together in the Americas, Asia, and Africa, protected areas denoted a type of territorial bounding to keep wildlife *in* and kick people *out*. Anthropologist Dan Brockington coined the term "fortress conservation" to describe how Western wilderness ideas operated as colonial mechanisms, removing Indigenous nations from ancestral homelands. Seeing parks as biological fortresses with segregated borders has been a longstanding tradition. In South Africa, the white Afrikaner government expelled

Black Africans to create the Sabi Sands and Singwitsi Game Reserves, which in 1926 became part of Kruger National Park. In effect, fortress conservation hardened park borders.[12]

Just as imperial expansion spurred the creation of national parks across the colonized world, the related overseas trade and travel brought scientists into contact with oceanic islands that in due time altered the perception of protected areas. Insular environments represented these self-contained units to study living things, most famously with Charles Darwin on the Galápagos in 1835 and Alfred Russel Wallace on the Malay Archipelago in 1854, both of whom separately developed a natural selection theory to explain the evolution of all life. The random processes of speciation and extinction stood out starkly as the British naturalists went from island to island, collecting many specimens of flora and fauna, like giant tortoises or birds of paradise, and observing slight variations among them. Understanding the evolutionary connections between species and space continued into the twentieth century with US entomologist Edward O. Wilson, who earned a PhD at Harvard and conducted field expeditions on New Guinea, Fiji, and Australia to survey ant populations. During a 1961 sabbatical, Wilson visited Trinidad and Tobago, a pair of islands just off the Venezuelan coastline, and noticed that insects were far more abundant on the larger landmass of the two.[13]

E. O. Wilson then teamed up with Princeton ecologist Robert MacArthur to write an influential 1967 book, *The Theory of Island Biogeography*. MacArthur and Wilson proposed a general rule: The smaller the island, the fewer species it contained. Building on the foundational observations of Darwin and Wallace, they developed a "species-area relationship" as a conceptual tool in which biological diversity, or the richness and variety of organisms, was stable over time when migration rates to an island matched extinction rates on that same island. The two scientists laid out its predictive power with Caribbean examples where bigger islands of the Greater Antilles, like Cuba, held more species of amphibians and reptiles than tinier islands of the Lesser Antilles. "We have provided some theoretical arguments, together with some documentation," they wrote, "which indicate that on very small islands the process of natural extinction is accelerated."[14] This new scientific framework, they suggested, could be applied from oceanic islands to other types of insularized environments: lakes, gallery forests, tide pools, and so on. MacArthur and Wilson even hinted that "island biogeography theory" might be relevant to parks and protected areas.

Between Wallace and Wilson, site-specific protections dictated US conservation policy from 1864 to 1964, from the Yosemite tract to the Wilderness Act.

At Rocky, park officials dealt with the multispecies saga of managing elk. Like beaver, elk populations had plummeted during the nineteenth century when market hunters scoured Colorado's Front Range for meat and hides to provision mining camps. In 1913, the first two shipments of wapiti arrived from Yellowstone National Park to restock the Estes Park region. Elk numbers grew under park oversight, guided more by tourist desires and less by ecological principles, but there were signs of trouble by the 1930s. In the absence of predation (wolves and bears had been eradicated locally) or human hunting, elk herds at Rocky increased exponentially, which resulted in their overbrowsing of aspen and willow. Without these vegetative building materials, beavers could not perform their ecosystem roles as hydrological engineers. The once-boggy Moraine Park converted into a drier grassland. During the 1940s and 1950s, park managers responded by culling elk to reduce herd sizes within the protected area and by purchasing nearby private grazing lands to redraw park boundaries along more biological lines.[15]

But, as MacArthur and Wilson might have predicted, park actions were limited by island effects. In the 1960s, Rocky adopted a new management philosophy called "natural regulation," a hands-off approach claiming that culling was unnecessary since limited forage, not predation, would eventually curb elk populations. Except that scenario never happened. Lacking top-down pressures on the food web, elk numbers exploded within the insular refuge, and their herbivory decimated riparian flora. To help beavers out, park managers built temporary fences around aspen and willow stands to shield them from overabundant elk. The policy became so untenable, with beaver populations declining by the mid-1990s, that Rocky decided to shift course. The result was the still-unfolding Elk and Vegetation Management Plan. Public stakeholders debated three possible strategies for controlling wapiti: human hunting, wildlife contraception, or wolf reintroduction. In 2007, park officials decided on the first option, partly because the protected area was deemed too small in size and too close to human communities for the restoration of wolves.[16]

When Rocky reinstated elk culling the next year, wolf tracks were identified at Moraine Park. It had been more than a half century since the apex predator last set foot in the national park. The wild canid was never spotted afterward, but it was easy to speculate where the animal came from. Using GPS technologies, scientists followed a lone wolf that had trekked roughly three thousand miles (4,825 kilometers) on a circuitous path from the Canadian line to Colorado. This event was one year into Ben Bobowski's tenure at Rocky. As a scientist who was trained to value empirical knowledge over personal feelings, Bobowski realized his tough guy image of reflector sunglasses and arrowhead logo uniform couldn't hide his pleasure knowing that if two wolves of the opposite sex moved in, they could reestablish a predator population at the park on their own. Despite Rocky's location in the center of the continental United States, ostensibly far away from its national borders, Bobowski couldn't avoid the conservation consequences of wolves moving in from the north or neotropical birds from the south.[17]

Protecting "mobile nature" like bird migrations has typically been addressed by state-to-state actions through international conservation politics. For example, Canada and the United States authorized the Migratory Bird Treaty of 1916 to provide legal safeguards for shared waterfowl species, which eventually was expanded to include Mexico in 1936, and to cover nongame birds like the yellow warbler. Additionally, global environmental issues have been tackled by international governing bodies like the United Nations. In 1976, Rocky joined the UNESCO's Man and the Biosphere program to recognize the park as a reservoir of biological diversity for its biome types (temperate forest and alpine tundra) and to coordinate long-term scientific monitoring among a worldwide network of protected sites. Biosphere reserves, encompassing wild lands and human communities in a single ecosystem, tried to reconcile biodiversity conservation with sustainable development. Scientists of different nationalities might collaborate on projects, but accounts of these top-down measures emphasize the work of government officials who usually approached the negotiating table to defend their own self-interests.[18]

Bobowski eventually came to work on breaching national borders for conservation, but years before he was dedicated to enforcing them for capitalism. Before joining the National Park Service's ranks in 1998, he served as a port security officer for the US Coast Guard Reserve. His duties, apart from performing search-and-rescue operations, involved inspecting the vessels of commercial ships for drugs and other illicit cargo on the Atlantic and Pacific Seaboards. By policing what goods moved in and out of the country, Bobowski personified how borderlands transform into borders. *Borderlands* are meeting places where no single polity reigns supreme; *borders* are demarcations where powerful nation-states attempt to define political membership according to their economic advantage. But ocean-going contraband, Bobowski admitted, still slipped by his patrols. He realized that, like the flow of goods, the movement of biota constantly undermined borders, rendering these international lines unstable. Hardening and softening national borders are central features of modern statecraft.[19]

When Bobowski entered graduate school in the mid-1990s, intellectual giants in ecology were debating the appropriate size of protected areas to safeguard against extinction. Prominent scientists like E. O. Wilson and Jared Diamond had proposed that if parks were terrestrial islands, then larger reserves functioned better at maintaining species' richness than smaller ones. To test "island biogeography theory" on charismatic mammals, biologist William Newark surveyed twenty-four western national parks in Canada and the United States, including Rocky, for their historical data on animal sightings—from bears and badgers to moose and mule deer. The theory held that local extinctions ("extirpations" in technical jargon) happened constantly, but these isolated losses could be offset to equilibrium by speciation, when new biota evolved to fill the void, or by immigration, when the same species from an adjacent area came to occupy the vacated habitat. However, as natural landscapes were shredded into tinier habitat patches by roads, fences, and other human-constructed barriers, wildlife could not enter to replenish the depleted population. In 1987, Newark confirmed in the high-profile journal *Nature* that smaller or isolated sites feature higher rates of extinction than bigger or well-connected ones. Bobowski learned the academic lesson of forging connections outside a park's administrative lines.[20]

"Protected areas are islands," Bobowski conceded, recognizing that most endangered species today face threats on surrounding nonfederal lands. Habitat fragmentation is one of the primary factors for why we have arrived at the sixth mass extinction, with species disappearing at a pace at least a hundred times greater than the rate shown in the fossil record. Our geological calendar has been punctuated by five major extinctions, such as when a giant asteroid collision wiped out the dinosaurs at the end of the Cretaceous period. The sixth major extinction event, though, will not be defined by a crashing object from space.[21]

By enacting policies that divide up territories, transforming once-intact large landscapes into ever-smaller tracts of land, modern settler states erode the biological base on which human flourishing rests. Over three hundred known species of terrestrial vertebrates have been annihilated since 1500, a biocidal turning point linked with the beginning of European imperial expansion. Because science has defined the extinction crisis, scientists have gained the authority to offer a solution: Build an integrated, transborder system of protected areas through large landscape conservation.[22]

Most prominently, in 2016, Wilson proposed a "Half-Earth" goal in which humanity works to conserve at least 50 percent of the world's lands and seas for nonhuman organisms. "I am convinced that only by setting aside half of the planet in reserve, or more, can we save the living part of the environment and achieve the stabilization required for our own survival," the scientist remarked. "Why one half? ... Because large plots, whether they already stand or can be created from corridors connecting smaller plots, harbor many more ecosystems and the species composing them at a sustainable level."[23]

Beyond Wilson's call, conserving half the Earth for other life-forms has developed from an idealistic proposal into a serious target. The World Database on Protected Areas, created by the United Nations in 1959, monitors the spatial coverage of parks and preserves across the globe to serve as a metric for evaluating conservation progress. An international push to enlarge protected areas hastened following the 1992 Rio Earth Summit, when 1,700 researchers issued the "World Scientists' Warning to Humanity" about species losses, and UN members signed the Convention on Biological Diversity (CBD). As a part of the agreement, biologists adopted the goal of tripling the extent of parks and preserves, which at the time was only 4 percent of the Earth's surface. In 2010, signatory countries gathered in Japan and agreed to Aichi Target 11, revising the CBD benchmark to protecting 17 percent of their lands and 10 percent of their waters by 2020. Realizing that geopolitical standards did not match the severity of extinctions, scientists began to question how much space was needed precisely for flora and fauna to persist in the future. At the 2022 United Nations Biodiversity Conference in Montreal, CBD signees agreed to place 30 percent of their lands and oceans under protection by 2030 (or "30 × 30" as a shorthand phrase) and reaching the 50 percent mark by 2050.[24]

Setting aside half for nature can be done, Bobowski told me, by observing Costa Rica's historic trajectory to achieve the milestone. Before 1970, the nation-state featured no single protected area and faced intense deforestation from land clearings for coffee, banana, and oil palm plantations. By 2010, the year of Bobowski's first visit, Costa Rica had designated almost 30 percent of its landmass as parks and preserves (twice as much as in the United States). This shift is illustrated by the Santa Elena Cloud Forest Reserve, which was intended originally to be subdivided as farms. During the late 1980s, university students at the Agricultural College of Santa Elena noticed that wildlife at a private forest reserve near Monteverde was attracting a steady flow of scientists and tourists to the region. Realizing the opportunities for community development, they asked the Ministry of Agriculture and Livestock in 1992 to incorporate the Santa Elena Cloud Forest into the Arenal Tempisque Conservation Area. Costa Rica has earned a reputation internationally as a "biodiversity hotspot" since the isthmus nation functions like a funnel between North America and South America, concentrating wildlife mobility to possess 5 percent of the world's known species. "They're at the forefront of conservation," Bobowski judged. "I call it Park Service 2.0 because protections necessarily included people."[25]

While expanding protected areas is promising, policy challenges remain in determining the most effective means to conserve wildlife in human-dominated landscapes. Using satellite imaging, we know that half the world's ice-free, terrestrial surfaces can now be categorized as anthropogenic biomes, or "anthromes" for short, places of intensive human use such as urban settlements or industrial agriculture. Extending the administrative lines of protected areas may not be a feasible option where the human footprint is too formidable or where protections impinge on disputed territorial claims. As an alternative, creating larger de facto parks and preserves by connecting core areas with wildlife corridors that transcend various jurisdictional boundaries may increase the likelihood of species survival. For this global change adaptation strategy to work, a research problem becomes understanding how and why past societies amended their national borders to help threatened biota.[26]

It would be impossible to focus this historical study on biological diversity, other than as a scientific or intellectual concept, so I have chosen to center the narrative on three of the most charismatic species in North America: gray wolves, monarch butterflies, and giant sequoias. Wolves, monarchs, and sequoias are "charismatic" because they became well-known crucibles of conservation. Each life-form faced periods of near collapse due to human activities—sequoias in the 1890s, wolves in the 1930s, monarchs in the 2010s—representing the bigger picture in which 34 percent of conifers, 26 percent of mammals, and 40 percent of insects are recognized today as "threatened with extinction."[27]

One side of the equation might be about eradication and elimination, but the reverse was about resistance and recovery. These iconic species persevered into the present because of the politics of ecological internationalism and the practice of large landscape conservation.

Following Nature Beyond the Nation

Drawing the administrative lines for parks, or the international lines for nation-states, has been political because it was (and still is) about who controls the land. Illustrating alternative geographies based on biology subverts the goals of governing space to satisfy the powerful. Border studies scholars Michiel Baud and Willem Van Schendel have suggested that any changes to these human-made demarcations can be understood through the "triangle of power relations" between centralized states, regional elites, and local people. For transnational conservation of the yellow warbler, centralized states were found in the governing halls of Washington and San José, regional elites were scientists like Ben Bobowski and Yaxine María Arias Núñez whose authority was derived from researching birds and speaking for biota, and local people came together through grassroots organizations such as Sister Cities. Narrating three case studies through this three-sided lens explains how and why different societies embraced biocentric borders after mapping nature beyond the nation.[28]

The story of gray wolves (*Canis lupus*) can be told on a series of escalating scales up to the binational.[29] The Montana-Alberta borderlands transformed into the US-Canada border in the context of policing trade in wolves and whiskey. Animals were historical actors on the western colonial frontier by exercising forms of territoriality on par with settler states: Wolves defended space by howling, scent marking, and fighting; cowhands defended space by cussing, building fences, and shooting firearms. During the late nineteenth century, cattle capitalists made habitat fragmentation real through a gridded landscape of barbed wire and steel traps, enforcing property lines by inflicting pain. Wolves became the targets of eradication campaigns in the western United States and Canada, but they were only extirpated south of the international line by the 1930s. The spatial difference boiled down to the role of centralized states: Washington turned wolf elimination into a federal issue by tapping the US Bureau of Biological Survey (BBS), while Ottawa was more reluctant to get involved in provincial matters. Nation-building followed different forms along the far western border.[30]

Wolves, however, kept subverting the international line with their dispersal movements as regional elites reframed their value. Biologists explored how predators made ecosystems whole just as environmentalists equated wolves with wilderness under siege. In the 1970s, University of Montana biologists started to utilize radio collars to track wolf dispersal at the same time these wild canids gained legal protections as endangered species to reclaim ancestral territories. Geolocating devices revealed that juvenile wolves moving from the Canadian to US Rockies, and vice versa, dispersed on ranges far exceeding the size of any park. Conservation activists, after learning to question the "national" in "national parks," proposed a new geography called Yellowstone to Yukon (Y2Y). This binational vision consisted of creating a series of wildlife corridors to link parks and protected areas allowing large carnivores and other mobile animals to travel in between. Large landscape conservation was the most successful where the Y2Y organization provided ranchers and other rural peoples with the resources they needed to maintain habitat connectivity.[31]

The story of monarch butterflies (*Danaus plexippus*) takes flight on another level as their migration operates on a trinational scale. Beginning in the spring, these winged wanderers spread out over two million square miles from Minnesota to Maine, Mississippi to Manitoba, as three to four generations follow blooming milkweed plants northward. In the fall, most butterflies east of the Rockies, though not all, fly southward to Mexico, overwintering at a mountainous location 0.015 percent of the area they occupied in the summer. Before 1975, this epic journey was basically a mystery. Canadians, Americans, and Mexicans saw plenty of monarchs coming and going over the twentieth century, often pondering where they originated. A pair of Canadian researchers at the University of Toronto established the Insect Migration Association to find out. The forty-year-long butterfly tagging effort was a contradiction in citizen science: It included thousands of lay participants, mostly from Canada and the United States, while it downplayed local Mexicans. The scientific discovery sparked a flurry of federal measures for overwintering monarchs, privileging outsiders over insiders, regional elites over *ejidos*, as the International Union for the Conservation of Nature (IUCN) pushed for establishing the Monarch Butterfly Biosphere Reserve.[32]

Transnational conservation then filtered through neoliberal economics in which capital moved across borders like butterflies. In 1986, the same year the biosphere reserve was created, Mexico entered the General Agreement on Tariffs and Trade. The federal government did not commit money to administering the newly protected area, relying instead on the environmental nonprofit Pro-Monarca. Founded by Mexico City professionals, Pro-Monarca obtained most of its financing though World Wildlife Fund, the IUCN's fundraising arm established by British aristocrats. Pro-Monarca tried to develop infrastructure and programming at the roosting sites for ejido communities to replace logging income with tourism revenue. During negotiations over the North American Free Trade Agreement, the Commission for Environmental Cooperation (CEC) was formed to coordinate top-down responses on any transboundary issues. That same decade, Monsanto introduced herbicide-tolerant corn and soybeans, called Roundup Ready crops, resulting in the spraying of milkweeds across the US Midwest. In response, Washington officials proposed wildlife corridors to bail out big agribusiness more than butterflies. Midwestern farmers in the United States and communal *ejidos* in Mexico felt abandoned by centralized states blaming one another for austerity or inaction. The CEC dispensed money to regional elites, but payments rarely led to economic stability for local communities. In the end, jobs and justice were more effective conservation tools than guns and guards.

The story of giant sequoias (*Sequoiadendron giganteum*) unfolded on subnational and supranational scales. In 1851, the US government signed treaties with Indigenous nations to placate violence between Natives and newcomers during the California Gold Rush. In the Sierra Nevada, for example, the Mono Nation agreed to lands mapped from the Kings River to Kaweah River, encompassing sequoia groves where they frequently burned plants to enhance cultural resources. As legal agreements between one sovereign nation and another, the treaties had to be ratified by the US Senate to become law, but they never were because of uproar from powerful California representatives. Indigenous forms of land tenure were then suppressed as the 1878 Timber and Stone Act privatized parcels of sequoia–mixed conifer forest. Timber capitalists incorporated redwood into a transpacific market; and logging operations cut down one-third of all sequoias by 1920. The cutover forest served as an inspiration both for bounding nature in national parks and for creating tree-ring science in which climatic forces transcended those very same boundaries.[33]

Under park oversight, the interplay of fire and climate drove resource managers at Sequoia–Kings Canyon to think about sequoias on a landscape scale over time. Influenced by scientific foresters and timber capitalists in national forests, the National Park Service adopted the management philosophy of total fire suppression during the early twentieth century. The so-called Indian way of forestry was snuffed out as militarized approaches to extinguish blazes took hold. Catastrophic wildfires, though, forced park staff to reevaluate their position. In the 1960s, fire ecologists at the University of California, Berkeley, rediscovered the role of the Native firestick in maintaining forest health. Prescribed fire became important as tree-ring science showed just how frequently the long-lived Big Trees burned before federal protection, especially during past warm periods. It was no coincidence, then, that the US agency began collaborating with the Mono Nation on burns during a global change–induced "hotter drought." In 2016, tribal elders reasserted the territorial claims of lost treaties when the Keepers of the Flame conducted "cultural burns" inside of and adjacent to the park's boundaries.

Taken together, living things became extensions of how scientists viewed the world and its borders, rubbing against the national desire to consolidate space.[34] Three new disciplines grew out of thinking through particular species. Corridor ecology, the study of connecting separated habitats for wildlife populations, developed by tracking wolves and other large carnivores. Chemical ecology, the study of chemically mediated interactions between organisms, derived from researching the toxic relationship of monarchs and milkweeds. And dendrochronology, or tree-ring dating, originated when analyzing long-lived sequoias to reconstruct past climates and fire records. These new fields of inquiry showed the active role of environments in shaping cognitive processes. Ecological knowledge, in turn, shaped the formation of transnational political identities.[35]

Large landscape conservation must still be approached with caution, as scientific framing on continental, hemispheric, or planetary scales was also used during the twentieth century to justify imperial and capitalist interests to expand state power. For example, US geologists saw veins of minerals that stretched beyond nation-states, claiming that these subterranean resources constituted "global commons" for private companies to exploit. US botanists set up permanent field stations across the circum-Caribbean from Cuba to Colombia to study tropical plants as a proxy for "global biodiversity." And US agronomists introduced new crop varieties to so-called third world countries, teaching input-intensive farming to address "world hunger" and act as an ideological shield against state communism.[36] Calls for embracing an "Anthropocene biosphere" during the twenty-first century, therefore, represents another dangerous trend of flattening the world's many peoples, nations, and ecologies into a single planetary force to hide the powerful, and their policies, who are responsible for the sixth mass extinction.[37]

But as the 2015 bilateral agreement between Rocky and Agua y Paz made evident, there has been an untold side to the story. The sister park relationship derived its political legitimacy from Sister Cities organizers. The case studies about wolves, monarchs, and sequoias that follow offer a bottom-up perspective of nonstate actors on the community level, making up what I define as "civil society," who united behind transboundary ecologies. The binational Yellowstone to Yukon Conservation Initiative created wildlife corridors for large carnivores and other animals; the trinational Insect Migration Association organized volunteers for tracking and protecting the annual flight of butterflies; and the multinational Keepers of the Flame brought Mono, Miwok, and Yokut Nations together to reclaim US-negotiated treaty lands in California's Sierra Nevada through cultural burning practices. By subscribing to cooperative management based on biological concerns, civil society disputed fundamental principles of state power and territorial sovereignty.[38]

The meaning of internationalism, broadly conceived as greater cooperation among nations, has never been stable. Rather, many variants of internationalism have emerged over the past two centuries. Socialist internationalism took root with the formation of the International Workingmen's Association in 1864, rallying behind the universal demand for an eight-hour workday at annual Labor Day (May Day) demonstrations. Liberal internationalism focused on maintaining the global capitalist order through networks of central banks in London and New York City, upholding the gold standard to remove monetary differences across national economies. Fascist internationalism, with the Japanese conquest of Manchuria in 1931, united Germany and Italy against the League of Nations in mediating peace in Asia. In 1977, a delegation of Indigenous peoples from Brazil, Canada, Chile, Guatemala, Mexico, Panama, and the United States arrived in Geneva to argue before the United Nations assembly that world leaders have repeatedly denied their status as nations. This display of Indigenous internationalism received support from third world nations of the Non-Aligned Movement in solidarity with parallel national liberation struggles from colonial powers. In 1980, the French Ministry of the Interior warned an "ecological international" that might spring up from antinuclear protests spanning multiple sites from energy production facilities in Germany and France to uranium mining centers in Niger and Togo. In sum, internationalism was a flexible concept over time.[39]

Reconciling state-imposed borders with biocentric ones was what I term "ecological internationalism," a political alternative to border-enforcing nationalism and border-expanding imperialism. It recognized that biota have their own forms of territoriality irrespective of nation-states that should be respected and protected. It involved creating bonds across geopolitical borders despite the differences in culture, language, and nationality. It challenged how international lines were the product of asymmetrical power relations between countries. Ecological because it was rooted in communities of living things. International because their function and fate was tied to other nation-states. If the separation of habitats led down the path to extinction, then connection established a road to recovery and repair.[40]

LIKE NEOTROPICAL BIRDS going from Colorado to Costa Rica, I follow the multispecies histories of wolves, monarchs, and sequoias across the nation-states of North America. Each chapter reflects how these threatened life-forms were pivotal to the making, unmaking, and remaking of national and administrative borders that become a layering process, a thick cartography, in written word. Each species will take us outside the international lines of modern maps, from the US-Canada border, to the US-Mexico border, to the treaty borders of Indigenous nations subsumed within the United States.

Science and technology helped people to understand these geographies of life in motion. Radio-collaring wolves revealed long-range dispersal movements when members left their packs in search of a mate, food, or territory. Tagging butterflies uncovered how those millions of monarchs, spread all over the continent for eight months out of the year, arrived at a unique overwintering destination. Tree-ring dating sequoias exposed that the mammoth plant had relied on biophysical processes in supranational climate and transboundary fire. These new scientific practices allowed us to see how nature operates beyond the nation.

By reconstructing places through the charismatic species that we share, three strands of civil society arrived at the political imaginary of ecological internationalism. They drew lines not by how nation-states operate but by how nature lives, softening national borders in the process to bring forth a biocentric vision of large landscape conservation. They paired up the need for improving habitat connectivity for nonhuman species with the obligation of repairing uneven geographies among human neighbors. Ecological internationalism offers us a version of the past and a vision for the future. As Bobowski reminded me with birds, and which applies to other biota, our choice has become solidarity or extinction.

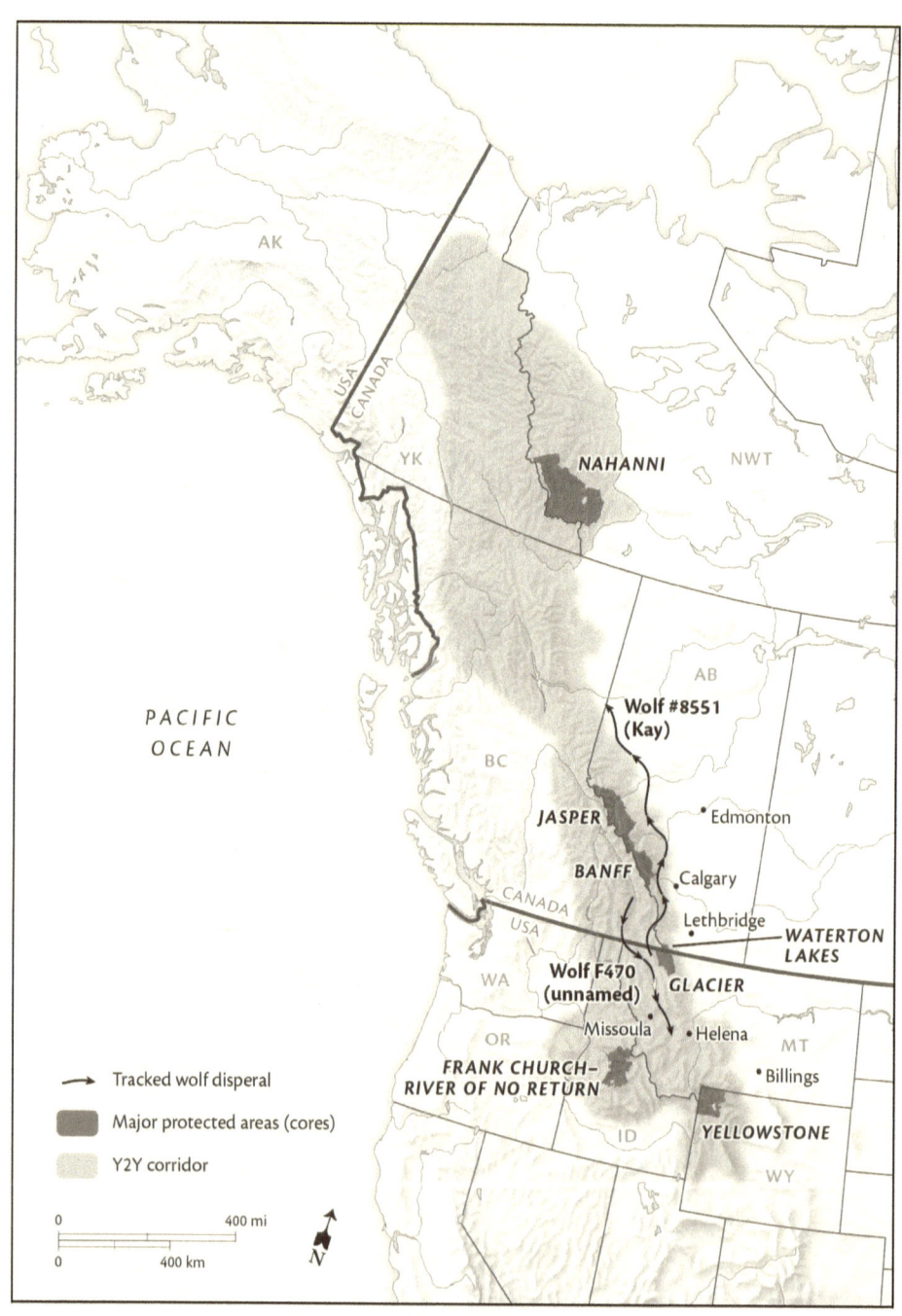

Wolves of the Yellowstone to Yukon corridor. Map by Erin Greb Cartography.

PART I
Gray Wolves
(Canis lupus)

CHAPTER ONE

Uneven Eradication

Driving down Central Avenue of Stanford, Montana, I couldn't help feeling transported back a century. Railroad tracks and grain elevators framed a single block of brick veneer buildings. Numbered streets no higher than Fifth bisected the downtown of a rural community whose census population has hovered around five hundred people since 1920. At one end of Main Street, the county museum exhibited a print collection from western artist Charles Marion "Charlie" Russell, who had worked as a cowhand on the O-H Ranch in the Judith Basin prior to a painting career. Russell illustrated frontier adventures in both Montana and Alberta, but his artwork was not Stanford's main attraction. At the opposite end of the street, I parked my vehicle facing a storefront with the block words "Basin Trading Post." Inside the building was what I really came to see: the display of a taxidermied animal named the White Wolf.

In the hallway, the creature with creamy-white fur stood in a glass case surrounded by pine boughs, deer antlers, and tree stumps. Its ears were pinned back, and its teeth were exposed as if the ferocious-looking beast was snarling at me. Pictures hung showing the wolf was taller from nose to tail than the man who had killed him; I could make out where the bullet had entered its skull near the eye socket. Newspaper clippings described how the old male renegade was first sighted in 1915 near the Little Belt Mountains, but the Ghost Wolf, as the canid was also called, often escaped notice after terrorizing Judith Basin ranchers. His notoriety grew when Stanford librarian Elva Wineman published accounts in the national press about how the predator regularly ate cows and sheep—valued at $35,000 in total—and routinely avoided every trap, poison, or bullet that hunters could muster. A plaque at its feet noted the White Wolf was a 2017 inductee in the Montana Cowboy Hall of Fame, the only animal I knew with such a distinction.

Stanford residents preserved the White Wolf as a town mascot because of its legendary, though mythologized, near and final deaths. One episode occurred in

The dead body of the White Wolf hangs from Al Close's horse in Stanford, Montana, 1930. Also known as the Ghost Wolf, the animal was one of the last wolves to roam the state of Montana before their extirpation. Courtesy of Montana Memory Project, Lewistown Public Library.

1926 when stockman Earl Neill clipped the wolf in its hind leg with a gunshot, crippling him, but the animal vanished into a snowdrift before it could be found. Another incident happened four years later when the wolf defended himself against five Russian wolfhounds while rancher Amby Cheney unsuccessfully tried to rope him with a lasso. On May 5, 1930, Neill spotted the White Wolf once more. Neill enlisted Al Close, who sent two canines to follow the carnivore's trail along skiffs of snow and stands of timber. In a patch of fir trees, the savvy dogs finally cornered the wolf. As the animals barked and clawed at one another, Close stepped out from behind a tree and fired his gun. The White Wolf collapsed. In Stanford, a crowd gathered on Main Street to see its dead body, and locals agreed to stuff the animal as a historic tribute. Gray wolves, unless mounted in a glass case, were gone from the state of Montana.[1]

Like a Russell painting, the White Wolf display obscured as much about the past as it revealed. A historical saga of biocidal campaigns for eradicating predators hid beneath a heroic struggle between male hunters and a cunning, yet doomed, wolf. Global processes tied to the twin engines of colonialism and capitalism lurked behind local people demonstrating their perseverance and prowess. Fraught history faded into frontier myth.[2]

Another story about "the last of the loners," as the famous Bureau of Biological Survey director Stanley Paul Young admiringly called these few remaining wolves, was more instructive about their disappearance from the western United States. In 1911, a mating pair established residence in Montana's Sweet Grass Hills, a group of forested buttes that rise from the plains near the Alberta border. The large carnivores took refuge among stands of timber during the day and then came down to nearby ranches during the night to be, in the words of one stockman, "handy to baby beef." With a reliable food supply of cows on the range, as well as deer and elk in the low-lying mountains, the alpha female had a litter of pups. The wolf pack's growing size alerted ranchers to their presence, and locals turned to the usual eradication tactics: poisoning, trapping, and denning, but mostly cussing. In 1926, after years of failed attempts at eliminating the creatures, stockgrowers did the only thing they could. They called for help from the federal government.[3]

The US Bureau of Biological Survey dispatched one of its professional trappers, John Oswood, to take care of the ranchers' problem. Oswood was a Norwegian immigrant who had become a "very efficient industrious" wolfer during his previous time as a cowhand. According to his supervisor, Oswood almost immediately found the wolves' den and "cleaned the country of all their young, but this last old pair were cunning, like many of our old Montana wolves."[4] To bring an end to these elusive animals, instead of lacing a cow carcass with poison, Oswood prepared different kinds of tallow lures with strychnine.

It worked, at least partially. The alpha female ingested a piece of deadly bait, but the alpha male got away, perhaps forewarned by his mate before swallowing the toxic substance. In a letter of praise to the Biological Survey, Thomas Strode of the Stirrup Ranch speculated on the creature's whereabouts: "It is uncertain whether the male wolf is a native of Canada or the U.S.," Strode remarked. "If he is an alien he should certainly be deported and if he is a native he should of course be hung."[5] Strode's harsh choice of words suggested that border-hardening nativism, as codified by the 1924 National Origins Act, could extend from people to other animals.[6]

Another stockgrower, Gomer Thomas, sent a letter of gratitude discussing the future of these border-crossing interlopers. "Last winter, and this winter so far, nothing has disturbed us. Of course, some stray wolves could stray in here now from Canada, but with the hunter working in the winter they will not be able to gain such a foothold again."[7] Thomas was right. Within the next four years, federal trappers had virtually eliminated gray wolves from Montana, but populations remained in Alberta and British Columbia.[8]

This spatial divergence got me thinking. If settler-colonial frontiers based on capitalist livestock production enveloped the Rocky Mountain interior of the United States and Canada, then why were wolves eliminated from one place but not the other?

This comparative history reveals how nation-building projects made a difference in the extirpation or conservation of our focal species, *Canis lupus*. During the late nineteenth century, livestock industries employed similar tactics to destroy the "wolf menace." Montana adopted its first bounty law to incentivize killing in 1883, and Alberta (a part of the North-West Territories, NWT) followed in 1897. On both sides of the international border, ranchers directed their eradication campaigns toward federal public lands, which were regarded as high-altitude refuges for large carnivores. However, over the early twentieth century, centralized states came to differ in their approaches to predatory animals.[9] Factors such as what level of government was involved, who was employed to do the killing, and how other economic players responded to wolfing were all in action. By the middle of the century, US-led extermination efforts created an ecological vacuum south of the 49th Parallel by wiping out wolves from the landscape. The uneven application of state power across the Montana-Alberta borderlands explained more about the absence of *Canis lupus* south of the international line than the embalmed White Wolf.

Capitalizing on Wolves

The near destruction of bison caused wolf numbers to surge. From 1865 to 1884, commercial buffalo hunters slaughtered millions of hoofed animals to force Indigenous peoples onto reservations, opening the grasslands for livestock. Wolves quickly learned that gunshots meant wasted bison meat, and following this food source allowed litter sizes to double. Increasing wolf populations along the Rockies impressed human observers. In 1865, hunter John McDougall was processing bison robes south of Edmonton when he reported, "I have seen great numbers of grey wolves, but never, I think, did I see them more numerous than at this time." McDougall continued, "When we were butchering the animals we had killed, they would form a circle around us."[10] Robe trader James Willard Schultz also recalled the ever-present carnivores on the Upper Missouri River near Fort Benton during the "buffalo days" of the 1870s. "And the wolves!" Schultz exclaimed, "There seemed to be thousands and thousands of the great shaggy fellows."[11] Extrapolating from current range densities, scientists have estimated North America's peak wolf population stood at a half million animals. In the Montana-Alberta borderlands, there were probably 22,500 gray wolves.[12]

Wolfing, or the practice of killing wolves for their fur, became a valued industry in a global capitalist market. The same clothing merchant firms that amassed riches shipping bison robes to New York City, Montreal, and London also exported wolf skins abroad. And since regional fur traders paid upward of two dollars for each pelt, itinerant hunters called "wolfers" could make a decent living by taking a few hundred animals each year. These mortalities did not cause wolf populations to collapse because their reproduction rates after mass die-offs, according to recent studies, can range between 200 and 400 percent. From 1871 to 1875, the steamboat freighting company T. C. Power & Brother—stationed at Fort Benton, Montana—purchased 34,000 wolf pelts from northern Montana and southern Alberta without totally destroying the species.[13] "Wolf skin overcoats becoming a part of the uniform of soldiers of portions of the Russian army, and the popularity of the wolf robe in all fur wearing countries, made the demand steady and profitable to the fur dealer and the wolf trapper," wolfer Joseph Henry Taylor explained, "so that new and more systematic ways were devised to destroy wolves for their fur value."[14]

Taylor was referring to the wolfer's application of strychnine. Hunters had used arsenic, mercury, and other poisons to harvest fur-bearing animals in the past, but these substances killed too slowly, allowing the targeted creatures to wander away before dying. Strychnine, in contrast, worked in minutes after ingesting small doses. Derived from the fruit seeds of an East Indian tree *Strychnos nux-vomica*, strychnine attacked the victim's nervous system, causing muscle spasms and severe cramps. Death resulted from asphyxiation due to convulsions of the respiratory tract. By the mid-nineteenth century, London-based pharmaceutical companies made strychnine widely available to wolfers. The lethal method was straightforward: Shoot down a buffalo, cut open its abdomen, and cover the carcass with granules of poison.[15]

The borderlands life of John George "Kootenai" Brown illustrates what it was like to become a wolfer during the era of the buffalo commons when transboundary movements across the 49th Parallel were prevalent. Born in Ireland in 1839, Brown was orphaned as a child, and at eighteen, he joined the British Army and was shipped to India to help suppress a massive anticolonial rebellion. After a stint in the military, Brown left home again in 1861 with a fellow veteran to join prospectors at the Cariboo Gold Rush in British Columbia. His placer diggings never panned out there, so Brown crossed the Rockies at South Kootenay Pass

An artistic rendition of a wolf pack consuming a strychnine-laced bison corpse. Note the dead wolf at bottom left. From "Poisoning Carnivorous Animals," in J. H. Batty, *How to Hunt and Trap* (New York: Albert Cogswell, 1878), 215.

in 1865 for a rumored mining boom near the Saskatchewan River. The eastward route did not bring Brown to gold, but he did find a chain of sapphire-colored pools known as Kootenay Lakes, later renamed Waterton Lakes, a beautiful place to which he vowed to one day return. "This is what I have seen in my dreams," Brown told travel companions. "This is the country for me."[16]

Brown found lucrative work farther east as a mail courier and whiskey trader, hauling loads back and forth across the international line between Fort Stevenson on the Missouri River and Fort Garry (Winnipeg) on the Red River. His very first encounter with the northern plains instilled a sense of wonder over the "one living moving mass of buffalo" pursued by "hundreds of wolves."[17]

In delivering letters and selling alcohol, Kootenai Brown developed contacts with Metis and Native peoples who reoriented his life. During snowy winters, Brown worked alongside Metis men who were hired by the US Army to drive mail-carrying dogsleds across the plains. During summers, Brown swindled Assiniboines, Crees, and Ojibweg by exchanging whiskey for animal furs. At the trading post of Pembina, Dakota Territory, Brown's brief courtship of a Métis woman of Cree and French ancestry, Olivia D'Lonais, led to their marriage on

September 26, 1869. D'Lonais belonged to the Plains Métis, a series of borderlands communities along the Red River Valley who traced their origins to relationships between Indigenous women and European men during the early fur trade. D'Lonais and Brown's nuptials represented both the strengthening ties between Natives and Euro–North Americans in a world system of capitalist trade and the weakening land base on which traditional livelihoods rested. The couple quickly moved west to escape the Red River Valley's growing political unrest, instigated by Louis Riel and other Métis nationalists who disputed territorial claims over Rupert's Land. In the same year as their wedding, the Dominion of Canada had purchased Rupert's Land from the Hudson's Bay Company, reorganizing it as the North-West Territories under the new confederation.[18]

Beginning in 1874, Brown and D'Lonais joined her Métis family every autumn to pursue roaming bison herds near the Milk River of Alberta and Montana. Once scouts located a buffalo herd, the Métis community settled down into camps. Brown and around three hundred of his male relatives galloped on horseback toward clusters of bison, firing volley after volley of their muzzle-loaders at close range to take down the fleeing animals. These chases resulted in hundreds of slain buffalo strewn across the prairies, leaving Métis women—D'Lonais included—with the work of picking up the bison on horse-drawn Red River carts and processing them back at camp. Unlike commercial buffalo hunters who were only after the robes, Brown remembered that "it was a rule in all half-breed hunting camps that every part of every animal must be used." Métis women butchered and prepared the bison: cutting off hides for coats and blankets; slicing meat into thin strips to be dried in the sun or over a fire; and mixing rendered tallow, dried fruits, and animal protein into pemmican. In these borderlands, Brown recalled that Métis "hunters didn't know where the [international] line was any more than the buffalo did."[19]

If buffalo herds and Métis peoples crossed the US-Canada border without exception, so did wolves. The Treaty of Ghent had set the 49th Parallel to separate western land claims in 1814, but national territories were not marked by Ottawa or Washington officials until sixty years later. A joint boundary survey commission piled mounds of dirt with a stake on top to demarcate the international line every mile. Wolves, bison, and people ignored them.[20]

Kootenai Brown turned to wolfing during the winters because, like whiskey trading, it promised generous returns. From camp, Brown rode a circuit of several miles where he would plant red flags near poisoned bison corpses. He returned the next day to find between twenty and eighty dead wolves around each. Equivalent to the gendered division of labor during bison hunts, D'Lonais and

A portrait of wolfer John George "Kootenai" Brown, circa 1910. Courtesy of the Glenbow Museum, Calgary.

their two daughters followed with Red River carts and skinned the carnivores. Brown described how money came easy: "We were living on buffalo meat which didn't cost us anything and using tents or cabins we built with our own hands, our only expense was ammunition we used for killing buffalo and the strychnine for poisoning wolves."[21] Brown purchased bottles of strychnine at five to seven dollars apiece, applying three to four of them per bison, but he killed nearly a thousand wolves every season, which roughly translated to a $2,000 payout. In comparison to an Appalachian coal miner's income, Brown made four times the earnings for less strenuous work. As friend and fellow hunter James McDevitt reported, he and other seasoned wolfers made "small fortunes at the business."[22]

The wolfing business, however, entailed both profits and perils. In April 1877, when steamboat navigation opened on the Missouri River, Brown brought his wolf pelts south to the trading hub of Fort Benton, much as he had in previous springs. Because commercial activities attracted a frenzied crowd of gamblers, soldiers, and traders, he and D'Lonais preferred to camp on the outskirts of town with her Métis community. Business transactions between wolfers and merchants involved a fair amount of drinking and bartering, and inebriated dealings sometimes erupted into violence. Despite arbitrating a dispute between two "half-breeds" over who should be paid for a wolf skin that both claimed, Brown himself was implicated in one such incident. A drunken Brown and trader Louis Ell were riding their horses between the Métis camp and Fort Benton when the Frenchman charged that Brown, who was known for bouts of rage while intoxicated, must turn over some wolf pelts to settle a debt. Brown refused, tempers flared, and the two men began fighting. During the brawl, according to a local newspaper, "Brown plunged a sharp edged butcher knife into Ell's abdomen."[23] Ell died from blood loss, and a Fort Benton sheriff tracked down Brown, who was trying to flee for Canada, and arrested him on the charge of murder.[24]

Five months later, local authorities transferred Brown from jail to the territorial capital of Helena to stand trial. Brown claimed the homicide was an act of self-defense. After several days of courtroom theatrics, the grand jury exonerated him for lack of evidence. Brown commented in his old age that his two favorite words in the English language were "not guilty," even if that was not the verdict. With a renewed perspective on life, Brown left Montana with his wife Olivia and their two children for a squatter's claim four miles north of the 49th Parallel. It was the Kootenay Lakes area.[25]

Frontier violence related to whiskey and wolves began to transform fluid borderlands into an international border. When Rupert's Land became the North-West Territories, Hudson's Bay Company traders lost their monopolistic hold on animal hides, allowing so-called free traders to barter unrestricted with Native peoples. Although the territorial government banned the sale of alcohol within western Canada in October 1870, free traders continued to smuggle whiskey on a trail from Fort Benton to Fort Whoop-Up because it was one of the primary goods for obtaining Native-procured bison robes and wolf pelts. The prodigious amounts of "white man's water" had such damaging effects on the Indigenous communities that Crowfoot, a prominent chief of the Siksika Blackfeet, stated, "The whiskey brought among us by the Traders is fast killing us off and we are powerless before the evil." Crowfoot observed that crossing the "medicine line" into the North-West Territories was no protection from Montana liquor: "We also are unable to pitch anywhere that the Trader cannot follow us."[26]

On top of loathing whiskey smugglers, Indigenous communities despised white wolfers for their widespread use of strychnine. Poisoned buffalo carcasses held unintended consequences by killing coyotes, foxes, birds, and other scavengers, but what enraged Native peoples the most was the loss of dogs that helped with everyday tasks such as hunting and hauling. They attacked Euro–North Americans who laced dead bison with strychnine to dissuade the practice. Tensions came to a flashpoint in June 1873 when Thomas Hardwick and a group of wolfers crossed into Canada to search for stolen horses. Hardwick's party slaughtered at least twenty Assiniboines, many of whom were drunk and defenseless, in the Cypress Hills. Once news of the massacre finally reached Ottawa, the Dominion government dispatched a new constabulary force called the North-West Mounted Police to the far western border.[27]

The Mounties restricted borderlands trade and, in the process, reassembled the social composition of those who killed wolves. Three hundred armed police, led by James MacLeod, marched west from Fort Garry, Manitoba, to clamp down on the whiskey smuggling that took place nearly 750 miles away along the Fort Benton to Fort Whoop-Up axis. In autumn 1874, they arrived at the Alberta post to find that most liquor dealers had been forewarned and vacated the premises. The Mounties then established Fort Macleod near the Oldman River, sending out regular patrols along the US-Canada border. Policing the plains not only solidified the Dominion's territorial claims north of the international line that had been surveyed recently, but federal presence also resulted in the separation of wolfers into "Yankees and half-breeds." On the US side, hunters mostly came from European ethnic backgrounds who saw their work as conquering wilderness; on the Canadian side, they were generally Indigenous or Metis trappers with spiritual traditions against eradicating animal kin. These social differences across the border would be significant in the one-sided extirpation of gray wolves.[28]

At Kootenay Lakes, Kootenai Brown constructed a general store where Indigenous customers traded furs for dry goods hauled in from Fort Macleod. According to Métis friend and historian Marie Rose Smith, Brown was more successful as a tourist guide for visitors "from the thickly populated eastern cities."[29] The Canadian Pacific Railway, a transcontinental line completed in 1885, brought a "stampede of speculators" to the region via a branch line after underground petroleum fields were discovered near Upper Kootenay Lake. While Alberta's first oil boom of 1890–91 turned bust when the wells failed to produce vast quantities of black gold, Euro-Canadians continued to flow into the mountains. Concerned that a general land rush would damage the wildlife, watershed, and scenery, Brown and rancher F. W. Godsal implored Ottawa officials to set aside the area as a "forest park" under a Dominion Lands Act amendment to halt further developments. On May 30, 1895, Canada's Department of the Interior withdrew fifty-five square miles from entry to create the Kootenay Lakes Forest Reserve.[30]

Brown was appointed as forest warden because, having been the longest Euro-Canadian settler at Kootenay Lakes, he intimately knew the flora and fauna. To encourage the growth of tourist-pleasing ungulate populations of deer, elk, and bighorn sheep, Brown kept up the practice of poisoning large carnivores, even though other duties preoccupied more of his time. Brown noted that "only an odd timber wolf is heard to howl and is rarely caught" as the wild canids became more wary of strychnine carcasses.[31]

The forest reserve took on a different meaning when other protected areas were created along the Rocky Mountains. To the north of the lakes, the Canadian Parliament established Banff National Park in 1885 and Jasper National Park in 1907; and to the south, US Congress founded Yellowstone National Park in 1872 and Glacier National Park in 1910. After Glacier was added across the international line, Brown and other local conservationists petitioned Canada's commissioner of national parks to alter the Kootenay Lakes Forest Reserve's size and purpose. "It seems advisable to greatly enlarge this park," Brown reported, "to have a [wildlife] preserve and breeding ground in conjunction with the United States Glacier Park."[32] In 1911, the same year the Dominion Parks Branch was created, Ottawa officials redesignated those lands as Waterton Lakes National Park (named for English naturalist Charles Waterton) with Brown serving as the first superintendent. Three years later, with Brown's urging in annual reports, Waterton was expanded to 423 square miles (one thousand square kilometers).[33]

National parks, for a brief time, sheltered wolves from the ecological transformations that came with privatization on the prairies. As ranchers worked to fence and enclose the open range, a patchwork of public lands offered remnant habitats for native predators. The conversion of the buffalo commons into gridded pastures of bovines and barbed wire would have consequences for the survival of wolves.

Beefing Up Bounties

The Smithsonian National Museum of Natural History in Washington, DC, gave me the impression that the space has become dedicated to exhibiting future extinctions. In the rotunda, a taxidermied African elephant towers over incoming visitors with the words "Elephants in Danger: Your Choices Make a Difference" etched into marble. The panel below explained the litany of human-caused factors leading to the rapid decline of these creatures: habitat loss, invasive species, population growth, pollution, and overhunting, or what a Conservation Biology 101 course would summarize with the acronym HIPPO.

I was reminded of the abbreviation after staring at a hippopotamus, posed with its jaws gaping open, within the Africa section of the Kenneth E. Behring Family Hall of Mammals. Ironically, Smithsonian donor Ken Behring might learn a thing or two from visiting given that he and other affluent sportsmen contributed to the annihilation of species. A real-estate developer who once made the Forbes list of wealthiest Americans, Behring earned notoriety as a trophy hunter and game poacher in 1997 when he pledged $20 million to the Smithsonian with strings attached. To receive the money, Behring demanded museum officials accept four Kara Tau argali that he had killed in central Asia. At the time, these bighorn sheep were one of the world's most endangered animals.[34]

Just as intriguing as the private donations that shape museum curation at Behring's Hall of Mammals was the proximity of two related canids. I saw a mounted gray wolf, like Montana's White Wolf, postured howling to mark the North America section, and not more than a few steps away, a dingo in the Australia section. The smaller carnivore with a rust-colored body and white paws stood in front of a gossamer screen. Pushing a button near the base of the animal, the stage lit up in the background to reveal a thylacine and display titled "An Extinction Story: What happened when these two predators faced off?" The panel suggested that the dingo, a species introduced to mainland Australia by Aboriginal peoples, was largely responsible for the thylacine's downfall some three thousand years ago. Missing from the display was any mention that a remnant population of thylacines remained on the island of Tasmania, sheltered by isolation, until the early 1900s when eradication efforts finally wiped them out.

The Hall of Mammals hinted at an interpretation that the destruction of native predators should not be seen as something isolated to North America but as one part of a larger extinction trend related to global settler frontiers of capitalist livestock production. Between 1822 and 1850, when wool became a major export for British colonists in Australia, sheep numbers grew more than a hundred times over, from 120,000 to 16 million. To curb the "dingo threat," livestock owners introduced strychnine in the 1840s. And after sheep investors overstated losses due to the thylacine, more commonly known as the marsupial wolf or Tasmanian tiger, the colonial administration passed a bounty law in 1888 to encourage killing them. Government payments for thylacine hides mirrored other bounty systems emerging during the late nineteenth century. In Japan, Meiji officials started offering small rewards in 1877 for turning over the ears of slain wolves. In South Africa, Cape authorities began providing money in 1889 for the tails of poisoned jackals. Taken together, the prospect of government-sponsored extinction arrived on the heels of imperial expansion.[35]

Like Britain in New South Wales or Japan on Hokkaido, Canada and the United States saw the North American West as colonies where the movements of Native peoples would be limited to allow for land expropriation by European immigrants and their biological package of cows, wheat, and other companion species. However, Indigenous nations did not easily fit into one settler state or the other's western land claims. In the late 1860s, Lieutenant Colonel Alfred Sully, the superintendent of Indian Affairs in Montana, observed that the Blackfoot Confederacy "claim[ed] in common a section of the country from the British line south some miles to the city of Helena, and north to the line of the Saskatchewan River."[36] The Blackfeet Nation was neither Canadian nor American.

Ottawa and Washington officials used reservation systems, backed by policies of forced starvation and military removal, to divide and confine "their" Indians onto discrete portions of land. In 1877, Pikani, Kainai, and Siksika leaders signed Treaty Seven, which ceded most of southern Alberta to the Dominion of Canada in exchange for annuities and three small reserves. An 1855 US treaty recognized the Blackfeet Indian Reservation as all territory north of the Musselshell River, a series of executive orders and acts by the federal government reduced tribal lands to the eastern slopes of the Rockies and later outside of Glacier National Park. One method to restrict the mobility of Native peoples was feeding them beef rations at government posts. In 1878, James MacLeod of the North-West Mounted Police suggested that Dominion officials buy cattle from Montana for those Blackfeet who stayed on treaty reserves.[37] For settler states, beef rations could move across the international border, Native peoples should not.

As wolf populations reached their numeric height, cows were replacing bison as a principal food source on the prairies. In 1884, a US Army officer at Fort Keogh near Miles City noticed that "when the great herds of domestic cattle succeeded to the stamping grounds of the native bison, there was a remarkable increase in the number of gray wolves on the Montana ranges."[38] Cowpunchers had driven large herds of Texas longhorns and other grazers on southern trails and eastern rails onto the northern plains. Scientific understandings reinforced the soldier's comment that wolves did not "object to beefsteak when buffalo hump was not to be had" for the adaptable predators can switch from one species to another based on the scarcity or abundance of prey animals. By 1890, Montana boasted 665,000 head of cattle and Alberta held 115,000.[39]

This biological invasion can be attributed to federal policies that opened western lands to a speculative boom of wealthy investors, who in turn pushed intensive animal production. In Canada, a Conservative Party government passed an 1881 law amending the Dominion Lands Act to allow large grazing leases on Crown lands. Applicants could secure leases of up to one hundred thousand acres over twenty-one years at one cent per acre per year, or they could purchase land outright for two dollars per acre. The bill was advanced by Senator Matthew Cochrane, a Montreal leather magnate, who had made friends with cabinet members of John A. Macdonald's administration. Four companies—Cochrane, Bar U, Oxley, and Walrond (the latter two were owned by the British aristocracy)—would in due time control nearly half of all leased land in Alberta, as well as an equivalent share of cattle.[40]

In the United States, Army general James Brisbin wrote an 1881 book titled *The Beef Bonanza; or, How to Get Rich on the Plains* praising the fortunes to be grown on the national estate. After purchasing a home range through existing land laws, such as 640 acres under the Desert Land Act of 1877 by irrigating a portion of it, cattle owners could ensure money on the hoof by grazing cows on the public domain at no charge. Even if some herds in Montana came from local cattle kings, as Conrad Kohrs made early profits selling beef to mining communities, most grazers arrived under the auspices of distant investment firms. By the mid-1880s, a special congressional investigation found that twenty-one million acres of US federal lands, or roughly the same size as the state of Maine, were controlled by foreign livestock entities. In Montana, the English-owned XIT Ranch and the Scottish-owned Matador Land and Cattle Company possessed vast ranges in Custer and Beaverhead counties.[41]

The visible hand of capitalism made wolf depredations more common. Although wolves killed cows just as they had bison, Texas longhorns' namesake biological feature made them less vulnerable to predators than other breeds. After 1883, when the Northern Pacific Railroad finally connected the cattle ranges of Montana to the industrial meatpacking plants of Chicago, railroad executives accumulated more wealth by transporting shorthorn varieties because the fixed costs of shipping remained the same no matter how many cows fit onto each railcar. As an illustration, twice as many English shorthorns could be packed onto a single train as Texas longhorns. This factor, along with changing livestock preferences after the Great Die-Up of 1886–87 when thousands of longhorn cattle perished on the Great Plains during a harsh winter, caused shorthorn breeds to become one of wolves' predominant prey species.[42]

Investors also received greater returns by pressuring cattle managers to maximize herd sizes and minimize labor costs. While small outfits might have five hundred to a thousand head managed by two or three cowhands, bigger operations tried to get away with as few people as possible. At its peak, Alberta's Cochrane Ranche hired a dozen cowboys to oversee fourteen thousand cows during roundups and kept only two foremen during winters, leaving the hoofed grazers to mostly fend for themselves. The lopsided bovine-to-buckaroo ratio on corporate spreads meant higher levels of wolf-related deaths.[43]

Because cattle capitalists viewed domesticated animals as investment property, or literally as "live stock," wolves changed from objects of earning money into agents of stealing money. The two biggest concerns for regional livestock organizations when they formed during the late nineteenth century were to prevent mavericking and to promote wolfing. At the inaugural meeting of the Montana Stockgrowers Association in April 1885, Conrad Kohrs of Deer Lodge testified that his business depended on maximizing the "calf-crop" and deterring "swindlers."[44] To accomplish both aims, the association registered branding irons and coordinated spring roundups so that fewer "mavericks," the name for unbranded—and thus unclaimed—calves on the open range, would be available to cattle thieves. At the first executive meeting of the Western Stock Growers' Association in 1897, ranch manager William Cochrane, son of Matthew Cochrane, resolved that the North-West Legislative Council require all mavericks be sold at auction to fund a new wolf bounty. Just as the extralegal practice of mavericking turned into the illegal act of rustling, commercial livestock producers recast wolf depredations as theft of private property.[45]

At the top rung of the economic ladder, corporate ranchers led the push for instituting wolf bounties. The leadership boards of both livestock organizations were mostly represented by big operations. In the Western Stock Growers' Association, some of the powerful cattle owners or managers included William Cochrane of the Cochrane Ranche, Fred Stimson of the Bar U, and Allan Macdonald of the Glengarry "44" Ranch. In 1885, Cochrane reported to his father that of the 356,000 acres in leased or owned land, wolves were most problematic near High River and Big Hill Springs areas for his cows numbering in the tens of thousands. In the Montana Stockgrowers Association, notable members were Granville Stuart of the DHS Ranch, Thomas C. Power of the Judith Cattle

Company, and Conrad Kohrs, who had some fifty thousand head of cattle and ten million acres of grazing pasture across the state. In 1888, Power testified to losing over a third of his calves "when the wolves are desperate & hungry," and the merchant-turned-banker agreed to serve as chair of the Committee on Bounties whose duty was pressuring legislators on a bounty bill.[46]

At the bottom rung, wolfing turned into labor for the poor and the colonized. Poisoning during the winter had become less productive, as wolves were fewer and learned to evade bait stations, so following mothers to den sites during the spring and setting up traplines during the summer augmented the usual duties. Both tasks were time-consuming work, and cattle barons formed racialized labor systems to do them. The Western Stock Growers' Association made its wolf bounty redeemable only by "Indians and Half Breeds" to persuade Native communities about the virtues of capitalist agriculture. In 1895, for example, Îyârhe Nakoda wolfers killed 120 animals and collected bounties on them over three months. When the livestock organization ended its exclusive offer in the early twentieth century, young cowhand W. H. McKay recalled that ranch owner James McGregor had hired Sun Child and twenty other Plains Cree to hunt wolves on his place west of Medicine Hat. After the Cree wolfers scoured McGregor's spread, he paid them $800 for turning over eight dead wolves. For McKay, the lesson of the private bounty was that "the government [of Alberta] some how never seemed to care whether wolves were done away with or not."[47] However, the out-of-pocket expense for McGregor was a small price to pay given Grand Fork Cattle Company's enormous herds of ten thousand cows.

Overstocking the range eventually drove stockmen to enclose the diminishing resources of grass and water. After the Massachusetts-based company Washburn and Moen bought Joseph Glidden's barbed wire patent in 1876, cheap fencing was made available to cattle companies on the prairies where traditional materials—wood, stone, or hedges—were scarce. The effectiveness of controlling space with barbed wire lay in its ability to inflict pain on animal flesh; cows learned to avoid sharp pieces of metal coiled around double-stranded steel. Intensifying competition between big operations and small outfits, as well as between ranchers and farmers, swayed many stockgrowers to fence "their" pasturage whether they had a title to it or not. In 1885, US General Land Office agents estimated that 4.5 million acres of illegal fences were erected in the western United States alone. By the turn of the century, some three-quarters of stockgrowers in Montana, and nearly as many in Alberta, had already fenced their properties with barbed wire. On settler-colonial frontiers, property lines were more than mere legal fictions when animals of enterprise demanded that the "range," meaning unbounded space, transform into the "ranch," an enclosed piece of land.[48]

LIVESTOCK INDUSTRIES on both sides of the border were responsible for introducing bounty systems to incentivize destroying wolves. In 1883, the Montana Legislature, known by Conrad Kohrs as the "cowboy legislature" for the high number of ranchers-turned-politicians in its ranks, passed its first bounty law on wolves and coyotes that would change the scope of wolf killing. Wolfers now slayed large carnivores for payments instead of pelts, turning seasonal occupations into full-time endeavors. Although various iterations of Montana's bounty system paid from one to eight dollars per wolf, bounty laws basically endured through statehood in 1889 until the 1920s when federal agents took over predator control.[49]

In Canada, neither the North-West Territories nor Alberta implemented long-standing bounty bills because ranching interests wielded less political power. The Western Stock Growers' Association, as well as private outfits, tried to fill the void left by the government by offering their own wolf bounties, but voluntary dues made funding unreliable and payments initially could only be redeemed in three locations across the whole territory: Calgary, Fort Macleod, or Maple Creek. The uneven application of these bounty systems meant that fewer wolves north of the international line met the same fate as those south of it.[50]

In Montana, the initial bounty law faced opposition because it became too expensive for legislative coffers. After adding prairie dogs and ground squirrels to the list of predatory animals for slaughter, bounty payments consumed two-thirds of the entire annual budget and bankrupted the treasury. In 1887, mining capitalists in the territorial government repealed the law and argued that passing another bounty bill on roaming wolves would be futile "unless [Canada's] Dominion government also puts a price upon the heads of the creatures."[51] Montana stockmen responded by inflating their losses to large carnivores; one Chouteau County rancher claimed to have 20 percent of calves taken in a single year. This political pressure worked. In 1891, the Montana Legislature approved a second bounty law on wolves and coyotes, amending it in subsequent years only to increase payments. Eight years later, the state government instituted a special tax on stockgrowers to ease financial constraints. For instance, a larger ranch with 2,500 cows would turn over about ten dollars toward a permanent bounty fund, which subsidized only two bounties but was still cheaper than paying wages to a private wolfer.[52]

One reason for the patchier results above the 49th Parallel was the Western Stock Growers' Association held less political influence in the North-West Territories because the organization encompassed a larger area. The NWT commissioner of agriculture, G. H. V. Bulyea, remarked that his aversion to backing a bounty law, in contrast to Montana, came from the northern lands where a lack of Euro-Canadian settlers would translate into a population reservoir of wolves. "The conditions of the state of Montana very much resemble those prevailing in the Territories," Bulyea observed. "But with regard to the extermination of the predatory animals, the state has the advantage of the North-West Territories in not having an enormous area of unsettled country adjacent to it."[53]

Another factor was that, unlike Montana's cattle barons who organized against mining capitalists to maintain the bounty system, Alberta's stockmen could not overcome the competing interests of dryland farmers. In 1899, the NWT Department of Agriculture finally agreed to split the costs of wolf bounties with the Western Stock Growers' Association. However, stock raisers near Medicine Hat argued that the cost-sharing duty revealed a preference for "more deserving" farmers in the eastern portion of the territory, who were closer to the government seat in Regina and did not have to put up any matching contributions for weed-control programs, even though wolves were as "noxious" as Russian thistle.[54]

The North-West Territories' reluctance to support a full-fledged bounty law meant that the number of state-funded wolf kills disappointed Canadian ranchers. From 1899 to 1905, the NWT government rewarded bounties on only 2,464 wolves. In contrast, the State of Montana paid ten times as many bounties in 1899 alone; between 1883 and 1918, the state government reported giving bounty payments on 80,730 wolves—costing the legislature $342,764 total. As a comparison, Montana paid the same number of bounties *every year* on average that the North-West Territories did in more than a *half decade*.[55]

Problems with administering the wolf bounty in western Canada continued during its transition from territorial to provincial status. In 1905, after an influx of white homesteaders to the prairie region from eastern Canada, Europe, and the United States, the Dominion government created the provinces of Alberta and Saskatchewan from the southwestern portion of the North-West Territories. Like the territorial precedent, the Alberta government shared the expenses of a bounty system with the Western Stock Growers' Association. However, Chief Bounty Inspector Benjamin Lawton commented that since the ranching

Cowhands with a dead wolf, Dorothy area, Alberta, 1905. Even when Alberta became a province in 1905, the ranching industry felt the provincial government did not do enough to help them out with predatory animals. Courtesy of the Glenbow Western Research Centre, University of Calgary.

industry "operated largely in that portion of the province lying south of the main line of the Canadian Pacific Railway, the effect of this regulation was that bounty was paid in only a portion of the province."[56] The Legislative Assembly of Alberta solved this issue in 1907 by passing the Game Act, which sanctioned the first bounty on wolves and coyotes to be paid exclusively by the provincial agriculture department. The law specified that bounty management would be split at the 55th parallel (which basically went by Edmonton): South of the line, inspectors appointed to larger towns would process bounty claims, and north of it, the Mounties would take them. The Alberta government also allocated $20,000 to be spent on bounty payments, stipulating that once funds were expended, no more would be paid for the year. This provision created yet another difficulty when the amount was exhausted by April 1908 and inspectors ceased all payments.[57]

In the same year that Ottawa officials established the province of Alberta, Washington bureaucrats initiated some reorganization that would lay the administrative groundwork for the total annihilation of wolves. The Division of

Economic Ornithology and Mammalogy, founded in the late nineteenth century, initially hired government personnel to study birds and other life-forms that might be useful in bolstering agricultural pursuits. Director C. Hart Merriam renamed it the Division of Biological Survey in 1896 to reflect the growing cadre of field naturalists who conducted research for economic applications. In 1905, the federal agency attained bureau status and solidified its place within the US Department of Agriculture by helping another entity that was charged with conserving the country's disappearing forests: the US Forest Service.[58]

Chief Forester Gifford Pinchot, who called American stockgrowers "the best-organized interest in the West," worried that purchasing grazing permits in the national forests would estrange ranchers from the new agency. To fend off political backlash from the livestock industry accustomed to grazing for free, Pinchot asked Bureau of Biological Survey staff to assist forest rangers in hunting down stock-killing wolves and other predatory animals. The federal presence south of the US-Canada border marked a crucial shift in the one-sided extirpation of a species.[59]

Federal Lands, Federal Agents

I strapped on my cross-country skis and prepared for a long winter day in Yellowstone National Park. From the steaming terraces at Mammoth Hot Springs, I followed wildlife technician Elise Loggers as she broke trail through fresh powder up to Fawn Creek. After eight miles of an uphill climb, we finally reached the snow-covered stream at the headwaters of the Gardner River. I was exhausted, panting to catch my breath. A river otter poked its head out of a hole where the creek wasn't frozen over as if to see how I was doing. "Are you coming?" Loggers asked me restlessly. "I see blood from a probable kill."

During the months of December and March, Loggers was working on a "cluster crew" for the park's winter field studies on wolves. Loggers and other technicians followed clusters of GPS coordinates on a map to the places where collared animals in each wolf pack had stayed put for at least an hour, which indicated either resting or eating. Cluster crews verified what species the wolves were killing, and how often, to determine predation rates. Though varying over time and season, nearly two-thirds of their winter diet in Yellowstone consisted of elk, one-eighth of bison, one-tenth of deer, and the rest from various species like river otter or pronghorn antelope. At the creek bottom, the snowy ground preserved a disorderly scene of darting tracks and spattered blood, which seemed to be evidence that the 8-Mile Pack had chased a mule deer. Loggers and I confirmed this hunch by finding what remained of the prey: a femur, a few strips of fur, and a mandible. Loggers wrote down the kill information on a sheet for inclusion in a computer database. We then skied down the mountain.

During this field season, the 8-Mile Pack mostly stayed within the northwest portion of Yellowstone. Before 2011, though, it was considered a "boundary pack" that ranged in and out of the park between Paradise Valley and the Mammoth area until it displaced another wolf pack that had collapsed. Wolf packs likely evolved to be territorial because the behavior comes with the benefits of protecting vulnerable pups from intruders and securing uncontested food sources. They defend home ranges through scent marking, howling, and at times, fighting. While territory size and shape depend on wolf populations, prey availability, and landscape features, most home ranges have core areas that remain stable but outer boundaries that are constantly shifting. This is where ecology becomes history. Scientific research indicates that human hunting can increase wolf mobility by leaving pack structures fragmented and former territories vacated. Veterinarian biologists found higher levels of stress hormones in those carnivores that were constantly hunted versus those that were not, concluding that human pressures created pack instability, especially when the mating pair was killed.[60]

Turn-of-the-century hunters, therefore, shaped ecological conditions on both sides of the international line that produced more transborder movements than ever before. As packs were destroyed and their territories loosened, Canada and the United States exposed fault lines in their response to wolf mobility: Washington responded to this transnational nature by applying more state power; Ottawa did not.[61]

Government officials took issue with wolves because their territories broke down legal divides between public and private lands. The US Forest Service commissioned Vernon Bailey, a Biological Survey naturalist, to research the location of wolves in national forests and the best tactics to eliminate them. In 1907, Bailey penned a report titled *Wolves in Relation to Stock, Game, and the National Forest Reserves*. Locating the den sites of twenty mating pairs in Wyoming's Green River and Bighorn River basins, Bailey concluded that wolves were breeding in foothill areas "mainly below the edge of the forest reserves," not directly in the national forests as previously assumed. However, Bailey noticed that wolves frequently moved onto federal lands to follow domesticated cattle and sheep up to mountain meadows. The Biological Survey also turned its attention to national parks. Between 1914 and 1916, Bailey trained Army scouts on the most effective means of hunting down native carnivores that were active "along the northern border" of Yellowstone National Park. On the advice of the Biological Survey, Yellowstone hired two special rangers who killed a dozen wolves and shipped their skulls to the Smithsonian.[62]

Based on the perception that public lands offered sanctuaries for wolves, American stockmen launched a political campaign demanding that the federal government fund their total destruction. On November 14, 1914, S. W. McClure of the National Wool Growers Association published an article in *The Country Gentleman*, titled "The Wolf at the Stockman's Door: Sheep and Cattle Killers Breed in the National Reserves," that epitomized the industry's message to politicians. Despite western states offering bounties, McClure claimed that stockgrowers still lost a cumulative total of $15 million in beef and mutton due to depredations, leaving the supply of the "Nation's meat" in limbo. McClure blamed the government for harboring "carnivorous wild animals" in national forests, national parks, and other public lands that would inevitably roam onto private ranches. "The Nation, therefore, should assume the obligations which the withdrawals impose, at least to the extent of keeping the lands from becoming a menace to the surrounding country," McClure asserted.[63] On behalf of the livestock associations, McClure asked US Congress to furnish $350,000 for the final solution to the wolf problem.

With successful lobbying, the charge of wolf eradication in the western United States shifted from state to federal actors. In 1915, Congress appropriated $125,000 to the Biological Survey for the purpose of "suppressing carnivorous wild animals destructive to live stock in the public-land States of the West." Citing McClure's article, Senator Francis Warren of Wyoming argued that federal agents should be held responsible for protecting the pasturage around federal lands by eliminating all predators. "The Government owns and controls over 200,000,000 acres of land carved out and segregated in [park and forest] reservations in the various States," Warren stated, and those public lands "practically furnish the breeding grounds of wild animals, which come from there out onto the lands of the settlers and destroy their stock."[64] With new funds, the Biological Survey hired about three hundred salaried trappers to be stationed in eight districts throughout the western United States. Each BBS district, headed by a predatory animal inspector, solicited "cooperative agreements" with federal, state, county, and private landholders to clear this landscape of carnivores more systematically.[65]

The difference in wolfing for salaries versus bounties was profound. Under the bounty system, wolfers only hunted when and where those animals were plentiful and easy to kill because their income relied on harvesting as many pelts as possible. Under the new system, fixed salaries meant that wolf hunting would occur whenever and wherever predators were reported because payment was no longer contingent on pelts. Montana rancher Wallis Huidekoper, vice president of the National Live Stock Association, stressed this point in a speech called "The Wolf Question and What the Government Is Doing to Help" at the 1916 convention of the Montana Stockgrowers Association. "Large bounties have been offered by cattle and horse companies, and by individuals, for the past thirty years, for wolves killed on their ranges; but what has been the result?" Huidekoper asked. "The cowboys kill a few, and the renegade wolves are temporarily run out of that particular country to some adjoining range where no bounty is offered." In contrast, the Biological Survey offered professionals operating in the entire West, or in the words of Huidekoper, "a more persistent and well-studied plan" for finally exterminating those "stock-killing individuals [that] are the most wary and cunning."[66] Huidekoper ended his address by inviting the Montana Stockgrowers Association to pass a resolution endorsing the Biological Survey's work.

In the Montana district, Predatory Animal Inspector Robert Bateman initially recruited fifteen hunters on the government payroll with a salary of seventy-five dollars a month. After receiving notice of wolf sightings or depredations, Bateman dispatched federal trappers wherever needed. Their success depended on coordinating with state agencies and locals to intercept wolves moving from core areas to other habitat patches. In 1919, for example, rancher Charles Marble of Paradise Valley wrote Bateman about "the depredations of wolves on domestic livestock and game; particularly among the elk ranging that part of the country and in and out of the National Park, the State Game Preserve, and the Gallatin National Forest."[67] BBS officials visited with Yellowstone's superintendent and Montana's fish and game commissioner to discuss working across multiple jurisdictions. The various parties agreed to post one government hunter who would oversee an eradication campaign spanning a national park, national forest, state preserve, and private grasslands. A year later, the wolf pack was dead.[68]

At roughly the same time that Montana began turning over predator eradication to federal agents, Alberta's bounty system was eroding. In 1916, the provincial department of agriculture suspended all bounty payments from January to May because of the challenge in differentiating between "timber wolves" (wolves) and "prairie wolves" (coyotes). When turning over pelts to bounty inspectors, trappers often convinced authorities that larger coyotes with thick winter coats were wolves to receive payments of ten dollars per wolf instead of one dollar per coyote. Another issue was enticing wolfers to stay in Alberta when higher bounties were offered in other provinces, such as fifteen dollars per wolf in British Columbia.[69] Newer regulations tried to address these issues, but recurrent frustrations with the system led the Alberta Legislature in 1920 to repeal its bounty law. Unlike ranching interests in the state of Montana, Walter Huckvale of the Western Stock Growers' Association commented, "since the formation of the Province has taken place we have had very little assistance from the Government in this way."[70] After 1920, provincial bounty laws came and went as stockmen's political power waxed and waned. National priorities in Canada and the United States left wolves to deal with nonstate actors above the international line and state actors below it.[71]

THE DIVERGENT APPROACHES to eradicating predators were demonstrated at the borderline parks. During July 1917, BBS naturalist Vernon Bailey surveyed Glacier National Park for its varied mammal life, reporting how lupine populations moved to and from Waterton Lakes National Park. For instance, a park ranger had killed two wolves that roamed on the Belly River Trail going through both parks, but he was unable to remove the entire transborder wolf pack. Bailey recommended these mobile animals "should be destroyed both for the protection of the game in the park and the stock outside."[72] A year later, Glacier entered a cooperative agreement with the Biological Survey to send three government hunters from Idaho who would train the ranger force. They distributed one hundred Newhouse No. 4 traps and supervised park rangers in setting up traplines along the valley bottoms and in locating den sites. With federal support, park staff wiped out a few dozen wolves over the next decade, eradicating the three to four resident packs.[73] In 1930, Bailey returned to Glacier to find the carnivores were virtually gone. "Gray wolves are now so scarce on both sides of Glacier Park," Bailey stated. He continued that the federal trapper "has not found signs of any for several years back. The last known were on the Belly River Country close to the Alberta line and a few may come across the line at any time from Alberta or British Columbia."[74]

In 1918, the same year that Glacier enlisted BBS hunters to help with predator control, Waterton recruited Indigenous peoples to trap wolves. In response to local ranchers complaining about wolf depredations in the vicinity of the park, Chief Warden Howard Sibbald went to David Bearspaw on the Stoney Reserve, drafting him and four other Îyârhe Nakoda men with wolfing experience to work temporarily for Waterton. While park officials were satisfied with the Nakoda wolfers for turning in several pups and adult males over the next three

David Bearspaw, holding two wolf skins, stands next to wife Peggy, daughter Elizabeth (*far right*) and son Sala (*left*) on O. C. Royal's Ranch near High River, Alberta, circa 1915. Like this Îyârhe Nakoda family, Native and Métis peoples were often employed as wolfers north of the medicine line. Courtesy of the Glenbow Western Research Centre, University of Calgary.

years, local stockmen were not. They grumbled to James Harkin, commissioner of the Dominion Parks Branch, that livestock attacks did not end because "the Indians invariably allowed the bitch wolf to get away, and some wolf experts believe that is done purposely so as not to kill the goose that lays the golden eggs."[75] Their criticism demonstrated a partial understanding of Nakoda tactics, for other contemporary sources indicate that some used "wolf ranching," which involved either raising wolf pups to adulthood for higher bounty payments or saving dens to cash out a new litter of pups. A Montana rancher near the international border remarked that "the [half-]Breeds in Canada catch wolves for the stockmen's bounty over there, but if they can possibly help it they won't harm an old female wolf."[76] By conserving gray wolves, Indigenous peoples conserved themselves on the fringes of a capitalist economy.[77]

But there was more to the aversion of slaying female wolves. These animals represented other-than-human kin with special powers for Indigenous peoples.

According to an oral tradition passed down to elder George Ear, a Nakoda warrior named Wolf-Come-Into-View owed his life to a wolf spirit for saving him during a near-death experience. When scouting for bison herds in the Bow Valley long ago, Wolf-Come-Into-View was spotted by a group of enemy warriors who charged toward him. Wolf-Come-Into-View ducked into nearby bushes where a guardian wolf visited him and then assisted his transformation into a fellow wolf. The lupine shapeshifter tricked the hostile warriors, who scoured the bushes for the man but could only find two wolves. Similarly, legends from the Blackfoot Confederacy emphasized how a wolf spirit, called Black Wolf or Wolf Chief, took pity on a starving family on the brink of death and brought a supply of buffalo meat to their lodge. In other stories, the first peoples learned how to hunt from wolves. This spiritual admiration for the animal explains why some Canadian agents insisted that Indigenous peoples held a "strong superstition" against predator eradication for most of them believed "that evil results will follow" those who unjustly kill wolves.[78]

Because the Canadian government incorporated nonstate actors in Native wolfers to hunt predatory animals, but the US Bureau of Biological Survey did not, wolves were more likely to persist north of the international border. The Dominion Parks Branch's Animal Division, which was charged with overseeing wildlife management, highlighted those institutionalized social divisions. Animal Division Chief Maxwell Graham surveyed for wolves, coyotes, and mountain lions in British Columbia and Alberta near the US-Canada border. For a government report, he interviewed wolfers, mostly "half-breeds" and "full blooded Crees," who at the time worked for private ranching spreads around Manyberries, Alberta, but traveled wherever bounty payments were highest. Despite receiving a fifty-dollar bounty for pups or adult males and up to one hundred dollars for adult females, Graham learned that "even this sum would seldom tempt them to destroy one of [the] latter."[79] In 1921, rancher Henry Riviere of Pincher Creek criticized Waterton Lakes National Park for employing "Stony Indians," telling Graham that the Dominion Parks Branch should either hire "expert hunters" from the Biological Survey or approve stockmen as "honourary" park wardens to hunt wolves. Superintendent George Bevan authorized Riviere and eight other stockmen to poison and trap within the park, but they pursued wolves only when depredations on their surrounding ranches became a nuisance.[80]

The challenges of wolfing across the administrative borders on public and private lands were also tested across national borders. Maxwell Graham of the Animal Division remarked that "many wolves follow[ed] game or stock from

Canada into the United States and vice versa" and their "considerable radius of action" undermined Waterton's management policies. For Graham, the Sweet Grass Hills and Milk River Valley, situated outside of park jurisdiction and in-between nation-states, were identified as wildlife corridors for carnivore species.[81]

Biocidal markers hinted at the range of wolf mobility. In 1905, the Montana Legislature authorized state veterinarian Dr. M. E. Knowles to infect captured wolves and coyotes with sarcoptic mange. This contagious disease, caused by an infestation of mites burrowing through the skin, triggers itching to the point that the animal's hair falls out. With fur loss during winter, canids die from exposure. In 1908, Knowles worked with a local game warden in Chouteau County, located northeast of Great Falls, to release six wolves and fourteen coyotes infected with mange. Three years later, state authorities received word of this skin disease among canid populations along the 49th Parallel and six years later, along the foothills of southern Alberta. Although the state program was discontinued when ranchers questioned its efficacy, mange continued to spread northward. During the 1930s, Hubert Green, chief park warden at Banff National Park, reported observing mange-afflicted wolves within the park, whose disease he surmised had come a few decades earlier from northern Montana.[82]

Transboundary movements of wolves forced US and Canadian officials to consider the possibility of collaborating across the international line. In 1915, J. S. Jackson, president of the National Cattle Ranchers Association, asked the US Secretary of State William Jennings Bryan to reach out to Ottawa to see if the Dominion government was interested in cooperating with the Biological Survey in wolf eradication. "As these predatory animals are of a migratory nature," Jackson argued, "it is necessary that the work of extermination be extended beyond our North-western and south-western borders, to some extent, in order to afford the herds and flocks of our farmers and ranchmen in those localities, protection from depredation by these destructive animals."[83] In response, BBS officials suggested dispatching federal trappers from the bordering state of Montana to the so-called Prairie Provinces of Alberta, Saskatchewan, and Manitoba. Canada's Department of External Affairs arranged for official communication to begin between the chief of the Biological Survey, Henry Henshaw, and the commissioner of Dominion Parks, James Harkin, on the idea of coordinating along a "boundary strip." Although local interest existed on both sides of the international border, no formal agreement was ever reached. These diplomatic negotiations, however, revealed how mobile species destabilized environmental governance bound by the nation-state.[84]

When international collaboration did not materialize, Ottawa bureaucrats took the opportunity to make comparisons regarding predator control. In 1919, Dominion zoologist Dr. C. Gordon Hewitt finished two years researching the shortcomings of the present bounty system in western Canada. Hewitt concluded that bounty incentives, despite spending more than $75,000 by the Alberta and British Columbia governments over the last decade, failed to eradicate wolves due to their uneven application. Provincial laws did not work together because they were passed, amended, or repealed at different times, allowing mobile predators to flee to areas where no bounty was offered. Alberta's Deputy Minister of Agriculture stated, "There is little use in one province trying to establish a [bounty] system to control these animals unless a similar effort is made by the surrounding provinces and states, as you are aware many of these animals move long distances in the course of a year."[85] In contrast, Hewitt remarked, the US Bureau of Biological Survey worked across state lines by "carrying on a systemic campaign in the infested regions." Hewitt recommended that the Dominion of Canada adopt a similar federal program.[86]

Another Canadian scientist, Norman Criddle, echoed calls for centralizing authority to wipe out wolves. In a 1925 report, the Dominion entomologist argued the Biological Survey provided "the most effective method of controlling predatory animals on a large scale" because its federal hunters operated across multiple jurisdictions. "The chief points of advantage in the above system," Criddle explained, "are in being able to concentrate upon any given point—the central organization, and the thorough co-operation possible throughout the infested territory." He added that federal power was particularly useful "in the case of countries where predatory animals occur on both sides of the boundary, as the operations on one side would be more or less abortive unless accompanied by similar efforts on the other."[87] Parks Commissioner J. B. Harkin reiterated the advantages of political centralization: "I may say that we have known of their [BBS] methods, which show that the United States considers the wolf menace a Federal matter, whereas, here it has largely been considered a question for the Provincial authorities."[88]

BBS methods on public lands were part of a larger campaign to eliminate wolves from the western United States. Washington bureaucrats defended this position by arguing that *Canis lupus* would continue to live in neighboring countries. In the 1925 annual report, Edward Nelson, chief of the Biological Survey, claimed that clearing the lupine presence from national parks and other wildlife preserves was "intelligent conservation" because "wolves will doubtless continue to exist indefinitely in the wilder parts of Canada and Mexico, where they now occur in large numbers."[89] Population reservoirs, though, also provided justification for maintaining federal trappers, even if domestic numbers were dwindling. Nelson remarked that wolves "will undoubtedly continue to invade the United States along the border for a long time in the future and only vigilant efforts will prevent them from reinfesting vast areas."[90] As long as Canada held wolves, the US Bureau of Biological Survey had a defining mission.

THE FINAL PUSH to eliminate lupine predators from the state of Montana involved the Biological Survey's pursuit of "loner" wolves. Because hunting destabilized pack structure and territoriality, most wolves at this point were solitary wanderers trying to find a mate. High levels of wolf mobility can be inferred in comments like one from a ranch manager in Cascade, Montana, who reported that a few roaming wolves took twelve calves in May 1918 and then left the herd alone for the rest of the year. BBS Inspector Robert Bateman stated that depredations of this type showed "the damages three or four wolves will do when passing through a country they do not use as their permanent range."[91] Government trappers across the western United States gave nicknames to the most cunning singletons, like the White Wolf of Montana's Judith Basin, whose white fur came with advanced age (black, brown, or gray pelage lightens over time) and experience in avoiding traps and poison.[92]

Federal effectiveness at killing the elusive wolves convinced the Montana Legislature that state funds were better spent on the Biological Survey. Inspector Bateman provided ranchers' testimonies from across Montana who "have expressed themselves in favor of the work by the Bureau, by stating that it was the only system from which they were receiving any protection or benefits."[93] In 1922, the Montana Stockgrowers Association adopted a resolution calling for the suspension of bounty payments and for livestock taxes to be given to the federal bureau. During the 1923 legislative session, the state assembly passed Senate Bill No. 41, which provided that twenty-five cents of each hunting and fishing license go toward predator eradication, and Senate Bill No. 72, which repealed the bounty law and redirected all funds to the Biological Survey. With state and federal governments now splitting costs, BBS hunters could "finish up our work in the extermination of the grey wolf in this State."[94]

In October 1930, Vernon Bailey, who now held the high-ranking position of chief field naturalist for the Biological Survey, arrived in Billings to meet Inspector Bateman. For the next month, the two BBS officials planned a statewide tour covering "the trap and poison lines of four of our best men"—Walter Standish around Lodge Grass, E. B. "Burt" Warren near the Big Belt Mountains, Chance

Bebe of the Glacier National Park region, and John Van Deutzen in the Flathead Valley. On November 2, Bailey and Bateman stayed with Warren in log cabins between White Sulphur Springs and Cascade. Burt showed them around, sharing stories about his work. Bailey learned that Warren began trapping for the Biological Survey back in 1922 when ranchers grumbled about offering private bounties of $150 per adult and $25 per pup "without any reduction in the [wolf] numbers." Warren set up a trapline seventy-five miles long, made up of two hundred Newhouse No. 4s; and within fifteen months, he killed nineteen wolves. Like barbed wire for cows, this steel barrier of traps made administrative lines real for large carnivores.[95]

The government trapper kept one pup, "Billy," alive and raised it as a family pet. Warren's two sons were wrestling with the wolf when Bailey tried to pat the animal's fur, but Billy "at once made it plain that he would eat me up if I got within reach of his chain." The Biological Survey had shipped the tame canid to state fairs and sportsmen's shows to build public support for the federal agency, Bailey noted, and so Billy "doesn't enjoy being put in a crate and taken away to be viewed by curious throngs. No wonder he was cross when I tried to make friends with him."[96] Bateman recalled taking the wolf to the 1926 State Fair in Helena, where he was approached by an Alberta rancher, J. B. Ririe, who "expressed a desire of joining with the stockmen along the boundary between Canada and the United States to cooperate with the Bureau."[97] Although Bateman was excited by the inquiry, he told Ririe that it was beyond US jurisdiction to be working in western Canada without the Dominion's consent. If Bateman had said yes and both governments had approved, the Bureau of Biological Survey would have made a more lasting impression on extinction in North America.

As federal agents killed the last few wolves, some developed feelings of sympathy for the animal, or even guilt for their actions. Despite all the popular mythologizing that wolves were vicious, bloodthirsty slayers, wolfers saw evidence to the contrary. Bailey reported, "Mr. Bateman and Mr. Warren have known of old wolves being cornered in the dens with pups and in such case they will not fight and will crawl into the farthest corner to hide." Bateman felt remorse after dragging the offspring out and butchering them. Warren struggled with similar emotions when he caught an old female wolf and seven pups, one of which had an injured back and could not walk, telling Bailey that "she moved her family several times and each time carried the poor little cripple to the next place, with the others who could run and keep up with her." Warren killed the wolves, but the memory still haunted him. These interactions foreshadow a more compassionate view of wolves that would take hold in a portion of the populace during the latter half of the twentieth century.[98]

Individual anxieties, however, did not stop the institutional push for total elimination in the western United States, as demonstrated by Montana's crashing wolf population. In 1924, the Bureau of Biological Survey reported killing seventy-three wolves across the state, followed by thirty-three wolves in 1925, twenty-eight wolves in 1926, and only one wolf in 1927. Federal hunter Barney Brannin poisoned a loner wolf called Snowslide that roamed across the Highwood Mountains southeast of Great Falls. "Everybody said that I never would get the job done," Brannin remembered, but after eight months of pursuit, a bait station of laced pork fat, dipped in horse tallow, finally did Snowslide in. Two years later, only five pups were taken from a den site. In July 1930, Inspector Bateman reported to Washington, DC: "The wolf situation in this state is under control. At this time we have no record of any wolves in the state."[99] Bailey confirmed their extirpation on his autumn trip, noting, "Big wolves have been practically eradicated."[100]

After the Biological Survey arrived in 1915, it took federal agents only fifteen years to destroy what few hundred wolves remained in Montana. Across the international line in western Canada, the bounty system targeted wild canids mostly in those areas where capitalist livestock production was heavily concentrated, which meant that a healthy population of wolves persisted north of the Athabasca River. South of the watercourse, provincial politics kept the application of bounties inconsistent at best, ineffective at worst, allowing remnant pockets of wolves to survive where traps and poison were scarce. The nation-building projects of Ottawa and Washington took similar trajectories into their western hinterlands, but differences in how to use government power translated into wolves' presence in Alberta—and absence from Montana—by the mid-twentieth century.[101]

I STROLLED INTO the Yellowstone Center for Resources, an 1897 building originally used as troop barracks that was repurposed in the 1990s to centralize park science and resource management. I was trying to find Doug Smith, who first came to Yellowstone in 1994 as project biologist for the park's wolf-restoration goals and now functioned as the senior wildlife biologist. I knew I had arrived at the Wildlife and Aquatic Resources Branch when I started seeing posters and photographs in the hallway—a trumpeter swan here, a cutthroat trout there.

I poked my head into each one of the offices until Smith, whose face was seasoned with wrinkles and a handle-bar mustache, appeared from behind a computer screen. "Can you wait a minute or two?" Smith asked. "I just need to finish sending an email."

"Sure," I responded, staring around the room to pass the time. A couple of things grabbed my attention. On the wall next to me hung a vintage leghold trap by its drag chain. "Can I hold it?" I asked Smith, who nodded.

The steel instrument, showing its age from a layer of rust, was not very heavy. I pressed two springs to release the jaws, which measured from the tip of my middle finger to base of my palm. Its circular pan on a dog lever served as the trigger mechanism and read: "S. Newhouse No. 4 Oneida Community N.Y." I held *the* industrial antique for capturing wolves. Back in the 1840s, blacksmith Sewell Newhouse hand-forged metal devices that were superior to other traps at the time in catching beaver and other fur-bearing animals. Newhouse was living in a religious commune turned corporation in central New York, known as the Oneida Community, which took over his original model for large-scale factory manufacturing. Ranging in size from the No. 0 for gophers to the No. 6 for bears, Newhouse traps became standard equipment for trapping animals. By 1910, the Oneida Community Ltd. reached its commercial zenith as the largest producer of steel traps in the world, exceeding seven million traps sold in that year alone, with buyers from Argentina to Australia. Unsurprisingly, its single biggest government customer was the US Bureau of Biological Survey.[102]

In Montana, the Biological Survey distributed at least 9,100 traps and 50,700 poisoned baits during one season to "cooperative" hunters who used them to defend administrative lines and property boundaries. Miles of traplines joined fences, roads, buildings, and other types of settler infrastructure that would impede animal movements. Borders created by nation-states, national parks, or new ranches were not only ideological lines drawn on paper maps; habitat fragmentation was made tangible to animals through disciplining technologies like barbed wire and steel traps. And the extension of federal agents into Montana, and not Alberta, was significant to the extirpation of *Canis lupus* south of the international line.[103]

I came to understand the ramifications of shredding up ecosystems in my conversation about wolves with Smith. During the 1970s, the biologist explained, wolves began moving across the US-Canada border into their ancestral homelands of northwestern Montana. Instead of being killed, they were legally protected under the Endangered Species Act of 1973. Human values were changing partly because scientists recognized that predators served a role in regulating prey populations, partly because rural traditions that cast wolves as sworn enemies had been lost with more and more Americans living in cities and suburbs, and partly because there was growing appreciation that wilderness areas in North America didn't exist without the presence of native carnivores. When the US government charged a group of wildlife biologists with developing a wolf recovery plan for the Rocky Mountains, they debated whether a breeding pair of wolves from Canada could make it back to Yellowstone National Park—or other suitable habitats—on their own. The recovery team decided that a "natural" migration scenario was unlikely and thus opted for "assisted" migration.[104]

Smith was hired by Yellowstone to oversee the translocation of wolves and to get them reestablished in the park. "I was one of those sappy childhood stories," Smith recalled. Born in 1960 in Ohio, Smith grew up on romantic tales about wolves that sparked a passion for them as a teenager after reading biologist Dave Mech's 1970 classic, *The Wolf: The Ecology and Behavior of an Endangered Species*. During his final year of high school, Smith wrote to four biologists, Dr. Mech included, to see if he could work on something related to wolves for a senior capstone project. All wrote back, but Durward Allen of Purdue University invited Smith to help in captive wolf research at his institution. Back in 1957, Allen had initiated what would become the world's longest continuous study of predator-prey dynamics by examining interactions between wolves and moose on Isle Royale National Park, an island on Lake Superior where wolves moved in by crossing an ice bridge from Canada. For nine summers, Smith worked at Isle Royale while earning college degrees: a bachelor's from the University of Idaho in 1985 and a master's in wildlife biology at Michigan Tech in 1988 under Rolf Peterson, who had taken over the wolf-moose study. In 1997, during his employment at Yellowstone, Smith completed a PhD in ecology, evolution, and conservation biology at the University of Nevada, Reno.[105]

Assisted migration turned out to be a wise management decision given the scientific data on wolf dispersal that reintroduction produced. In January 1995, land managers captured thirty-three gray wolves, shot with immobilizing darts from helicopters, east of Jasper National Park in Alberta. Later that same month, they released fourteen wolves in Yellowstone and the rest in the central Idaho wilderness complex. In the following year, after receiving more Canadian transplants Smith and the recovery team worked toward the goal of reestablishing one hundred wolves in the park. With geolocating devices modernizing wildlife research in the biological sciences, Smith remarked that the Yellowstone study was—and still is—unique in that *all* wolf packs were radio-collared from the very beginning.

The textbook answer for recovery was that dispersing wolves, at least one per generation, would be able to move between the three core areas of northwestern Montana, central Idaho, and Yellowstone to prevent inbreeding. But some twenty-five years of tracking lupine movements have revealed that too many barriers still stand in the way. Park scientists have documented only one migrant per every two generations successfully reaching central Idaho, or what Smith said was "minimally acceptable" for maintaining genetic diversity, and none to northwestern Montana. "We are on an island," Smith remarked, because wolves "have a lot of corridor gaps to jump over. They can get shot on ranches in eastern Idaho or get hit by vehicles on I-15, and they don't make it through."[106]

On the wall behind where Smith was seated, a laminated topographical map of the region was attached with markings drawn all over it: Red lines outlined the park's boundaries and black lines delineated wolf packs' territories. The lines crossed each other like many interlocking Venn diagrams; only half of the ten packs, radio collars had shown, lived totally within the park. In the 1990s, conservationists in Canada and the United States took some lessons about wolf dispersal to heart, scrapped the map, and redrew it with new lines. Yellowstone to Yukon, as they called the geographic space, represented a transnational vision that humans and other animals could move beyond national borders for biological well-being once again. In other words, Y2Y was a version of ecological internationalism.

CHAPTER TWO

(De)stabilizing Technologies

In Canmore, Alberta, the gateway town for Banff National Park, I sat in the lobby headquarters of the Yellowstone to Yukon Conservation Initiative. I was waiting to meet Y2Y president Jodi Hilty, a scientist who wrote the manual on "corridor ecology," detailing for conservation practitioners the needs of living things so they can successfully move across separated habitats. The utility of linking landscapes was powerfully illustrated by a piece of technology resting on a table near me. I picked up a black nylon collar with a silver aluminum box attached and, upon examining it, noticed "Telonics" inscribed on an adhesive tag. Telonics, an engineering company with ties to the Cold War aerospace industry, was the maker of this early satellite transmitting system used to track animal movements. A bullet had obviously pierced the box for sending and receiving signals, but, as a small placard explained, not before this apparatus captured the famous journey of a lone female wolf.[1]

In the late 1980s, biologist Paul Paquet established the Central Rockies Wolf Project to study wolves that were returning to areas around Banff. During the first couple years of working for the Canadian Park Service, his team of researchers used radio telemetry, locating the collared canids by picking up signals from antenna on a vehicle or plane. If wolves left the study area, Paquet had no choice but to ignore them due to technical constraints. On a drizzly day in June 1991, however, scientists trapped a wolf that they had named Pluie (the French word for "rain") and outfitted her with one of the first generations of satellite transmitters. Even though Pluie was collared in Alberta's Kananaskis region just outside the park, she could now be followed on a continental scale. For months, Pluie stayed within the territory of a resident wolf pack in Banff until, one day, she wandered off and disappeared. When Paquet started receiving Pluie's collar signals near Browning, Montana, he could not believe it. "We thought she was on a pickup truck for a while," Paquet recalled. "She was moving so fast."[2] The next locations were from around Spokane, Washington, then Bonners Ferry, Idaho, and finally Fernie, British Columbia. Over the nine-month journey,

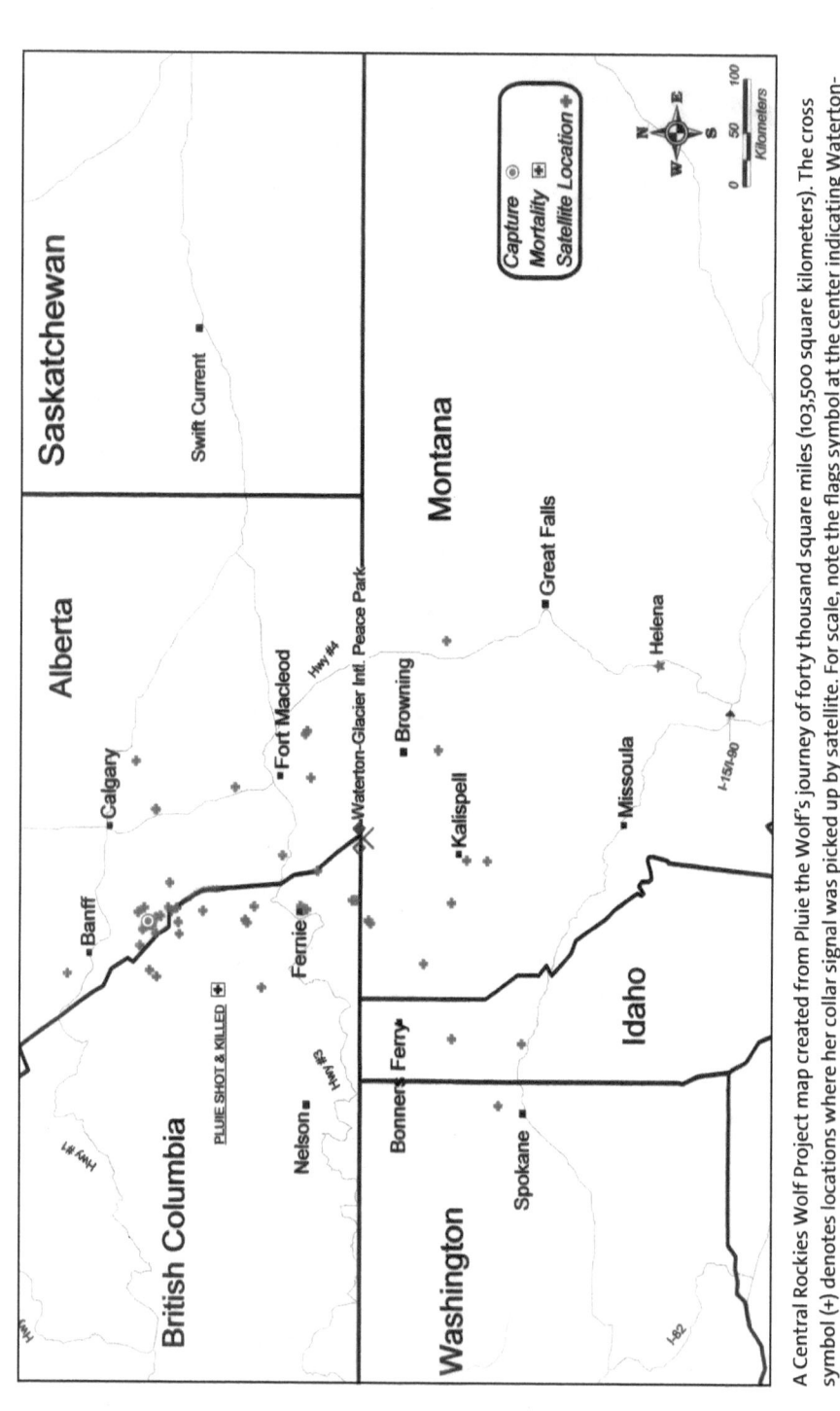

A Central Rockies Wolf Project map created from Pluie the Wolf's journey of forty thousand square miles (103,500 square kilometers). The cross symbol (+) denotes locations where her collar signal was picked up by satellite. For scale, note the flags symbol at the center indicating Waterton-Glacier International Peace Park. Courtesy of Paul C. Paquet.

Pluie traveled some forty thousand square miles, an area fifteen times larger than Banff National Park and ten times larger than Yellowstone National Park.[3]

When Hilty walked into the lobby to greet me, I asked her, with collar in hand, what lessons Pluie's journey had taught conservationists. Hilty took a deep breath and gave a rapid-fire response. First, she replied, tracking studies revealed that parks and protected areas are too small to safeguard against extinction because most preserves didn't offer mobile animals like wolves, bears, or elk enough room to roam. Second, the survival of itinerant species depends on softening of administrative borders to achieve what ecologists call "landscape permeability," or a living thing's ability to travel from one suitable habitat patch to another. Over a nine-month journey, Pluie trekked across two countries, five states and provinces, and more than thirty different political jurisdictions with differing—and at times conflicting—priorities for managing wildlife. Wandering outside of the park was dangerous, as evidenced by the bullet hole through the transmitter; Pluie was legally killed by a hunter in December 1995. And third, transnational cooperation would be vital to preserving native fauna. "At that time, conservation was happening in Canada, conservation was happening in the United States, but there was no collaboration," Hilty stated, until Pluie's "audacious movements brought people together."[4]

This was a strand of ecological internationalism. The Y2Y organization, whose mission was protecting and connecting the spine of the North American continent for the health of people and wildlife, started as an outgrowth of wolf-tracking studies that demonstrated the arbitrariness of national borders for conservation. Wolves were almost exterminated from Alberta by the mid-twentieth century during a rabies scare. After the population sank to an all-time low in the Rocky Mountains, wolves rebounded in the 1960s and 1970s as scientists grasped how these carnivores made ecosystems whole and as environmentalists transformed wolves into symbols of wilderness. When habitats along the northern Rockies were more saturated with wolves, they began moving southward toward the international line. Under a federal mandate to reestablish populations of endangered species, biologists at the University of Montana tracked wolves that were reclaiming ancestral territories south of the 49th Parallel around Glacier National Park.

Radio-collaring opened new scientific windows into the lives and movements of wolves, giving conservationists reason to question core protected areas and to think in terms of habitat corridors. Not only were these spatial technologies used to justify the creation of a transnational political identity, as demonstrated by Yellowstone to Yukon efforts, but they could also be employed to enforce administrative lines. As wolves recovered in western Canada and returned to the western United States, scientific practices were deployed for the softening and hardening of borders.[5]

Recovering North of the Line

While national parks on both sides of the international border were initially hostile places for wolves, they later became the spatial anchors of wolf conservation and regeneration. This roller coaster pattern of a rising and falling population was demonstrated at Banff National Park. When overhunting drove elk toward extinction within the preserve by the turn of the century, wolves soon vanished due to the absence of prey animals and the attacks of park wardens. Like at Rocky Mountain National Park three years earlier, wapiti were reintroduced to Banff from Yellowstone herds in 1917, but wolves took longer to come back. The outbreak of World War II redirected men and munitions from killing lupine pests to battling soldiers in Europe and Asia. As a result of human neglect, wolves returned. "If there is under way a natural immigration of wolves into Banff National Park," Superintendent J. A. Hutchison wrote, "we may as well recognise that we shall not be able to keep them out, even though we should desire to do so."[6] At the war's end, park wardens reported seeing wolves in the Bow Valley; three years later, they estimated that thirty wolves, split into four packs, resided within the park.[7]

As wolves recovered, hunting organizations advocated for their elimination from Banff based on the contention that predators kept ungulate populations at intolerably low levels. W. C. Fisher, president of the Alberta Fish and Game Association, argued that park wardens should destroy carnivores like wolves, coyotes, and mountain lions because they were guilty of causing big game animals to disappear. In response, the Canadian Parks Service commissioned zoologist Ian McTaggert Cowan of the University of British Columbia to study the distribution of wolves in national parks and to assess the influence of their food habits on prey species, particularly elk. From 1943 to 1946, Cowan surveyed five parks situated along the Canadian Rockies, including Banff, by drawing on warden observations of wolf behavior and by analyzing the composition of wolf scat. Cowan recommended that park officials refrain from "more intensive control" efforts unless ungulate populations crashed. "Dr. Cowan estimates that the average density of wolves in the park is approximately one per 100 square miles," wrote a Dominion Wildlife Service official. "This information does not suggest that the park is overrun by wolves."[8] Cowan's study epitomized a larger bureaucratic trend over which scientific understandings began to challenge popular attitudes.[9]

At Banff National Park, Warden Hubert Green, writing under the alias Tony Lascelles, penned articles to educate public audiences about a shifting predator policy. When wildlife conservation was professionalizing with university programs, Green proved the exception holding no college degree but learning on the job as a sort of naturalist-in-residence. Green, writing as Lascelles, defended trained biologists who saw population explosions of Kaibab deer, Yellowstone elk, and Isle Royale moose as mounting evidence for predator eradication campaigns gone wrong. In a 1943 issue of *Hunting and Fishing in Canada*, Lascelles charged Fisher and other sportsmen with relying on wolf myths to criticize park management: "The opinions of laymen, when such opinions invade the field of science of which biology is part, must give way to the knowledge of those better equipped to understand the machinations of the ever-turning wheel of life."[10] Green argued that wolves and other large carnivores should remain in the park because they restored the "see-saw balance of nature" between predator and prey species.

However oversimplified Green's writings about food webs may have been, they reflected scientific principles about wildlife conservation at the time. Back in 1933, Aldo Leopold published the textbook *Game Management*, a revolutionary manual that redefined what predation meant to Green and other conservation practitioners. In his youth, Leopold, who had studied forestry at Yale University, had accepted a utilitarian worldview, which posited that the only animals that were worth protecting were those useful to humankind. In 1909, he began working for the US Forest Service in New Mexico and Arizona, helping the Biological Survey to destroy wolves. Leopold had a change of heart about predators after he shot a female wolf and, upon reaching the fallen animal, saw "the fierce green fire dying in her eyes." In 1931, Leopold also learned that thousands of deer had starved to death on the Kaibab Plateau along the North Rim of the Grand Canyon because, lacking carnivores, ungulates had denuded the range. For Leopold, as detailed in *Game Management*, "biotic communities" should have both predator and prey species to keep wildlife populations within the "carrying capacity" of any given habitat. From an ecological standpoint, wolves held functional value as apex predators—preventing elk, deer, and other herbivores from getting too large in number and overgrazing.[11]

New knowledge about wolves percolated into the Canadian parks system through the *Wildlife Bulletin*, a monthly newsletter edited by Hubert Green and distributed to all wardens. Created to educate the rank-and-file in the biological sciences, the publication became a voice for institutional change among park managers. In an April 1956 issue, Green discussed how the Kaibab deer "afford[ed] a striking example of the ill effect that can result from the killing of predatory animals." The following year, he echoed Leopold's ideas about carrying capacity in a circular. "One important role that any predator plays is that of counteracting the regular animal increase that occurs in any species," Green wrote. "There must be a top limit to the total number of those plant-eating animals that any one area can carry."[12] Despite educating wardens about apex predators, Green faced an uphill battle at Banff National Park given that previous officials had permitted park staff to augment low wages by killing wolves and selling their pelts.

For all this newfound evidence, the rabies crisis tested the limits of predator tolerance, bringing wolves to the brink of total collapse. In 1952, after scientists confirmed that rabid Arctic foxes had infected sled dogs near Wood Buffalo National Park in northern Alberta, disease pools were known to be spreading among wild carnivores. With anxieties that rabid creatures would bring deadly "hydrophobia" infections to human populations, provincial authorities responded to viral threats by initiating leash laws and vaccination for dogs, as well as implementing a wildlife "depopulation" campaign.[13]

Like the repression of leftists during the Red Scare, Canadian government officials hunted down wolves during the rabies scare. Directed by chief veterinarian Dr. A. A. Ballantyne, Alberta's forestry branch hired about 170 trappers to set up barriers of leghold traps to stop the disease from diffusing southward. By 1953, they had laid out five thousand miles (eight thousand kilometers) of trapline that, if connected from end to end, would have stretched three-quarters of the width of Canada. Within two years, the provincial government reported killing some 54,000 foxes, 45,000 coyotes, 9,850 lynx, 5,000 wolves, 3,440 bears, and 64 cougars in what amounted to creating a biological desert across the middle of Alberta devoid of predators.[14]

In Banff National Park, provincial authorities instructed park wardens to shoot any wolf, coyote, or stray dog on sight due to concerns about rabies. By 1955, *Canis lupus* was basically gone from the preserve, despite autopsies later showing that no slain wolf within the park had contracted the virus.[15] Once the panic died down, Herb Snowden of the Calgary Fish and Game Association insisted that predator extermination should continue because of management problems related to wolf mobility. "It's all very well leaving them unchecked in the park," Snowden wrote. "But they can't be kept there. They roam."[16] Confronting transboundary movements was an ongoing sticking point.

ECOLOGY THE SCIENCE proved less effective at protecting wolves than ecology-based movements. Grassroots politics started in the backdrop of the Cold War era, when visceral attention to both human destruction of the environment through nuclear bombs and the interconnectedness of all life-forms through nuclear radiation could not be ignored. That was why US Fish and Wildlife Service biologist Rachel Carson, in her now-classic 1962 book, *Silent Spring*, chose to compare radioactive materials to synthetic chemicals. "Strontium 90, released through nuclear explosions into the air, comes to earth in rain or drifts down as fallout, lodges in soil, enters into the grass or corn or wheat grown there, and in time takes up its abode in the bones of a human being, there to remain until his death," Carson wrote. "Similarly, chemicals sprayed on croplands or forests or gardens lie long in soil, entering into living organisms, passing from one to another in a chain of poisoning."[17] Documenting the indiscriminate use of an insecticide known as DDT, Carson's work sold a half million copies in hardcover and was translated into at least twenty-four different languages. There was enough momentum behind popular environmentalism amassed during the weeks near April 22, 1970, the first Earth Day "teach-in," that about twenty million Americans declared their commitment to ecological issues by planting trees, cleaning up garbage, or picketing with signs outside of polluting companies.[18]

For the white middle class living in affluent first world countries like Canada and the United States, a political premium was placed on preserving the outdoors and open spaces as another set of goods to be consumed. This phenomenon can be exemplified by the formation of the National and Provincial Parks Association of Canada in 1963 (later renamed the Canadian Parks and Wilderness Society), the World Wildlife Fund Canada in 1967, and the Canadian branch of the Sierra Club in 1970. In this context, saving wolves and saving wilderness became parallel endeavors.[19]

A shifting consciousness regarding wolves was best exemplified by the reception of Farley Mowat's 1963 novel, *Never Cry Wolf*. Mowat had joined the Canadian Wildlife Service as an intern about fifteen years earlier on a project investigating wolves and caribou in northern Manitoba. Rejecting conventional wisdom that wolves must be killed to protect herds, he wandered away from the study to make his own field observations. Mowat later drew on interactions with a wolf pack at Nueltin Lake to write a fictional account about a government biologist studying wolves in the far north. As a literary plea, Mowat depicted his wolf characters, such as Angeline, George, and Uncle Albert, in sympathetic—rather than adversarial—terms: They howled with human-like feelings, feasted only on sick or aging caribou, and cared for pups as an extended family. Mowat penned,

"Inescapably, the realization was being borne in upon my preconditioned mind that the centuries-old and universally accepted human concept of wolf character was a palpable lie."[20] After the book's release, professional biologists criticized Mowat's work for presenting his fabricated imaginings as authentic experiences and for blending rational science with emotional appeals.

But his detractors did not turn the rising tide of interest. The Canadian Wildlife Service received a flood of letters from concerned citizens who had read *Never Cry Wolf* and opposed the bureaucratic assault of wolves. As an example, Lorna Roberts of Kamloops, British Columbia, asked government agents, "What is the present policy on the preservation of the wolf? I do hope it is more humane and sensible policy than one had been employed in the 1950's."[21] Like social movements that pushed for the recognition of universal human rights, some organizations began advocating for animal rights. Farley Mowat's *Never Cry Wolf* captured a major realignment of mainstream attitudes about wolves much like Rachel Carson's *Silent Spring* did about industrial chemicals.

In concert with wildlife storytelling, conserving what few wolves remained in western Canada received a boost from grassroots environmentalism. In 1968, artist Robert Guest founded Canadian Wolf Defenders dedicated, according to its by-laws, "to promote conservation of and develop appreciation for one of Canada's most unusual and magnificent wild animals, the Wolf." Guest became enamored with *Canis lupus* as a child growing up in the Peace River country and hearing them howl on cold winter nights. Later, as an adult living near Edmonton, Alberta, Guest saw wolves as biological emissaries for remote wild places that were vanishing under the ravages of industrial civilization. Two years into creating the organization, Guest had recruited 350 people to join Canadian Wolf Defenders. One member, Mrs. B. Twigg of British Columbia, joined to learn more about wildlife advocacy after reading Mowat's book. Besides countering the "old wolf image" through education, Canadian Wolf Defenders' goals were organizing fellow citizens to abolish the government bounty systems, ban the use of poisons on public lands, and outlaw the practice of hunting wolves from aircraft.[22]

The Canadian Wolf Defenders, among other environmental organizations, pressed the provincial governments to alter their predator policies. In 1971, Alberta's Department of Lands and Forests enacted a new rule limiting wolf control only to verified cases of livestock damage or public endangerment. Gordon Kerr, director of the Fish and Wildlife Division, commented that his agency "has attempted to develop a policy and wolf management program which will fit the needs and desires of the entire public while at the same time exhibit some compassion for the wolves themselves."[23] In 1975, British Columbia restricted all poison use—strychnine baits, cyanide guns, and compound 1080—to government personnel who would only apply the toxic substances when dealing with "problem wolves." Even though wolves were still subject to trapping because they were categorized as a "furbearer," BC authorities could stop harvesting in certain regions if lupine populations dropped too low. With stronger protections in Alberta and British Columbia, wolves recovered from the rabies scare and began moving along the Canadian Rockies toward the international line.[24]

In the United States, a similar groundswell of popular environmental concern led to federal legislation aimed at saving imperiled animals, wolves included. The Endangered Species Act received bipartisan support with votes of 92 to 0 in the US Senate and 390 to 12 in the House of Representatives, and President Richard Nixon signed the bill into law on December 28, 1973. Aptly called the "bill of rights for nonhumans," the Endangered Species Act required that private enterprise—mining, timber, agriculture, and other industries—prove its actions would not harmfully impact species teetering on the edge of extinction. It mandated that protecting "critical habitat" would be essential to conserving threatened biota and, wherever possible, returning those species to what remained of their former habitats. Lastly, it directed all federal agencies to cooperate with the US Fish and Wildlife Service (the bureaucratic descendant of the Biological Survey) in adhering to the law's stipulations. After the law was enacted, federal officials listed the western gray wolf as an endangered species.[25]

In sum, environmental movements in both countries laid the political groundwork for wolves to begin crossing the international line. Civil society, rather than centralized states, were responsible for softening the border. Wildlife advocates, a subset comprising mostly of white middle-class suburbanites, saw native predators as relics of wild places under siege. Robert Guest of the Canadian Wolf Defenders laid out this sentiment: "The wolf is the true symbol of Canada's wilderness—and I don't even want to be around anymore if there's no wilderness and no wildlife."[26] Environmentalists pressured government officials to stop harmful practices and to write new laws. For the most part, their actions succeeded. In 1972, Canadian biologist Douglas Pimlott observed that "within the past twenty years people in Canada and the US have begun to make a determined effort to resist the seemingly inexorable forces that were sweeping the wolf to extinction."[27] But as Cold War anxieties ignited popular ecological awareness, Cold War technologies made the shortcomings of site-specific conservation much clearer too.

Returning South of the Line

Beep... beep... beep. Maddy Jackson adjusted a knob on the handheld radio monitor, switching between frequencies to try to pick up a stronger signal from one of the five collared wolves. I sat in the seat next to Jackson, both of us listening to the faint sounds that came in from an antenna on the top of the vehicle. Outside the car with vinyl wolf decals stood Jack Rabe, peering out through a tripod-anchored spotting scope at a snowy ridgeline. From early dawn until now at midday, we were struggling to see the Junction Butte Pack in the Lamar Valley, even though we knew from signals that they were bedded down somewhere in the nearby conifer forest. Jackson and Rabe were wildlife technicians on a ground crew for Yellowstone National Park's winter field studies, and their job consisted of driving on the road between Tower Junction and Cooke City to locate and observe wolves. Jackson and Rabe used radio-tracking equipment to find the animals, a technology that had revolutionized the discipline of wildlife biology only a few decades prior.

BEEP... BEEP... BEEP. The sound intensified from the radio signal of wolf #6048 as Jackson turned toward me and cheerfully explained, "It means they're moving." Sure enough, Rabe spotted the Junction Butte Pack darting out from the tree line. To the naked eye, the wolves looked like a dozen black spots, barely distinguishable from rock outcroppings, moving across the white backdrop. With the help of binoculars, I could see that pack members were encircling a bison cow and calf that had drifted away from the herd. For hours, we watched the wolves try to separate them, one charging at the adult buffalo while another swiped at her offspring, but their persistence ended with no payback. At dusk, the Junction Butte Pack stopped hunting. With little visibility left in the day, we packed up and drove away.

Beep... beep... beep. Diane Boyd scanned the landscape from the sky, riding in a single-propeller airplane with antenna mounted on its fixed wings. Boyd was soaring over the North Fork of the Flathead River, a remote drainage running along the western edge of Glacier National Park, to find a lone wolf dubbed Kishinena. In April 1979, a trapper had captured and collared the silver-colored female (her name was derived from nearby Kishinena Creek) seven miles north of the Canadian line. Five months after the radio-tagging event, Boyd started following Kishinena. During June 1980, the wildlife biologist reported flying over "the Flathead headwaters & couldn't find her there. On the way back to Moose City at >9000 feet I picked up her signal & found her along the ridge between Elder and Kish[i]nena Creeks."[28] On the next flight, Boyd confirmed that the wandering wolf had slipped into the United States. After plotting all these locations on a map, Boyd showed that Kishinena was a transborder animal. Wolves, simply put, were returning to Montana.[29]

Boyd's scientific investigation began under the direction of Robert Ream, a cosmopolitan professor who established the Wolf Ecology Project (WEP) at the University of Montana. Born in 1936, Ream spent his teenage years living in Thailand and the Philippines because his father worked as a scientist for the US Agency for International Development. Ream attended boarding school in India, where he became enamored with studying the natural world after collecting a variety of ferns for a biology class. Returning to his native Wisconsin, Ream graduated with a PhD in ecology in 1963. Ream joined a team of US Forest Service ecologists to conduct a botanical survey of the Boundary Waters Canoe Area Wilderness in northern Minnesota.[30]

There, Ream met renowned wolf expert L. David Mech, who was studying the only remaining population of gray wolves in the Lower 48. In November 1968, he and Mech worked together to trap and outfit a wolf with a radio collar, marking a significant moment in *Canis lupus* research as the first instance of deploying this technology on the species. Akin to Ream's own migrant childhood, the collaring of four more wolves exposed their border-crossing movements from Ontario's Quetico Provincial Park into Boundary Waters. The following spring, Bob Ream left the job to accept a faculty position at the University of Montana's School of Forestry and Conservation.[31]

While teaching in Missoula, Ream heard from colleagues of solitary wolves showing up across the state. Although Ream believed the reports were probably sightings of dogs or coyotes falsely mistaken as wolves, his curiosity piqued when museum staff showed him two wolf skulls that had been collected in the 1960s from Glacier National Park and near Lincoln, Montana. In 1973, the same year that the Endangered Species Act gave legal protections to wolf populations, Ream created the Wolf Ecology Project to research some basic questions for the federally mandated recovery efforts: Did wolves, especially breeding pairs, exist in Montana? If so, how many were there? And where were they coming from?[32]

Ream enlisted graduate students to conduct fieldwork for answers. In 1979, he hired Diane Boyd, who over her long career would earn a reputation as the "Jane Goodall of wolves." Raised in the Twin Cities area of Minnesota, Boyd fell in love with the outdoors as a young suburbanite, playing in a marshy area near her home that she affectionately called "the Swamp" and hunting with her father. Boyd attended the University of Minnesota as an undergraduate, majoring in preveterinary medicine, until she met biology professor David Mech and begged him for fieldwork. "I was like a good parasite—persist, persist, persist," Boyd recalled. Mech gave her an internship in 1977 at the International Wolf Center in Eley, Minnesota, and Boyd's interactions with wolves convinced her to change majors to wildlife management.[33]

After graduating, Boyd worked in Alaska before accepting a trapping job for the US Fish and Wildlife Service back in her home state of Minnesota. During summer 1979, the twenty-four-year-old woman adapted to the gendered expectations of working in a farming community to capture wolves that were depredating livestock. "The game warden would call me in, and the farmer would say, 'Well, where's the trapper?' Because all he saw was this blonde babe," Boyd recollected. "They always went slack-jawed. It pissed me off." Boyd, who had gained trapping experience in a male-dominated field, arrived at the University of Montana in fall 1979 to begin graduate school and work for Ream's Wolf Ecology Project.[34]

Boyd had no sooner reached Missoula than she left for the international line. She and a fellow graduate student, Mike Fairchild, headed to the North Fork of the Flathead River, where they would be stationed at Moose City, a former homestead settlement located a quarter mile from the US-Canada border. Sparsely populated, the collection of cabins lacked modern amenities—no electricity, no telephone, and no running water. From her rudimentary accommodations, Boyd passed by a customs house into British Columbia to monitor wolf #114, better known as Kishinena, from a safe distance that would not disrupt her natural behavior. Boyd spent months tracking the shy animal's radio-collar signal, on foot or in her pickup truck during the summer, skiing or snowmobiling during the winter, or flying in airplanes. For her master's thesis, Boyd documented Kishinena's food habits, interactions with coyotes, and most importantly, home range. Monitoring revealed that Kishinena occupied thirty miles of the North Fork drainage in two different countries.[35]

Two years' worth of fieldwork verified that although no breeding pairs lived in the state of Montana, a consistent number of dispersing wolves were entering from Alberta and British Columbia. Biologists have observed that the social structure of a wolf pack consists of a single male and female that mate, while the rest of the animals—usually their offspring—assume a subordinate role and do not reproduce. For wolves, "dispersal" can be described as exploratory movements that go far beyond their place of birth, which are typically juvenile pack members that strike out on their own to find a mate, an exclusive area, and food resources. If two dispersers pair up and habitat conditions are favorable, they can settle down to form a new pack and thus repopulate a territory. Dispersal maintains the evolutionary health of populations as individual wolves reproduce with others from distant groups, preventing inbred animals that are susceptible to reduced fertility, genetic disorders, and immunity problems.[36]

With knowledge about dispersal, graduate student Ursula Mattson proposed that wolves must be following a wilderness corridor somewhere between Banff National Park and Glacier National Park. The WEP study extended farther into southern Canada seeking people who could help identify these migration routes. While Mattson wrote letters of inquiry to Canadian trappers, Ream recruited Canadian resource managers to create the Border Wolf Technical Committee. At the first meeting, members supported developing a unified management plan across the international border, which, Ream reported, would "help [them] achieve [their] U.S. goal of recovery by protecting a narrow corridor along the continental divide so that dispersing wolves might have an easier chance of making it into Montana."[37] In brief, corridors facilitated wolf dispersal, and dispersal facilitated wolf recovery.

Biologists realized that large carnivores like wolves needed landscape connectivity for repopulating areas. On the US side, the Endangered Species Act stated that conserving "critical habitat," even places not currently occupied by near-extinct biota, was a tool for rehabilitating listed species. Bob Ream used this legal framework when persuading elected officials to vote for a new wilderness designation under the 1964 Wilderness Act. After a five-day backpacking trip to the proposed Great Bear Wilderness, Ream penned Representative Max Baucus in September 1978, explaining how the undeveloped land "provide[d] a vital link between the Bob Marshall [Wilderness] and Glacier Park and Canadian Wildlands to the north." Ream also emphasized to the Democratic congressman that because "wolves are very far-ranging animals," only a few canids dispersing into Montana would offer "the best opportunity anywhere for recovering the Northern Rocky Mountain Wolf to viable population levels."[38] With Baucus's support, Congress approved the Great Bear Wilderness that same year. On the Canadian side, with the Border Wolf Technical Committee's advocacy, the British Columbia government placed a moratorium on hunting and trapping wolves in 1979 for the southeastern portion of the province. Assembling a contiguous habitat across the spine of the continent acted as a biotic conveyor belt for wolves, bears, and other animals.[39]

Facilitating wolf movements was needed so that Kishinena could find a mate at a crucial juncture in the Wolf Ecology Project. Sixteen months into using the geolocating device, Boyd stopped receiving signals from Kishinena's collar because the transmitter failed. When the only radio-tagged animal in the study disappeared by the winter 1980–81, WEP funding dried up. Boyd could have returned to Missoula's campus, but she was too committed to the project and decided to stay with no pay. Boyd earned money chopping wood, working at a fire lookout, selling her wildlife paintings, and serving as caretaker of Moose City. "I lived on fresh air," she later joked. Boyd received descriptions of a lone silver-furred wolf that fit Kishinena's profile, even if she could never get visual confirmation. In fall 1981, however, Boyd discovered another set of tracks with a wolf's three-toed paw; in addition, a bear biologist reported seeing another wolf. A year later, the two wolves mated and denned just a rifle's shot from the US-Canada border. The Magic Pack, as their litter came to be known, formed the nucleus of wolf recovery in northwestern Montana.[40]

The Magic Pack received its namesake because tracking the young canids proved quite difficult in the absence of geolocating technologies. "When [WEP] biologists followed these wolves on snowshoes and skis without the help of radio telemetry," Ursula Mattson explained, "the pack would magically show up in an area and just as magically disappear."[41] After a bear snare hooked and killed the alpha male, Kishinena was left to raise the seven pups on her own. But she too went missing. Boyd and Fairchild set up their traplines, furnishing them with modified jaws to minimize the chances of injury, and they hoped to catch the offspring. In October 1984, Kishinena's offspring Sage (wolf #8401) was captured and outfitted with a radio collar. Boyd learned that Sage—a disperser from the Magic Pack—had covered some two thousand square miles during the first year, which at the time was the largest known home range of any wolf in North America.[42]

In May 1985, Phyllis (wolf #8850) was accidentally caught in a bear snare. During collaring, Boyd found out that Phyllis had become the alpha female since the wolf was lactating. Phyllis and the twelve other wolves that made up the Magic Pack could now be monitored together. In October, Boyd was alerted by circling ravens at a site in British Columbia of a decapitated wolf carcass. Although harvesting wolves was illegal in this part of the province, a big game hunter probably killed the animal. Reacting to hunting pressure, the Magic Pack extended its territory southward a month later into Glacier National Park. There, Phyllis gave birth the following spring to more whelps, marking the first documented wolf den in the western United States in more than fifty years.[43]

While the Magic Pack splintered as juvenile members left, park officials authorized the Wolf Ecology Project in 1986 to trap and collar wolves within Glacier National Park. As part of what would eventually become a PhD in biology at the University of Montana, Boyd outfitted three more wolves with

Biologist Diane Boyd using a jabstick to drug a trapped wolf with sedatives, circa 1980s. The wolf was radio-collared on the Flathead drainage, a few miles north of Glacier National Park in British Columbia. Courtesy of Diane K. Boyd.

radio-tags—including wolf #8551, or Kay—for researching this recolonizing population. The young wolf lingered in the northwest corner of the park until she left the study area. Boyd lost her signal. Six months later, in July 1987, Boyd was astonished to recover the collar after the disperser turned up dead near Dawson Creek, British Columbia, two-thirds of the way to the Yukon Territory. The rancher, who legally killed the wolf, observed that she had assimilated into a new wolf pack that straddled the Alberta-British Columbia border along the Peace River.[44] Ream was surprised, telling the Glacier superintendent, "The dramatic movement of W8551 nearly 600 miles to the north is very interesting scientifically, but also has major recovery implications." To put this distance into perspective, Ream underscored, "If she had gone south instead, she would be well past Yellowstone NP and nearly to Rocky Mountain NP."[45] Like Kay, the vast movements of dispersers helped scientists to think beyond national and jurisdictional borders.

For this reason, Ream, Boyd, and other WEP personnel took a minority stance in debates over US wolf recovery. Back in 1975, the Department of the Interior appointed the Northern Rocky Mountain Wolf Recovery Team to create a plan

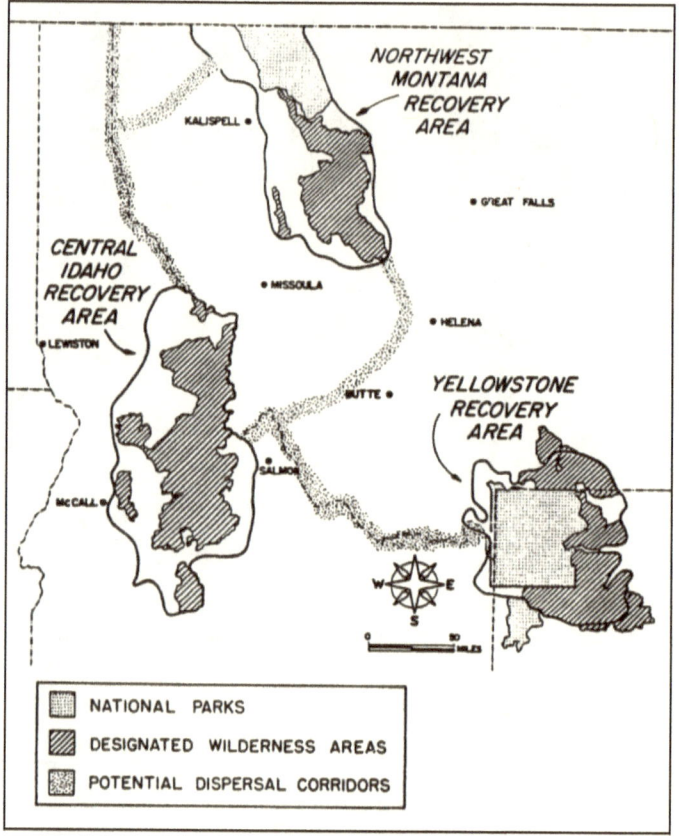

The Northern Rocky Mountain Wolf Recovery Team, which consisted of biologist Robert Ream and ten other members, developed a map of the three recovery areas and the potential dispersal corridors. US Fish and Wildlife Service, *Northern Rocky Mountain Wolf Recovery Plan* (US Fish and Wildlife Service, 1987).

for restoring the species to its former habitat and to coordinate among various state and federal agencies in its implementation. Composed of one rancher, one environmentalist, and nine biologists, including Ream, team members agreed to the goal of reestablishing ten breeding pairs of wolves to three core zones: northwestern Montana, central Idaho, and the Greater Yellowstone area. Most participants also reasoned that due to the geographic isolation of the latter two places from "seed" populations in Canada, human-assisted translocation would be the only viable recovery option. However, Ream opposed reintroduction because he believed that "natural recolonization" through dispersal would repopulate those areas and result in less backlash against federal agents.[46] Ursula Mattson of the Wolf Ecology Project remarked, "Politically there is a big difference between wolves being trucked across the Canadian/U.S. border or walking across on their own power even though the end result is biologically the same."[47] Despite identifying a legitimate concern about local resentment, other biologists won the argument stating that Ream had overestimated the potency of wolf dispersal. The Northern Rocky Mountain Wolf Recovery Team published a preliminary draft in 1980, followed by a final plan in 1987, recommending the majority position of reintroduction.

In contrast with wolves that quietly moved to the western side of Glacier, those on the eastern side of the park caused an uproar. In summer 1987, a wolf pack took down cows as their prey on the Blackfeet Indian Reservation near Browning, Montana. The depredations were reported to the US Fish and Wildlife Service's Animal Damage Control program, which deployed government trappers to collar and track the wolves. When a white rancher named Dan Greer saw wolves attacking his sheep on leased tribal lands, federal agents swooped in and euthanized the wolves. While Greer felt relief that the carnivores were dead, livestock owners were still dissatisfied. These first losses from wolves in fifty years ignited criticism about establishing different management zones for wolf recovery.[48] "They don't understand boundaries, they don't understand zones," stated Stuart Doggett of the Montana Stockgrowers Association. "You may talk about zone three, but it's not as if you can tell Mr. Wolf, 'Here's your boundary, please don't go any farther, we have cattle here.'"[49] Because parks and protected areas could never possibly contain dispersing wolves, human tolerance outside of those legal boundaries would determine their future.

WOLF RECOVERY challenged the administrative lines of the public-private interface, renewing calls for lupine exclusion from the livestock industry and hunting organizations. Sheep rancher Joe Helle from Dillon, Montana, a chairman of the National Woolgrowers Association, declared, "The livestock industries of Wyoming, Montana, and Idaho are totally against it [reintroduction]. And the reasons we are against it is because we realize that the wolves won't stay in the park."[50] Big game outfitter Jim Zumbo, who served as editor of *Outdoor Life* magazine, made a similar case. "One of the problems with introducing wolves into the park is the basic fact that there are no fences around Yellowstone," Zumbo commented. "And the wolves could roam at will to other areas around the park."[51] Overcoming these objections to wolf reintroduction became a monumental task for conservationists.

One significant method to curb dissent came from utilizing a revision of the Endangered Species Act. Under a 1982 amendment, Section 10(j) allowed for what was called "nonessential-experimental populations," permitting captured specimens of a listed species to be reintroduced to former habitats and subject to control activity so long as their loss did not imperil the species' continued existence. Simply put, the federal government lifted legal protections on any wolf that routinely preyed on livestock. Biologist John Weaver with the Northern Rocky Mountain Wolf Recovery Team explained, "If a wolf wandered out of Yellowstone and into an area that we recognize cannot realistically support wolves, [ESA] provisions allow federal or state agents to either trap and relocate the animal, or kill it."[52] Between 1987 and 2004, they killed nearly three hundred wolves in the tristate recovery area to resolve conflicts with livestock, including 166 wolves in the state of Montana. Nonessential status gave resource managers more flexibility in the control measures taken for wolves habituated to killing cows and sheep.[53]

Still, passions flared when receiving input for the environmental impact statement. At a formal hearing in Helena, Montana, in August 1992, citizens from near and far gave testimonies for and against returning wolves to the park. Jack Gladstone of the Blackfeet Nation read aloud a tribal council resolution endorsing restoration because his people viewed the wolf "as a sacred medicine animal worthy of respect." Rancher Dave Witt of Jordan saw the project as a veiled attack on resource westerners with out-of-state environmentalists chipping away at "our private property rights until we have no more left." Catherine Bushway of Potomac believed fears were overblown, commenting, "There will be no immediate hordes of wolves leaving Greater Yellowstone to prey on domestic cattle. There is sufficient space and prey to keep them in the ecosystem." Hunting guide Chad Shearer of Great Falls joked, "What is going to happen when the elk and other wildlife start to migrate out of the Park? Are the wolves going to stay in the Park and feed on trees and shrubs?" A telling statistic was that the environmental impact statement collected the most written comments for any federal project to date: one hundred thousand for reintroduction and sixty thousand against it. In the end, a two-to-one ratio of respondents supported wolf recovery.[54]

Another important tactic to ease criticism was creating a wolf compensation fund for ranchers. After the Browning fiasco, conservationist Hank Fischer remembered asking himself: "Did stockmen instinctively despise wolves, or did economics foster the hatred?"[55] Settling on the latter position, Fischer persuaded the organization he worked for, Defenders of Wildlife, to develop a private account that would pay fair market value for any livestock killed by wolves. In what Fischer termed "supply-side environmentalism," adapted from President Ronald Reagan's "supply-side economics" phrase about federal tax cuts, the new fund shifted the financial burden of depredations from ranchers to members of the public. By 1992, Defenders of Wildlife raised $100,000 to establish a permanent fund. Defenders of Wildlife paid more than $15,000 to ranchers between 1987 and 1995 on twenty-two verified cases of livestock losses in northwestern Montana. For the most part, payments put many small livestock producers at ease.[56]

The final environmental impact statement, written by US Fish and Wildlife biologist Ed Bangs, recommended that if a viable population of wolves couldn't reach central Idaho and Yellowstone on their own by October 1994, they should be reintroduced to those areas. When no breeding pairs could be documented, federal agents started an assisted migration program, called Operation Wolfstock, to transplant wolves from Canada to the United States. The American Farm Bureau Federation and Mountain States Legal Foundation sued to halt any reintroduction, but federal courts struck down the legal injunction for a lack of evidence that ranchers surrounding the park would be irreparably harmed by the return of wolves. By January 1995, US Fish and Wildlife Service personnel captured and collared thirty-three wolves east of Alberta's Jasper National Park, notable as the point of origin where Kay had dispersed to, for transport to Yellowstone and to Idaho. After spending about a month in acclimation pens, on March 21, 1995, the first group of wolves were released into their former ecosystem. A year later, another group of thirty were reintroduced. The apex predator was back.[57]

Wolf populations started the long recovery process in national parks along the Rocky Mountains. By 2000, there were 117 wolves in Yellowstone, 64 in Glacier, 40 in Banff, and 80 in Jasper. But a handful of scientists realized that without protective measures on a larger spatial scale, restoring native predators to small pockets of habitat would be meaningless in the bigger picture of mass extinction.[58]

CONSERVATION BIOLOGY, which formed as a new scientific field in the 1980s, redefined wolf recovery in the US and Canadian Rockies. In 1985, renowned biologist Michael Soulé wrote an article titled "What Is Conservation Biology?" calling on scientists in the "crisis discipline" to urgently develop applied knowledge to maintain a diversity of organisms. A year later, Soulé and other scientists cofounded the Society for Conservation Biology; and the professional organization swelled to over five thousand members in its first half decade. Due to the inadequacies of site- and species-specific conservation, Soulé also joined a team of scientists, activists, and capitalists in 1991 to form the Wildlands Project (now Wildlands Network), which converged around the idea of landscape-size "rewilding." Early members included Dr. Reed Noss, a panther specialist; Harvey Locke, a lawyer and president of the Canadian Parks and Wilderness Society; and Doug Tompkins, owner of the North Face outdoor clothing company who bankrolled the new endeavor with his corporate profits. Noss and Soulé distilled the rewilding concept down to the Three Cs: cores, corridors, and carnivores.[59]

The radio- to satellite-tracking of wolves brought rewilding from experiment to action. "Cores" were theorized as blocks of habitat minimally impacted by extractive activities, such as national parks or wilderness areas, and these protected tracts formed the spatial anchors of conservation. To keep them healthy, "corridors" were envisioned as multiple-use zones where special management practices allowed species to successfully move from one core area to the next. "Carnivores" were important to rewilding because they served as "umbrella species," a term used to indicate that conservation measures protecting a charismatic species like wolves would indirectly protect other species. Applying these ideas on the ground relied on GPS, or global positioning system, using multiple satellites deployed during the Cold War to pinpoint the latitude and longitude of signals coming from a collared species. By the late 1980s, the space agencies in France and the United States had equipped Earth-orbiting machines with new observation tools, simplified by the acronym ARGOS, to track environmental data from near space. Tagging wolves, in this case, allowed biologists like Paul Paquet and Diane Boyd to develop a carnivore's-eye view of landscape connectivity.[60]

Paquet thought about ethics as much as biology when it came to wolves. Maybe, he later suggested, this tendency to ask deep questions came from his Catholic education. Born in 1948, Paquet attended Jesuit High School near Portland, Oregon, and majored in moral philosophy at Santa Clara University in California. Growing up on fictionalized accounts of wolves like Jack London's book *White Fang*, Paquet's first in-person encounter with a gray wolf came during a trip to see his relatives in Italy's Apennine Mountains. This subspecies was nearly extinct on the European peninsula, causing Paquet to wonder why the carnivore was persecuted all over the world. In 1971, Paquet started a master's project on coyote biology at Arizona State University, where he learned about radio telemetry from fellow student Stan Tomkiewicz, an engineer who later founded the wildlife-tracking company Telonics. Paquet resisted the institutional pressure to adopt the new technology when pursuing a PhD in zoology at the University of Alberta, using his dissertation to demonstrate that "natural history" methods were just as effective for studying wolves and coyotes in Canada's Riding Mountain National Park. After finishing up in 1988, Paquet worked as a biologist for the Canadian Wildlife Service and World Wildlife Fund Canada.[61]

Upon graduation, Paquet initiated the Central Rockies Wolf Project to help the Canadian Park Service understand the ecological implications of a growing wolf population around Banff National Park. His assumption was that most wolves had come from the north in Jasper National Park because the two preserves shared a common border to form one contiguous habitat, or core area. Paquet's research team, which included Boyd, reluctantly outfitted these carnivores with very high frequency (VHF) collars, but tracking lupine movements was arduous given the limited signal range and the expense of flyovers. "We started looking at what other technologies we could use to monitor," Paquet recalled. In 1989, he contacted Stan Tomkiewicz at Telonics, who recommended trying a GPS-tracking device stocked with ARGOS.[62]

After securing a grant from the Alberta government, Paquet purchased three satellite-transmitting collars at $3,000 apiece (conventional radio collars were $250 each) for the wolf study. Paquet remembered encountering issues of spatial inaccuracy because the US military was distorting signals up to three miles for "national security reasons." However, he got around the problem by triangulating the satellite position with multiple radio positions. With expanded coverage, Paquet found out his hypothesis was wrong: Just as many wolves had been arriving from a different core area to the south. As recounted at the beginning of this chapter, Paquet's team collared Pluie, a disperser wolf that basically wandered away from Banff to make a complete loop around Waterton-Glacier International Peace Park.[63]

Boyd's doctoral study confirmed a shared population. Between 1985 and 1997, Boyd recorded that over half of the fifty-eight tagged wolves left Glacier National Park for some other distant area. "Locating wolves along their dispersal routes with our VHF radio collars was extremely difficult because of the rapidity and long distance that wolves moved," she wrote. Boyd gave the animals' radio frequencies to other field biologists working in the Rocky Mountains so that signals could hopefully be picked up in other places. Boyd recalled that collaboration with Paquet was mutually beneficial: "We shared data because, heck, we shared wolves that crossed an artificial boundary."[64] Some wolves moved south into the United States. One adult female, collared as F470 near Banff National

Park, dispersed three hundred miles to live near Deer Lodge, Montana. But most of them moved north into Canada where the likelihood of running into a mate was higher. Two Glacier juveniles, male #8703 and female #8857, trekked over one hundred miles to join a wolf pack around the Highwood River in Alberta. The regularity of transborder activities convinced Boyd that there was no such thing as a "foreign" wolf and, as followed, that conservationists should be thinking in terms of "landscape linkages."[65]

The case was taken up by Harvey Locke. The Calgary-based attorney, who had recently lost an election bid for the Alberta Legislative Assembly, attended two conferences that set the stage for the Yellowstone to Yukon idea. In 1993, the WWF Canada sponsored a meeting in Banff on large carnivore conservation. Wolf specialists Boyd and Paquet gave a joint presentation on the implications of their recent wildlife-tracking studies. While their specific words were lost, Locke and his fellow participants probably heard something like this: "The frequency and distance of these movements strongly suggests that we are dealing with a single wolf population between Jasper and Yellowstone national parks that spans four states and two provinces," Boyd asserted in the mid-1990s. "Our new understanding of the vast range of this wolf population and the high rate of human-caused mortality has convinced me that it is time to undertake an *international ecosystem approach*."[66]

That same year, Reed Noss of the Wildlands Project hosted a "vision mapping" workshop at the Kananaskis Field Station in Alberta where participants charted out what a system of connected preserves would look like. One mapped area, Locke and others discussed, ran up the mountainous spine of the continent "from the Yucatán to the Yukon." Locke preferred the phrase "Yellowstone to Yukon" because it linked two powerful historical cachets: Yellowstone, celebrated as the world's first national park by Americans, and the Yukon, seen as a primeval wilderness by Canadians. Whatever the two anchor points, they were making a biological case for large landscape conservation.[67]

In between the two conferences, Locke went on two backpacking trips in the mountains located northwest of Jasper National Park: one in Alberta's Willmore Wilderness Park and another in British Columbia's Muskwa-Kechika Range. Locke recalled feeling disoriented by his new environment because "a lot of people thought, and probably still do, that the Rockies kind of ended in Jasper."[68] Sitting with outfitter Wayne Sawchuck around a campfire, Locke wrote an essay on the fringes of a map laying out why the Yellowstone to Yukon vison was needed.

"The first step," Locke asserted, "is to ignore where the political boundaries have been drawn. They are meaningless to the biota." After invoking Boyd and Paquet's biological studies, Locke cited how "one of the wolves Boyd had collared in the Montana Flathead had travelled all the way up the Canadian Rockies to Dawson Creek, BC—Mile 0 on the Alaska Highway.... The implications were clear—the Canadian Rockies are one gigantic linear ecosystem." Not only did all North America's major river systems—the Missouri, Columbia, Saskatchewan, Fraser, Peace, Athabasca, and Mackenzie—originate in the Rockies, Locke remarked, but it was the only place across the whole continent where the entire collection of native carnivores remained intact. For Locke, this large landscape formed "the natural heart of Western North America."[69] New scientific practices inspired Locke's argument for ecological internationalism.

"How do you begin to create a framework to manage so vast an area?" Locke asked. The answer came from "the Wildlands Project and Dr. Reed Noss's concept of matrix, core, corridor, interconn[ected] to [form the] Yellowstone to Yukon Biodiversity Strategy." Locke envisioned the four major core areas going south-to-north as the Greater Yellowstone Ecosystem, the Crown of the Continent Ecosystem (Waterton-Glacier International Peace Park and the Bob Marshall Wilderness Area), the Rocky Mountain Parks complex (Banff, Jasper, Kootenay, and Yoho national parks), and finally where he was backpacking in British Columbia's northern Rockies. The challenge, Locke offered, was to ensure these cores were connected.[70]

Connecting in Between

For most of her life, Jodi Hilty's scientific work has been about corridors. If isolated pieces of habitat led species down the dangerous path of extinction, the objective becomes, as Hilty put it, "How do you connect the places where wildlife live so that populations can be sustained over the long term?" Hilty spent quite a bit of time pondering that question prior to stepping into her leadership role at the Yellowstone to Yukon Conservation Initiative in 2016. For a PhD in conservation biology at the University of California, Berkeley, in the late 1990s, Hilty turned her attention to the creeks of wine-growing regions in Sonoma County. Hilty measured the width of riparian vegetation near stream beds—then classified them as denuded, narrow, and wide—to determine the corridor quality for a suite of carnivores moving through vineyards. Using motion-triggered cameras, Hilty documented that species such as mountain lions, gray foxes, coyotes, and bobcats were eleven times more likely to travel in dense sprawling cover than in trimmed grape vines. The results shifted local conservationists' focus from the "cores" of mixed oak forest to riparian "corridors" in agricultural lands.[71]

Later as executive director of the Wildlife Conservation Society, Hilty's team researched pronghorn antelope movements in Wyoming. Scientists recommended protection of a millennia-old migration between Grand Teton National Park and the Green River Basin, leading to the first US federally designated wildlife corridor, aptly called the Path of the Pronghorn, in 2008. Her experiences reinforced a long-standing Y2Y goal: Corridors can deliver a proactive antidote to extinction.[72]

At the end of our meeting in Canmore, Alberta, I asked Hilty if she could point me in the direction of a project related to Y2Y's efforts. She told me to look no further than catching a ride on public transportation to Lake Louise. After boarding a shuttle bus and heading up the highway toward Banff National Park, I was pleased not to be in the driver's seat. Cars and trucks zoomed past us on the Trans-Canada Highway, a wide four-lane road channeling millions of passengers each year through the mountainous valley. Tucked in-between two walls of jagged, snow-capped peaks, the highway shares a narrow path with the Bow River. A dozen or so miles after passing by the Banff townsite, I gazed out my window at what looked like an exit ramp. A gray cement bridge framed two semicircle tunnels for automobiles to pass through. Everything looked like a typical highway feature, except for pine trees growing on it. I looked again, overhearing a fellow bus rider say it was the Red Earth Wildlife Overpass. I soon noticed large metal culverts buried beneath the roadway that offered more wildlife-crossing structures. Animals were passing over our roof and under our wheels at that very moment. This was a concrete step, literally, toward improving landscape connectivity.

The institutional origins of the Y2Y Initiative arrived in 1993 when a group of scientists and conservationists met to formally establish a binational partnership. Y2Y's mission was to ensure the survival of biological richness, as well as life-supporting processes, along the Rocky Mountains by linking protected areas through habitat corridors. Founding members spoke to the power of developing connections among conservation groups in Canada and the United States—forming transnational political solidarity. Wendy Francis of the Canadian Parks and Wilderness Society observed, "We weren't even talking to each other across the 49th parallel before Y2Y, and now we sit down and meet two or three times a year."[73] Gary Tabor, a US-based veterinarian, remarked that "Y2Y helps North Americans to view their common landscape not as several countries but as one continent."[74] The attendees of early meetings—conservation activists drawn mostly from Alberta, British Columbia, Idaho, and Montana—labeled themselves as a loose "network" rather than a distinct "organization" because they wanted to bring their disparate work together instead of supplant it.[75]

For that reason, Y2Y members reframed local actions with continental significance. At the second gathering, they agreed to develop a Yellowstone to Yukon "atlas" that would serve as the shared vision of the large landscape they were trying to protect. In October 1997, they also hosted a conference called Connections at Waterton Lakes National Park, workshopping the proposed maps and texts with 350 attendees. The whole process, Harvey Locke commented, was dedicated to "how the landscape lives, not how we draw lines."[76] Louisa Willcox, field director for the Greater Yellowstone Coalition in Montana, led the charge finishing up the atlas project. Willcox was motivated to participate after listening to a presentation a decade ago on the "island effects" of site-specific conservation given by prominent biologists Thomas Lovejoy and Jared Diamond, who told conservationists like herself that "they weren't thinking big enough" with the Greater Yellowstone Ecosystem. Their words resulted in a paradigm shift. "This was as big as it gets," Willcox remembered. "Nobody wanted to hear what Diamond and Lovejoy had to say, including me." She concluded, "If we're going to take care of animals such as grizzly bears and wolves, then we've got to think broader than the scraps of habitat we've got left."[77] In other words, Willcox's take on island biogeography translated to working toward large landscape conservation.

In April 1998, Willcox and other Y2Y advocates published *A Sense of Place*, the atlas assembled to define an "ecoregion," a geographic unit not commensurate with human-made polities. Delineated by biological communities on similar terrain, the landscape extended along the Rocky Mountains two thousand miles long north-to-south and two hundred miles wide east-to-west. Willcox noted a stark contrast in habitat fragmentation spatially. "Whereas the northern, Canadian area consisted of wild areas punctuated with islands of development," Willcox told an interviewer, "the southern portion—mostly in the United States—was an 'inverse template' of islands of wildlands in the midst of a 'sea of development.'"[78] The text reiterated that parks and protected areas alone were not big enough habitats to maintain viable populations of wolves and other carnivores. National Park Service and Parks Canada officials signed a memorandum of understanding a month later recognizing the "Yellowstone to

One of the first maps of the Yellowstone to Yukon ecoregion, created in 1996, to define the large landscape. Note the high level of habitat fragmentation as represented in "major roads" in the southern half of the region and how small the core areas in "national parks" and "wilderness areas" were. Courtesy of the Whyte Museum of the Canadian Rockies Archives.

Yukon Corridor" as a high priority area and thus affirming a mutual interest in "national parks close or contiguous with the border for the purpose of conserving shared ecosystems."[79]

One justification for Y2Y's work was a shift in international economics. As president of the Alberta Liberal Party during the 1990s, Locke lived the contradictions of capitalism. At one of the early Y2Y meetings, Locke told members, "We have moved on a socio-economic plane toward free trade. And we need to be talking about free trade in large carnivores. And we need to be defending those sorts of things with the same kind of vigor that the business interests defend free trade with."[80] Just as the North American Free Trade Agreement—signed by Canada, Mexico, and US governments in 1993—further opened the three countries to flows of corporate commerce, Locke argued that mobile animals like wolves and grizzlies should be afforded the same top-level privileges. "Paradoxically," Locke observed in the 2000s, "the United States and Canada, each other's largest trading partners, have organized themselves to ensure the free flow of investment capital, goods and energy across the border, but have not thought of the free movement of wildlife in the same way."[81] Organizing for a more connected landscape paralleled the capitalist push to structure a more connected economy.

This paradox came out as the Yellowstone to Yukon Initiative struggled with a bottom-up versus top-down organizational structure. To avoid a "command-and-control" hierarchy, they set up a democratic architecture radiating outward from the coordinating committee to council. Although Bart Robinson, founding editor of the naturalist magazine *Equinox*, was hired to manage day-to-day operations, the real power came from the 120 council members who adopted consensus decision-making. They agreed that Y2Y would not accept any funding from the "primary extractive" industries, such as Weyerhaeuser Timber, and instead seek out a few charitable organizations. The two major contributors were the Henry P. Kendall Foundation, a Massachusetts-based trust that derived its wealth from textile manufacturing, and the Wilburforce Foundation, which was founded by former Microsoft computer engineers. In 1998, Y2Y members dedicated money toward a "mini-grants" program, initiating scientific research to identify bottlenecks and barriers within the large landscape.[82]

An immediate challenge to wildlife connectivity was the Trans-Canada Highway. Back in 1958, Canadian authorities finished constructing the new road through Banff National Park; and in the early 1980s, they proposed widening it from two to four lanes to expand tourism. Locke of Y2Y dubbed the Trans-Canada Highway and Canadian Pacific Railway, both of which go through the

A car passes through the Red Earth Wildlife Overpass over the Trans-Canada Highway in Banff National Park. Photo by Josh Whetzel. Courtesy of the Yellowstone to Yukon Conservation Initiative.

narrow Bow Valley, the "Berlin Wall of Biodiversity" because of the high rate of wildlife deaths. From 1987 to 1996, automobile and train collisions accounted for 80 percent of all wolf fatalities in and around Banff, not to mention a large proportion of grizzly, deer, and elk deaths. In effect, animals either never crossed the infrastructure or ran the risk of being struck if they did. Much like breaking down the Berlin Wall to reunite Germans at the end of the Cold War, Y2Y advocates thought this transportation barrier must come down to protect a single population of wildlife.[83]

As part of environmental mitigation process for the Trans-Canada Highway, wolf expert Paul Paquet, who also served on Y2Y's science advisory board, recommended that an elevated thoroughfare or crossing structures be constructed. With pressure from the Y2Y Initiative, the Ottawa government allocated $3.3 million in the mid-1990s to build two wildlife overpasses and twenty-two underpasses. The structures were an easy sell after a cost-benefit analysis showed that drivers were paying hundreds of thousands of dollars every year in vehicle damage from hitting wildlife. The Banff Wildlife Crossings Project, a long-term research effort sponsored by Y2Y, has monitored the willingness of animals to

Trail cameras capture wildlife use of overpasses on the Trans-Canada Highway in Banff National Park, 2005–6. *Clockwise from upper left*: Moose, grizzly bear, gray wolf, and elk. Courtesy of Tony Clevenger, Western Transportation Institute.

cross those structures basically since they were built. Using hair snares, sand patches, and trail cameras, scientists documented one hundred thousand crossings from ten species of large mammals, with nearly seven thousand coming from wolves alone. On top of that, wildlife collisions were reduced by 80 percent. More underpasses and overpasses were soon constructed because of this evidence. By the mid-2000s, the forty-four total structures between the national park's eastern entrance and the continental divide at the BC-Alberta border were, as of this writing, the most for any single span of highway in the world. These transportation redesigns worked in making roads more permeable to animal movements.[84]

Not everyone was happy about these ideas, and most hostility to large landscape conservation originated from corporate influencers. In 1999, a spokesperson of the Competitive Enterprise Institute—a Washington, DC–based think tank funded by billionaires Charles and David Koch—stoked fears that Y2Y's goals represented environmentalists' desire to create one giant park devoid of resource extraction. "Once people zero in on what this is about," a libertarian pundit asserted, "that it's another massive layer of land-use control, they'll pay a lot more attention." The BC (British Columbia) Forest Alliance, an advocacy group for timber corporations, forecasted that implementing Y2Y would result in the loss of billions of dollars in revenues and thousands of jobs. "It's amazing to me," the BC Forest Alliance chair remarked, "that U.S.-based environmentalists have the nerve to come up here and propose that half our province be locked up in the so-called Y2Y corridor." Parallel to claims that Y2Y was really a cover-up for American interests on Canadian soil, others argued it was a ploy of the United Nations. In 2001, William Grigg of the far-right John Birch Society insisted that Y2Y's "binational land grab" meant that "roughly two-thirds of Idaho and nearly half of Montana would be subsumed into the bioregion, which would eventually be administered by a UN-approved 'bioregional council.'" Corporate messaging with nationalist undercurrents offered guidance to conservative political allies.[85]

Y2Y members counteracted negative publicity in a variety of ways. Some neutralized extreme critiques with humor. Lance Craighead of American Wildlands wrote a local newspaper editorial, requesting, "If anyone out there is part of this ecoterrorist paramilitary arm of the United Nations, or is otherwise concerned about environmental issues, please send us a (tax deductible) check to improve our web site."[86] Others provided educational outreach. From June 1998 to September 1999, Parks Canada employee Karsten Heuer completed what was called a Y2Y hike, walking the Rockies from south to north to "symbolically trace the dispersal of a wolf" and stopping at towns along the way to deliver presentations about the large landscape vision. After giving a hundred public talks and twice as many media interviews, Heuer noted that "regardless of their background or occupation, people who understand that Y2Y is *not* a huge park proposal are supportive of the idea and are often inspired enough to want to help."[87] Still others tapped into popular culture. In the first season of the television show *The West Wing*, Pluie the Wolf and Y2Y were dramatized when White House staff participated in Big Block of Cheese Day, a fictional workday in which they met with special interest groups who pitched ideas for government funding.[88] Whatever the platform, Y2Y representatives tried to offer an inspiring counternarrative of their internationalist mission.

Y2Y advocates realized the landscape vision would mean nothing on the ground without a base of grassroots support, from naturalist guides to hunting organizations, labor unions to civic leaders, religious institutions to Indigenous nations. In 2002, Y2Y hosted a meeting with Native Americans and First Nations groups to, in the words of Wendy Francis, "find ways to support your traditional work and lifestyles."[89] A year later, the Dehcho First Nations—a confederation of Dene and Metis peoples from the so-called Yukon and Northwest Territories—began negotiations with Parks Canada officials to expand Nahanni National Park. Herb Norwegian, Grand Chief of the Dehcho Dene, argued that large landscape conservation made sense for Indigenous nations because big contiguous habitats ensured that Native foodways could be preserved. Landscape species like woodland caribou, Dall sheep, and grizzly bears migrated across park boundaries that many scientists considered to be too small. The shared goal of park enlargement brought Indigenous leaders and conservation activists together.[90]

The proposal was contested on two fronts. Canadian Parliament, under majority control of the Conservative Party, wanted to keep mineral prospects open as a tungsten mine and a lead-silver-zinc mine operated in the region. Dehcho representatives, in contrast, wanted guarantees that any expansion of park lands under historic treaty rights would be subject to comanagement between sovereign governments. In 2009, Dehcho First Nations and Parks Canada agreed to the sixfold increase of Nahanni National Park and Reserve, now twelve thousand square miles in size. The designation turned a new page in park administration by incorporating traditional ecological knowledge into management decisions, allowing for subsistence uses like hunting and trapping, and supporting Dene culture in education and interpretation. The bigger park also created a new core area to serve as a northern anchor in the Y2Y landscape.[91]

Overall, the Yellowstone to Yukon Initiative has made progress in conserving a large landscape. Back in 1993, when the group started, only 11 percent of the region's land was protected—34.7 million acres, equivalent in size to New York state. This territory was represented by "cores" of national parks and wilderness areas. By 2013, protections nearly doubled to 21 percent, including an additional 31 percent with some sort of conservation status—166.9 million acres, roughly equal in size to the state of Texas. The latter area was mostly "corridors," a mishmash of political jurisdictions with special management practices for maintaining connectivity. In only twenty years, the Y2Y landscape reached a 50 percent threshold, as advocated by famous biologist E. O. Wilson, to safeguard biological diversity.[92]

Although the Yellowstone to Yukon Conservation Initiative has expanded its focus to connecting avian and aquatic ecosystems (birds and fish cross political borders too), the freedom of movement for charismatic mammals, especially carnivores, remains central to its work. At the first meeting in Kananaskis, Alberta, Harvey Locke told Y2Y members that wolf movements "blew the doors off of conventional thinking about what boundaries are for your ecological integrity." He concluded that the Y2Y landscape "is the continental seed stock for large carnivores. If we lose this, we lose them. Period."[93] Wolves had recovered inside cores, but the story remained to be told if these animals would be accepted moving through corridors.

AS A CORE AREA, Yellowstone National Park reached a point of saturation. For the ten years after wolves were first reintroduced in 1995, the lupine population steadily climbed almost every year. By 2007, biologist Doug Smith and Yellowstone Wolf Project members tallied 171 wolves in as many as fifteen packs. But numbers then dropped over the following half decade, stabilizing at a hundred wolves in ten packs as the ecosystem achieved what Smith called its "natural density." The delicate equilibrium between the number of wolves and the number of elk, deer, and other prey to eat would remain stable so long as other environmental conditions held. Pack territories had filled up the park. "From here on," Smith expected, "more wolves will disperse from their families."[94] Living with wolves inside the park was one thing, allowing them outside the park was quite another.

Except for those staying under park jurisdiction, wolf management was turned over from the federal government to state governments after removing Endangered Species Act protections. The delisting process got started back in 2002 when US Fish and Wildlife Service personnel documented thirty breeding pairs in the three recovery zones. Political battles dragged things out. Unlike Idaho and Montana, the Wyoming delegation could not put together a state management plan with adequate protections. For approval, they had to demonstrate how they would monitor and maintain a viable population—that is, a minimum of fifteen breeding pairs. Besides a small area near Yellowstone designated for trophy hunts, Wyoming officials classified wolves as "predatory animals" throughout the rest of the state, meaning they could be legally shot at will. Environmental and animal rights organizations filed lawsuits against potential reclassification. In the 2008 case *Defenders of Wildlife v. Hall*, plaintiffs argued that if rules were accepted under the proposed state plans, wolves in Yellowstone would be cut off from those in central Idaho and northwestern Montana, risking the population's long-term survival. A year later, in January 2009, the US Fish and Wildlife Service issued a rule delisting the western gray wolf in Montana and Idaho.[95]

But a legal injunction temporarily placed hunting and trapping activities on hold. After seventy-two wolves were harvested in Montana during the first season, US District Judge Don Molloy ruled in August 2010 that the Fish and Wildlife Service had "acted arbitrarily" when delisting the gray wolf. Molloy agreed with the argument that removing federal protections in only two states, and "without any evidence of genetic exchange" among the three recovery zones, had violated the Endangered Species Act because a regional plan needed to be developed to maintain a regional population. Simply put, state plans failed to ensure that wolves in Yellowstone could move and breed with wolves in central Idaho and northwestern Montana. After putting wolves back on the endangered species list, cooperation never happened. Instead, normal procedures were bypassed in April 2011 when President Barack Obama signed an appropriations bill into law with a special rider delisting the gray wolf in Idaho, Montana, and Wyoming. The three states then issued hunting and trapping permits.[96]

While conservationists used wolf-tracking studies as justification for softening political boundaries, politicians employed those same geolocating technologies to enforce them. In 2011, a Republican-majority Montana legislature approved a wolf management bill, with the support of Democratic Governor Brian Schweitzer, mandating a radio or GPS collar "be attached to at least one wolf in each wolf pack that is active near livestock or near a population center in areas where depredations are chronic or likely."[97] Given the wide ranges of wolves and the prevalence of livestock, the new law required that almost every wolf pack be monitored by the government. The legislation was ostensibly crafted to ascertain wolf populations, but other censusing methods have proven more effective, such as a one-time blood sampling to establish pedigrees. However, collaring these social animals allowed for the creation of what have derisively been called "Judas wolves." If these carnivores preyed on livestock, the signals of a Judas wolf betrayed the pack's location to government officials who would lethally remove them. The use of geolocating technologies held contradictions between border making and unmaking.[98]

Although wolves caused less than a half percent of overall livestock losses, individual ranchers still felt the economic pinch in places where predators habitually roamed. In 2008, the Montana Livestock Loss Board, a panel created by the state legislature with governor-appointed members, took over the compensation program from the Defenders of Wildlife. Between 2009 and 2015, payments for confirmed losses averaged $96,245 annually, with Beaverhead County near Yellowstone experiencing more depredations than any other county across the state. But wolves even squeezed the bottom line for ranchers among their surviving cattle. Examining the sales records from eighteen ranching operations across western Montana over a decade and a half, University of Montana researchers found that when a cattle herd suffered a wolf kill, the average calf weight in that specific herd fell about twenty-two pounds. This translated to a $6,679 loss when calves were sold. The study held a promising flipside: After mapping herd locations and wolf movements, investigators also discovered that simply having wolves in the vicinity had no tangible effect on calf weight. If equipped with labor and tools, ranchers themselves might be able to chart a way forward.[99]

Preventing wolf-livestock conflicts became one of the chief objectives for a conservation group called People and Carnivores. Founded in 1993, the organization has undergone multiple name changes (originally called the Predator Project) that reflect a shift in focus from influencing policy to deterring predation. "There was a lack of preparation, socially and technically, for what came to pass," stated People and Carnivores director Lisa Upson, referring to the surge of wolf numbers during the late 2000s. In response to increasing depredations, a dedicated corps of field staff—who lived in the rural communities—began introducing a bevy of coexistence methods to ranchers. Here, the Y2Y vision took hold with civil society.[100]

One of the most successful was a range riding program. In the Blackfoot River Valley of west-central Montana, a local landowners cooperative called the Blackfoot Challenge sought to modify animal husbandry practices to adapt to the presence of wolves. The Blackfoot Challenge had prior experience dealing with large carnivores, building electrified fencing around calving and lambing pens to deter grizzly bears. When a wolf population became evident in 2008 with confirmed case of livestock losses, Blackfoot Challenge members responded by hiring their first range rider from a prominent ranching family to monitor wolves. Range riding consists of watching the cows, particularly at dawn and dusk, for signs of predator duress and using horseback maneuvers to revive the herding instinct as protection comes from animals grazing together. It also involves identifying dead carcasses and removing the attractants so that a wolf pack does not become accustomed to preying on livestock. Initial success led to employing more range riders. Seth Wilson, a People and Carnivores field representative and Blackfoot Challenge program coordinator, trained them in radio telemetry to track wolves and then move herds away.[101]

Even if a cause-and-effect relationship cannot be established, preventative measures produced results. Between 2008 and 2015, the Blackfoot Challenge—comprising fifty ranches grazed by sixteen thousand head of livestock—averaged two and a half depredations per year. Meanwhile, fewer than three wolves each year had been lethally removed during the same period as the wolf population in the region increased from one pack to twelve packs.[102] A small measure of victory joined local action to continental vision as the Yellowstone to Yukon Initiative served as a funding partner for People and Carnivores, supporting them with range-riding salaries and predator-deterrence equipment. To protect wolves traveling through corridors, conservationists committed resources to preserving rural livelihoods.[103] Ecological internationalism took hold when and where it led to investments in civil society.

IT WAS A COLD spring morning on the Anderson-Pope Ranch. This livestock operation sits on the upper Tom Miner Basin, located a few miles from the northwest corner of Yellowstone National Park. The grassy rolling hills and forested mountain slopes serve as a shared habitat for people, cows, and all kinds of wildlife, such as elk, grizzlies, and wolves. It was also about calving time. Near a fenced pasture, I met third-generation rancher Malou Anderson-Ramirez to grab a spool of fladry—rectangular strips of red nylon fabric sewn to a polymer wire—out of the bed of her pickup truck. She had invited me, as well as a dozen high school students from a local FFA chapter, to help set up fladry fencing around calving pens. We pounded metal stakes into the ground, hung up a single strand around the square-shaped perimeter, and electrified it after connecting it to a portable solar-charging station. For the next month, this novel stimulus would deter carnivores that frequent the area from feasting on livestock's offspring. Anderson-Ramirez told me that for the eight seasons since her family had started using fladry in 2011, the ranch has not lost a single calf. And her rural community has taken notice of this nonlethal alternative.[104]

But it wasn't always that way. Anderson-Ramirez explained that the ranch was purchased in the late 1950s by her grandparents, who needed a respite because her grandfather was a prisoner of war during World War II. With little experience working in agriculture, they learned about conventional ranching from neighbors, passing on lessons to their four sons. Her father, Hannibal Anderson, and her uncles felt a "unique responsibility" to care for the larger ecosystem being so close to Yellowstone, but as Malou observed, "I grew up with a bunch of ranching men who lived in their challenges and never shared anything." They struggled, and to pay the bills, her parents lived part of the year in Livingston, where Hannibal worked as a history teacher and her mother, Julie, as a nurse. Born in 1981, Anderson-Ramirez recalled that seeing life outside of the family ranch opened her eyes to different ideas. So did education. Malou earned a degree in psychology at Montana State University and worked as a social worker in animal-assisted therapies in California. She later returned home to the Tom Miner Basin to take a position at the B Bar Ranch where she studied holistic range management. By the late 2000s, wolves showed up more often and most ranchers instinctively grabbed their guns.

Malou and sister-in-law Hilary Zaranek-Anderson, a Detroit native who studied wildlife biology at Montana State University and volunteered for a time as a field technician with Doug Smith's Yellowstone Wolf Project, tried everything else to protect livestock. They fired shotguns into the air every hour, they camped out all night long near the herd, and they played radio music on three sides of a pasture. "We joke that wolves really dislike eighties hits," Malou stated. They still lost two steers per day. Inspired by observations of bison herds, Hilary suggested they nudge the cows together and teach them to stand their ground en masse, much like their wild relatives did as a predator defense. It worked. In 2014, when grizzlies started killing livestock at nearby ranches, the two women cofounded the Tom Miner Basin Association as an endeavor to share coexistence tactics.[105]

The association's efforts took form through four programs: range riding, fladry fencing, carcass removal, and wildlife tracking. Malou and Hilary talked with neighbors and anyone who'd listen. At first, they got responses varying from "Bless your heart" to "There's one thing that'll stop a wolf and that's a bullet." Nonetheless, those stances changed over time when their results proved the techniques worked. Attitude turned out to be a significant barrier, but so was money. Malou and Hilary received grants from Defenders of Wildlife and People and Carnivores to pay for salaries, liability insurance, and equipment. In 2016, the Yellowstone to Yukon Initiative joined the other conservation organizations as a fiscal sponsor. Outside investment opened people up to the possibilities; the rhetorical question "What have we got to lose?" became a more common reaction. With backing, the Tom Miner Basin Association employed up to three range riders for patrolling ranches within the watershed, as well as offering fladry to neighbors. Community outreach included the FFA students, whom I labored alongside, adding predator prevention tools to ranching toolkits.[106]

As we walked around the fladry barrier to ensure it was eighteen inches off the ground, I was overcome by the comforting smell of sage grass and the chilling sound of wolf howls from a resident pack, which I was told were frequent sounds. "We've always looked at the landscape as something to be shared," Anderson-Ramirez commented, referring to her family's land ethic. Coexistence methods demonstrate that neither livestock raisers nor large carnivores harbor innate animosity toward each other, but that an economy and society forged by the twin ideologies of settler colonialism and global capitalism have more to do with present challenges to ranching. "The bigger picture is that this is a wildlife corridor that is one of the most unique on the entire globe," Malou continued, so the bigger perspective of rural communities "being more tolerant" to native predators "creates a feeling of being part of the system." Anderson-Ramirez thinks that ranchers, like wolves, are an endangered species. "If they don't survive, at some point we won't survive."[107]

Across fence lines dividing one ranch from another, one block of protected habitat from the next, cooperative endeavors like the Tom Miner Basin Association were trying to heal wounds on the larger landscape. Radio-collaring wolves led to the Yellowstone to Yukon vision, but this transnational political identity became anchored in women like Malou and Hilary Anderson only when it improved their ranching livelihoods.

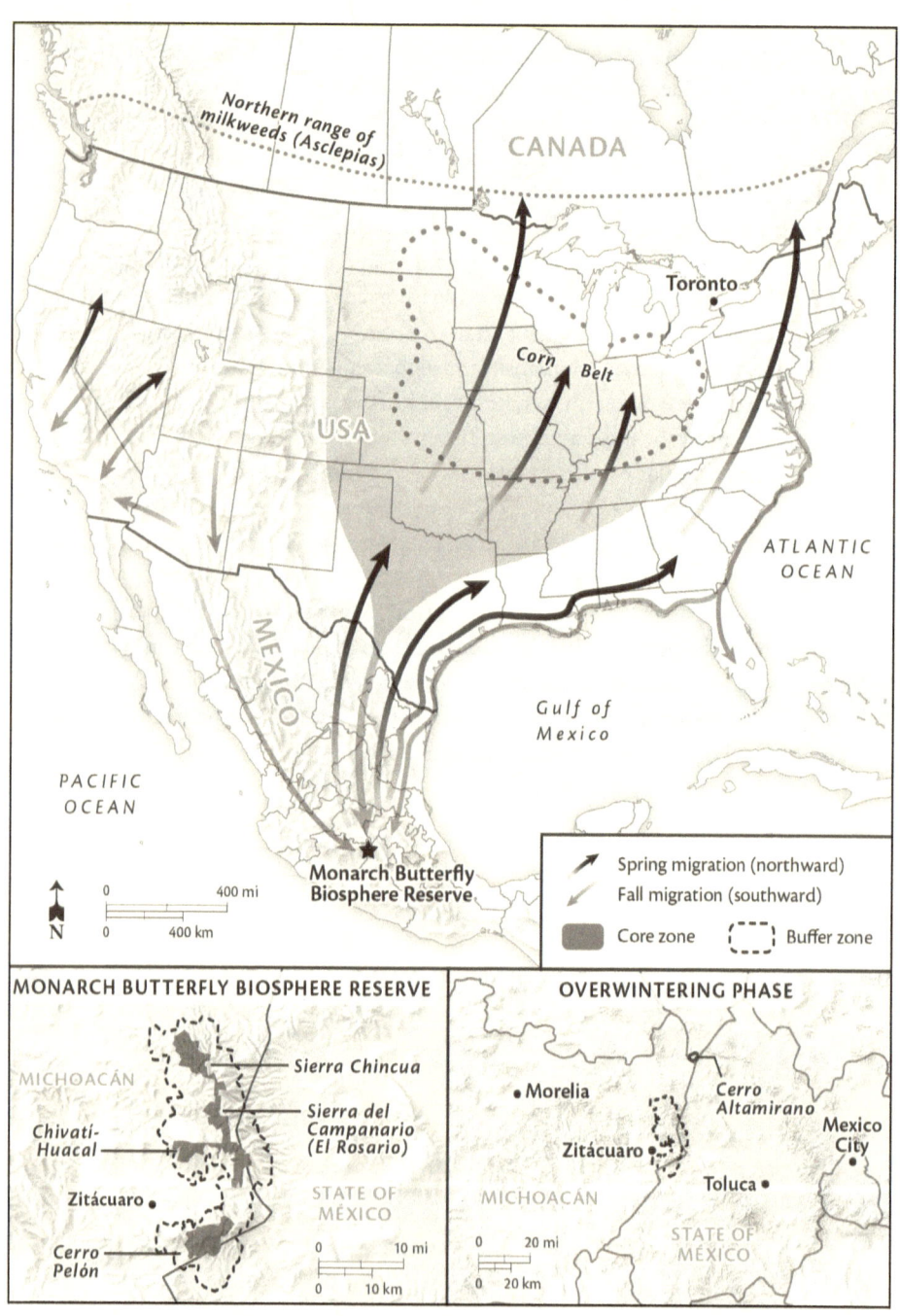

The Monarch Butterfly Migration and Overwintering Sites. Map by Erin Greb Cartography.

PART II
Monarch Butterflies
(Danaus plexippus)

CHAPTER THREE

Tracking Flight

At the Mexican village of Macheros, I climbed on a horse to ride to the monarch butterfly sanctuary of Cerro Pelón. A teenager led his family mare up a steep trail along with a string of other local *guías* who supplemented community income by escorting tourists from Mexico and abroad to the roosting sites. While ascending the mountainous terrain through conifer forest, I saw pine trees tapped for resin every few hundred feet, the sticky substance oozing into bottles nailed to trunks. My horse handler Adan told me they sold resin for making glue and medicine, adding, "El bosque nos apoya" (The forest supports us). I knew what he meant given my meal from the night before had been cooked in a wood-burning stove. Although a steady stream of butterflies flittered past us, they became orange-and-black-colored clouds floating on each gust of wind. When we reached our destination, a meadow named Llano de Tres Gobernadores, we got off the horses and hiked a short distance along the forest edge.

It was hard to describe my feelings when I spotted a stand of mature trees covered with millions of monarch butterflies. The winged creature averages a half gram, equal in weight to a paper clip, and yet so many monarchs clung onto the oyamel fir and Montezuma pine that the branches sagged toward the ground. Slivers of sunlight beamed down onto the clusters of insects, heating them up, until monarchs eventually burst into the air like a star exploding in outer space. Posted signs instructed us to *guarda silencio*, or stay quiet, to limit our disturbance of the monarchs. However, I thought that we kept our voices low to appreciate the faint clapping sound coming from the flapping of thousands of butterfly wings. They flew all around us. I stood in awe among the three forest rangers who kept watch.

One of them, Patricio "Pato" Moreno Rojas, approached me with a smile, reached into a pocket of his blue jeans, and extended his hand with something in it. Born in 1985, Moreno was raised in Macheros, a town straddling the border between the states of Michoacán and México, as well as at the entry of Reserva de la Biosfera Mariposa Monarca. Moreno and his brothers visited the butterfly colonies as local guides. He remembered, "We pulled the horses during the tourist season from a very young age."[1] His father, Melquiades Moreno de Jesús, had worked as a forest ranger for the State Commission of Natural Parks and Wildlife, patrolling Cerro Pelón on the México state side for over thirty-five years. Not only did his monitoring have a positive impact on forest health, which was visibly more intact than on the Michoacán side, but a steady government paycheck pulled his family out of poverty. In 2014, Melquiades turned over the conservation job to his son Pato.

A circle-shaped sticker, which read "Monarch Watch—AAUJ 492," rested on Pato's extended palm. Moreno explained to me that the adhesive label was a *marca* (tag) taken off a butterfly's forewing. He did not know where the invertebrate had been tagged yet, but this unique identifier would be forwarded to Monarch Watch, a conservation organization at the University of Kansas, so that scientists could track its flight. Moreno, among other local Mexicans who find *marcas*, receive about five US dollars per sticker as a financial incentive for the time and energy spent locating them. While only a small fraction is recovered from year to year, the tags sketch the contours of an epic migration across three countries.

The *marca* connected Moreno to a much longer history of citizen science, international conservation, and landscape use. Beginning in the 1930s, a team of Canadian researchers at the University of Toronto, Fred and Norah Urquhart, developed a postage stamp–sized tag to find out where monarch butterflies traveled. The Urquharts investigated accounts from the late nineteenth and early twentieth centuries about "plagues" of these invertebrates, but no one had proven any pattern of movement. During the same time, land reforms in post-Revolution Mexico created the ejido system based on communal forest ownership. With scientific oversight, rural campesinos collectively managed timber resources to support community needs. Events on opposite ends of the continent were brought together though a citizen-science brigade organized by the Urquharts, called the Insect Migration Association, which recruited laypersons from Canada, Mexico, and the United States to tag monarchs and forward recoveries. Citizen science laid the groundwork for ecological internationalism.

Although making scientific knowledge more participatory solved the mystery of the journey, inclusion did not extend to everyone along the migratory path. International experts drowned out local concerns when pressing Mexico's federal government to protect the roosting grounds. Nation-states shared a unique biological phenomenon, but they failed to share power when managing transnational nature.

Seeing the Butterflies

Every September from 1903 to 1911, Lillie George watched monarch butterflies gather on a tree in her backyard. George lived in Saint Paul, Minnesota, in a modest Prairie style house with a low-pitch hip roof, overhanging eaves, and a front porch. All were common architectural features for new dwellings across the Midwestern region. George, a homemaker, spent a great deal of time gardening outside since her husband, William, traveled as a salesman for a paper company. One autumn day at three o'clock in the afternoon, Lillie spotted hundreds of these "milkweed butterflies," as she and her neighbors called them, flying toward a box elder where they gathered on a sturdy limb to rest for the night. The next morning when she woke, they were gone. A local newspaper reported that "Mrs. George, who has watched them for years, and has seen thousands come and thousands go, offered no explanation for the ever-recurring visit."[2]

In hindsight, an explanation for these ever-recurring visits can be found in Lillie George's slang nomenclature and in William George's paper business. Monarchs really were "milkweed butterflies," because during their larval stage, caterpillars only eat a family of milkweeds known as *Asclepias*. Gardeners and scientists alike wondered why young monarchs dedicated so much of their time to munching on a single species and whether adult monarchs followed their bloom. Milkweed plants thrived in the open spaces created from cutovers in the north woods. Between 1860 and 1890, fifty million acres of old-growth forest, mostly white pine, were cut down in the Great Lakes region. Trees arrived by water or rail to commercial metropolises like Chicago and Minneapolis where pulp mills transformed them into paper and other commodities.[3] Clearing the forest vastly increased the abundance of a single type of milkweed, *Asclepias syriaca*, which, in turn, expanded where monarchs could lay eggs across eastern North America. But the landscape changes featured a spatial paradox: They ushered in the *expansion* of the monarch range in Canada and the United States while bringing the *contraction* of it in Mexico.[4]

Anna Botsford Comstock, who lived on the eastern side of the Great Lakes, epitomized the robust debate about where monarch butterflies traveled. Born in 1854, Botsford grew up as an only child on a family farm in upstate New York, where her Quaker mother encouraged watching wildflowers, insects, and birds.

Twenty-year-old Anna enrolled at a nearby land-grant institution in Ithaca, Cornell University. She attended for two years but left to marry a young professor and entomologist named John Henry Comstock, who nurtured her interest in illustrating insects. With little formal training in art, Anna created images to accompany her husband's lectures and publications. For example, while John briefly served as chief entomologist for the US Department of Agriculture, Anna prepared drawings of a pest known to the California citrus industry as red scale. In 1885, she reentered Cornell to finish a degree in natural history and developed into a prolific illustrator of and writer about an order of insects called lepidoptera—moths and butterflies.[5]

Comstock joined the Cornell faculty as its first female instructor in 1897, teaching new courses in nature study, although she was denied full professorship until late in her career. The nature-study movement emphasized that the best learning occurred outdoors. Comstock agreed with Swiss geologist Louis Agassiz who urged young students to "study nature, not books," a progressive-minded philosophy that swept across public schools in the English-speaking world as a remedy for the societal ills of industrialization. Anna herself defined nature-study as "the comprehension of the Individual life of the bird, insect or plant that is nearest at hand."[6] She developed curriculum for rural extension programs recommending to not begin an abstract theory but start with a material object close at hand.[7]

As an example, Comstock launched into the concept of biological imitation by examining monarchs. In the popular insect-teaching manual *Ways of the Six-Footed*, Anna recalled, "When I was a child I disturbed a flock of caterpillars resting together on the lower side of a leaf of milkweed, and I still remember the creepy fascination with which I gazed at the black and yellow ringed creatures."[8] Monarch larvae showed alternating bands of color across their bodies as a warning to birds, Comstock suggested. The same basic principle also held true for monarch adults. Anna penned a 1902 article for the magazine *Country Life in America*, observing that "the monarch is immune from bird enemies; the callow birdling take a bite from it, wipes his beak in disgust, and forever after connects the noisome taste with orange wings."[9]

Unlike other butterflies or moths that used camouflage for protection, Anna and John speculated that monarch resistance to avian predators came from a "disagreeable odor." They were wrong. But using the idiom "Eat or be eaten," they were right about another hypothesis: Viceroy butterflies look like monarch butterflies for protection. "There is another butterfly, called the viceroy, who is a pleasant morsel for birds," Anna wrote. "But he mimics the coloring of the monarch, and it takes a practiced eye to tell the difference. The birds are fooled, and many a viceroy flies to safety from under the beaks of hungry feathered folk."[10]

The Comstocks entered a larger dialogue among naturalists by theorizing that monarchs could be found anywhere milkweed grew. In the 1904 book *How to Know Butterflies*, Anna and John offered that "the mother butterfly follows the spring northward as it advances as far as she finds milkweed sprouting; there she deposits her eggs, from which hatch individuals that carry on the journey, and in turn lay their eggs as far north as possible." The Comstocks assumed that an annual migration might occur as far northward as Canada's Hudson Bay, but its southern limit was anybody's guess, since "no hibernating specimen has ever been found."[11]

Uncertainty about a migration lingered because, even though an overwintering site had been found in 1881 near Monterey, California, most observers during the early twentieth century thought that at least some butterflies hibernated elsewhere. H. V. Andrews saw orange-colored swarms from his hometown of Toronto gliding across Lake Ontario, noting that "their flight was rapid, as if they intended reaching the U.S.A. or wherever they were going in as short time as possible."[12] Jennie Brooks witnessed the yearly return of butterflies near Lawrence, Kansas, commenting, "All along the Canada line east and west the mighty winged host of monarchs advances, when instinct stirs, straight down across the states, to Mexico."[13] Some naturalists even posited that monarch butterflies, whose evolutionary history points to tropical origins, might go as far southward as Central America.[14]

This discussion attracted foreign scientists who tried to synthesize observations. British entomologist C. B. Williams published *The Migration of Butterflies*, which dedicated an entire chapter to monarchs. Compiling field reports between 1862 and 1924, Williams noted the month, location, and direction of every published sighting. Although monarchs could be found from Minnesota to Maine, Mississippi to Manitoba, the majority were seen traveling southward during autumn. None were found in the northeastern United States and southern Canada during the winter. With few accounts from the Rocky Mountains, Williams resolved public confusion by using the mountain barrier to separate monarchs into two migrating groups: a western population traveling to the California coastline and an eastern population possibly heading for the Florida peninsula.[15] German entomologist Adalbert Seitz suggested the driving force behind migrating northward in spring was to follow milkweed, particularly where it was abundant near cutovers. Seitz reported that monarchs "are fond of the open-country, accompanying cultivation further and further into the primeval forests as soon as a few clearings have been formed where the foodplants of the larvae, especially of *Asclepias*, can get a foothold."[16]

Seen by many, understood by few, the monarch range expanded across eastern North America as forests converted to fields and farms—a landscape of full sunlight ideal for milkweeds. Canadians and Americans delighted in the butterflies they saw every year, but no one knew precisely from where they came. Mexicans, who also noticed them, had their own explanations for *la Monarca*.

ACCORDING TO Leonel Moreno Espinoza, the people of Macheros believed the origins of monarchs to be earthly and local. "Some older folks said that butterflies were born from the oyamel seeds," Moreno recalled. "Others said that there was a cave in Cerro Pelón and that butterflies came from there."[17] They were right to connect monarchs with the forested mountaintop, one among a chain along the Trans-Mexican Volcanic Belt. His community came to possess a portion of those lands in 1937, a year before Leonel was born, when his father, uncles, and other land-reform beneficiaries called *ejidatarios* established the Ejido El Capulín. As a child, Moreno remembered herding cattle on communal plots and playing with the butterflies that came to drink at irrigation canals. He lived in the aftermath of the Mexican Revolution, which unleashed impulses for social justice transforming the management of woods and wildlife.[18]

Leonel's village of Macheros received its place name during the late nineteenth-century hacienda period when one-fifth of the national territory was held by wealthy owners and foreign investors with ties to the dictatorship of Porfirio Díaz. Before 1910, Cerro Pelón was controlled by the Hacienda San Bartolo, a large estate that helped to construct railroad tracks deep into the forest to cut down timber. As a comparison, the Michoacán y Pacífico Railroad, which extended to the nearby city of Zitácuaro, received permission from the regime to take whatever resources it needed to build and maintain the line. Constructing a thirty-five-mile-long section required fifty thousand downed trees for ties, trestles, and fuel. What trees the laborers of Hacienda San Bartolo could not reach by train, they dragged down the mountain with horses. Macheros was where they housed beasts of burden; *machero* means "corral" or "stable." "We are in Macheros," Moreno remarked, "because the *macheros* of the horses gave rise to the name of Macheros, and thus we are recognized worldwide."[19]

The Mexican Revolution of 1910–17 reinvented landscape use across the country, with rural campesinos seen as its rightful recipients. Military leader

Emiliano Zapata issued the Plan of Ayala in 1911, which rooted peasant anger in denying natural resources for the benefit of the many. "In virtue of the fact that the immense majority of Mexican pueblos and citizens are owners of no more than the land they walk on, suffering the horrors of poverty . . . because lands, timber, and water are monopolized in a few hands," the leftist manifesto declared that land redistribution to the working class would ensure "prosperity and well-being may improve in all and for all."[20] The aspiration was preserved in Article 27 of the 1917 Constitution, which stated that land and natural resources ultimately belonged to the nation and could be given to ordinary citizens on behalf of the public trust.

Between 1934 and 1940, during the center-left presidency of Lázaro Cárdenas, land reform increased. Under "restitution" (*restauración*), the federal government restored territories originally held by Indigenous communities if they could provide evidence that an outside party had illegally taken their land. More commonly, under "endowment" (*dotación*), the national government conferred a parcel of land with permanent usufruct rights, known as an *ejido*, to be communally managed by a group of applicants.[21] "My father and my uncle and other people were the initiators of the ejido, they were among those who fought for them to give them their land, they said that they had suffered a lot," Leonel recalled. "They spent days walking to Mexico City or Toluca for the efforts of the ejido . . . but yes, thank God, they managed to get the government to give them the land, and since then we stayed here."[22] The federal government in due time conferred ejidal rights to over fifty communities, including El Capulín, on lands that comprised the Trans-Mexican Volcanic Belt. The creation of a more equitable system meant that local people could reap the benefits of their land and labor.

Campesinos across the country started extracting timber on ejido lands after establishing cooperatives under the Forestry Code of 1926. The law articulated a vision of forest management combining rational science and social justice. Although subject to oversight by professional foresters, natural resource use was managed by local communities. In 1938, for example, four hundred ejidatarios in Contepec formed the Emiliano Zapata Forestry Cooperative to jointly cut down oak and pine trees around the mountain of Cerro Altomirano for charcoal production, with annual quotas set by state foresters. After Jaime Chaparro and Santos Pérez of the Department of Forestry, Game, and Fish demanded the ejidatarios pay two thousand pesos to continue felling the forest under their allowance, ejido commissioner Juan Correa asked for the removal of these two corrupt men from their positions. In response to the petition, the Cárdenas administration sent a federal agent from Mexico City who resolved the dispute on behalf of the ejidatarios. The Department of Forestry, Game, and Fish approved 866 such forestry collectives nationwide by 1940.[23]

At Cerro Pelón, Leonel's cousin Elidió Moreno de Jesús remembered working for a local cooperative that made charcoal, firewood, and axe handles from the trees they felled. They sold most products at Zitácuaro, Michoacán. "We got a whole bunch there," Elidió explained, dealing wooden goods like "corn vendors" in the city streets. During his 1940s childhood, Moreno saw monarch butterflies in the mixed pine-fir forest on Ejido El Capulín. "The branches were loaded with them," Elidió recalled, but as an adult who inherited rights to the forest, he never witnessed nor logged where any large clusters of butterflies were. On the other side of the mountain, the federal government issued a forest closure (*veda*) in 1950 for the state of Michoacán that drove many ejidatarios to work as tree-tappers in the resin industry. This decision was consistent with national goals that viewed the conservation of natural resources and their development by small producers as compatible.[24]

Unlike the Romantic intellectual tradition of unpeopled wilderness that undergirded US national parks, the revolutionary desires for social justice forged Mexico's nature protection. Between 1935 and 1940, the Cárdenas administration created forty national parks, more than ever were established before or after that five-year period. These protected areas outlawed commercial production, but subsistence uses remained within those places. At Popocatépetl-Iztaccíhuatl National Park, two extinct volcanoes located on the eastern side of México state, local communities still gathered firewood, tapped pine trees for resin, and processed timber into paper products. People and parks cohabitated through the social politics of the age.[25]

While living and working in the forests of the Trans-Mexican Volcanic Belt, local communities took notice of *la Monarca*. For Native peoples, some referred to these insects as "the butterflies that pass in November," since they arrived at beginning of that month every year, while others called them "the reapers" (*cosechadoras*) because their presence coincided with the maize harvest. Homero Aridjis, an eight-year-old boy in 1948 returning home from his first holy communion at a nearby church, fondly remembered, "Thousands of monarch butterflies were crossing the village. The air, like a river, carried currents of butterflies. Through the streets, above the houses, between the trees and people, they passed by going south."[26] Observations of this swarming behavior likely caused Indigenous Otomí of northeastern Michoacán to call them "doves" (*palomas*), since monarchs' clustering during flight resembled a flock of birds. Even though names for the monarch butterfly varied from place to place, they became sojourners for many paths of life.[27]

An accident altered the direction of one path. Born in 1940, Aridjis grew up in Contepec, Michoacán, a market town at the base of Cerro Altomirano, where his parents ran a general store. During his youth, Homero trekked with friends up to Llano de la Mula ("Plain of the Mule"), a meadow located on the forested mountaintop, to picnic and watch butterflies clinging to stands of oyamel fir. He recalled seeing woodcutters drag those same trees down the trail with mules. A near-death experience, however, transformed the adventurous boy into an introspective man. In January 1951, Aridjis woke up in a hospital bed after unintentionally shooting himself. He had taken a shotgun from his older brothers without permission, and while aiming the firearm at a bird in the sky, he dropped it, discharging a shot of pellets into his stomach. Aridjis credited the incident for his becoming a poet and, later, an environmentalist.[28]

Using the influence of his rising literary fame in adulthood, Aridjis would have much to say about those remaining woods, especially some thirty locations where butterfly colonies could still be found. Jerzy Rzedowski, a Polish-born botanist whose Jewish family had relocated to Mexico in 1946 after surviving Nazi concentration camps, suggested the monarchs had migrated there. While Rzedowski and a lab assistant were hiking in the Sierra Madre Oriental near Ciudad del Maíz, San Luis Potosí, in October 1956, they saw a "constant stream" of butterflies, four to ten every minute, flying southward. Rzedowski then saw several thousand monarchs resting on a mesquite tree, only to find them gone the next day. "It remains to be elucidated," Rzedowski reported in the journal *Acta Zoológica Mexicana*, "whether this is a migratory movement of local importance or whether the migratory origin or destination is located at great distance from the place of observation."[29] Convincing the national government to establish a protected area for monarchs meant that scientific and political elites like Rzedowski or Aridjis would have to confront the social justice tradition of national parks. Ejidatarios never forgot who gained revolutionary rights to the land. When asked who owns the forest where monarchs can be found, Elidió's brother Melquiades Moreno de Jesús responded without hesitation, "We are the owners."[30]

SOMEONE VERY INFLUENTIAL to Aridjis, and to the international community involved in monarch conservation, was Lincoln Brower. The US lepidopterist developed a new scientific field called chemical ecology to study the reciprocal adaptations between monarchs and milkweed, a path of laboratory research that brought him into the fold of protecting the migratory phenomenon.

Brower cultivated a passion for butterflies at an early age. Born in 1931, the Great Depression haunted his childhood in western New Jersey. His grandfather struggled in the greenhouse business, but despite never receiving a college education, he taught Brower about heredity through propagating a "sport" of roses in which like produced like. When not working on the family dairy farm, Brower explored a wetlands area leftover from the last Ice Age known as the Great Swamp. There, Brower met an old butterfly collector named Charles Rummel who mentored him in capturing and raising caterpillars, a skill that he would use for the rest of his life. However, it was harder to find specimens after World War II as real estate developers bulldozed the countryside to construct tract houses. For Brower, suburban sprawl offered no tranquility, causing the destruction of butterfly-loving open spaces.[31]

Brower got more serious about butterflies in college. A promising student, he attended Princeton University, majoring in biology and continuing the practice of rearing caterpillars, this time in his dorm room. Brower was captivated by lectures from leading geneticists on evolution through natural selection. Before graduating in 1953, Brower and his soon-to-be-wife, Jane van Zandt, attended an annual meeting of the Lepidopterists' Society in New York City. During the collectors' conference, Lincoln and Jane met an entomology professor named Charles Remington who recruited them both to join the doctoral program at Yale.[32]

While Lincoln worked on a dissertation about the evolutionary origins of swallowtail butterflies, Jane sought to demonstrate that viceroy buteries were mimicking monarchs. At the Archbold Biological Station near Lake Placid, Florida, the Browers devised an experimental test in April 1956 by capturing eight scrub jays, placing them in large cages, and breeding butterflies to feed them. They released the "model" monarch and "mimic" viceroy into the cages at the same time repeatedly until birds learned that the model species was distasteful. Then, when they released two viceroy butterflies, the preconditioned jays avoided them on sight. Jane proved the mimicry hypothesis, and Lincoln's help testing cultivated an obsession with monarch butterflies.[33]

Teaching at Amherst College, Lincoln Brower decided to take the experiment one step further. He wanted to answer a follow-up question: Why were monarchs unpalatable to birds in the first place? A hypothesis came to him when he was helping with Jane's study. "In the process of my raising Monarchs," Brower recalled, "I tasted one of these milkweed plants. It just about knocked me over. It was so foul tasting. I drooled and I almost vomited."[34] This response caused the lepidopterist to begin talking with pharmacologists about the chemical properties of *Asclepias*, a genus name that Brower learned was derived from Asklepius, the Greek god of medicine. Through these discussions, he found out that milkweeds possess a chemical substance called cardenolides, referred to as "cardiac glycosides" in medicine because similar compounds were used to treat irregular heart rhythms. Some plants generated toxins to repel herbivores from eating them, which had explained why ingesting milkweed made Brower feel sick. If these plants were poisonous, could the monarch butterflies sequester the poison to fend off birds?[35]

Lacking the lab equipment for chemical analysis, Brower and his colleagues designed a different trial to see if monarchs internalized cardenolides. Much like during his youth, Brower took thousands of monarch eggs and attempted to raise the caterpillars on cabbage, a host plant without toxins. Most died of starvation since they did not eat the unfamiliar food, but after five generations of selecting those that did, he had nurtured a few monarchs into adulthood that only ate cabbage. For the experiment, Brower and his Amherst students fed wild-caught blue jays a planned diet to assess their gastric responses. First, they introduced mealworms to the birds as a negative control because there was no prospect of retching the delicious food. As expected, no vomiting was induced. Next, they fed the jays a regimen of cabbage-reared monarchs until the birds grew accustomed to eating the butterflies. Still no vomiting. Then, they offered monarchs that had been raised on a tropical milkweed species, *Asclepias curassavica*, as a positive control. Vomiting occurred every time. After the findings of "Brower's barfing blue jay" were published in 1967, this predictable avian behavior was validated by chemists who measured cardenolides in monarchs during their larval and adult stages.[36]

But there was a surprise. Brower discovered that not all milkweed plants—and thus not all monarch butterflies—possessed the same level of toxicity. This fact became evident because Brower had also reared caterpillars on a milkweed vine called *Gonolobus* and anticipated that the jays, during feeding experiments, would puke the monarchs up. Astonishingly, the birds didn't. When Brower sent a sample of *Gonolobus* leaves to a chemist for testing, the results came back measuring no cardiac glycosides. "I liked experimental biology," Brower remarked, because the laboratory "allows you efficiently to answer questions—more so than by just muddling around in the field."[37] Since there was great diversity in cardenolide production among milkweeds, as the caterpillars only food source, it meant a wide range of palatability among butterflies. In 1969, Brower wrote an article titled "Ecological Chemistry" for *Scientific American*, explaining the relationships between monarchs and milkweeds and, more importantly, announcing the possibilities of what a new scientific discipline had to offer.[38]

Outfitted with a lab toolkit, Brower was now fascinated with another aspect of monarch biology. Milkweeds, having varied levels of cardenolide toxicity over the one hundred different species, were found in distinct regions throughout North America. Brower wanted to study the geographic distribution of these plants in relation to the flight of monarch butterflies, which led him to the ongoing scientific work of Canadian researchers Fred and Norah Urquhart. "Basically, Urquhart's and my interests converged," Brower remembered. A problem was that despite the parallel pursuits, the lepidopterists despised each other.[39]

Searching for Butterflies

On the campus grounds of the University of Toronto, Scarborough, I was looking for a garden and a guide. The Urquhart Memorial Garden stood out among the greenery with bursting color: yellow black-eyed Susans, pink Autumn Fire stonecrop, and orange butterfly weed (*Asclepias tuberosa*, part of the milkweed family). This blend of flowers attracted monarch butterflies, many of which were feeding on plant nectar to build up their fat reserves for a long flight southward. My guide, Don Davis, who stood out among the young students as an old man with a wiry frame and a wiry mustache, has been a lifelong supporter of Fred and Norah Urquhart. The Urquharts developed an early example of "citizen science," enlisting thousands of volunteers, Davis included, to help find answers to their research question: Where did monarchs they saw in Canada every summer go during the winter? For the university's fiftieth anniversary in 2014, Davis arranged for this butterfly garden to be planted in their honor for organizing many people of different nationalities to solve the migration mystery.

In 1967, at seventeen years old, Davis joined the Insect Migration Association. He originally had wanted to help Ontario naturalists with banding birds, but the capture-and-release process seemed too complicated. Instead, Davis connected with the citizen-science organization led by Fred and Norah Urquhart. He and other "research associates," as the Urquharts called their contributors, tagged monarchs wherever they lived. Davis remembered waiting eagerly to receive the annual newsletter and see if any of his tags had been found. He never stopped tagging, still marking thousands of monarchs every year. Davis bragged to me that his name can be found in the *Guinness Book of World Records* for tagging a butterfly at Presqu'ile Provincial Park in 1989 with the longest recorded flight, 2,880 miles. "The Urquharts made us all feel like one big family," Davis remarked. "It changed the trajectory of my life." His path speaks to the power of grassroots participation in monarch science.[40]

A wellspring of the Insect Migration Association can be traced back to a side hobby in a marsh. Born in Toronto in 1911 to Scottish immigrants, Frederick Urquhart was consistently in trouble for skipping classes. It wasn't his fault, Fred told his parents, because in order get to school he had to walk past this boggy area along the railroad tracks, which was the perfect place to catch insects. Finally, his biology teacher, Mr. Pollard, caught the truant in the wetland. Instead of bringing Urquhart to the principal's office, Pollard showed him how to find cocoons and then asked him to write up a report on one after metamorphosis. A jade-green chrysalis turned out to be a monarch butterfly. Urquhart underwent a metamorphosis of his own, becoming a dedicated student and deciding he wanted to be a biology teacher. As a teenager in the Great Depression, Urquhart stayed close to home for his undergraduate training at the University of Toronto. Some professors kindled his enthusiasm for natural history on course excursions, studying living things through observation rather than experiments. Other professors bored him to tears by striving to reduce biology down to laws of mathematics, chemistry, and physics. Drawn to the field over the lab, Urquhart came from the opposite intellectual tradition as Brower.[41]

After finishing a degree in 1935, Urquhart couldn't find a teaching position due to the challenging economic times and thus joined the University of Toronto's graduate program. For income during his studies, Fred applied for a fellowship that required part-time service as a tour docent at the Royal Ontario Museum. Urquhart added samples to the museum's insect collection through his dissertation project about grasshoppers on the Canadian side of Lake Erie at Point Pelee. Museum staff saw value in Urquhart's work and eventually hired him as curator of invertebrates for cataloguing new species and producing exhibits. Point Pelee was also chock-full of monarchs. Outside of his college tasks and museum duties, Fred devoted any spare moment trying to solve a riddle that had dogged him since adolescence: Where did these butterflies migrate to?[42]

Through trial and error, Urquhart developed a tagging system to track monarchs' flight south from Toronto. To begin with, he tried staining the wings with paint spots to identify each butterfly. He collected no results. Urquhart then tested gluing a postage stamp–sized piece of paper with return information to the wings. Again, no results—the tag hindered movement and fell off whenever rain washed the paste away. Fred achieved a measure of success with what he called the "alar tag," which consisted of making a hole in the front-right wing with a paper punch, folding a paper sheet over the wing's edge, and gluing the tag to itself. These quarter-inch-by-half-inch tags read "Return to Museum of Toronto, Ont." and held a unique serial number to establish who did the tagging, when it was done, and where the recovery was made. Fred remembered, "I had to put something on the wing so that if you caught it, you'd know what to do with it, like bird banding." His study of monarchs was interrupted in 1940 when, doctorate finally in hand, he joined the meteorological service for the Royal Canadian Air Force, training British pilots during World War II on how to read weather patterns.[43]

After the Allied victory over fascism, a significant person for the butterfly investigation entered the picture: Fred's spouse, Norah Patterson. Urquhart had started dating the Toronto-born social worker, seven years his junior, during graduate school. In 1945, the two married. They became complementary partners in the monarch-tagging project: Whereas Fred held the biological training to interpret butterfly behaviors, Norah brought the organizational skills to manage huge volumes of incoming data. In 1949, Fred became director of the Royal Ontario Museum with a cross-appointment to teach at the University of Toronto. Meanwhile, Norah improved tag function with a new product from the Avery Adhesive Label Corporation, a pressure-sensitive adhesive that stayed put without having to punch a hole through the butterfly's wing. Increasingly, tag recoveries now showed monarch movements from Toronto to other parts of Ontario and even to New York State. But the Urquharts soon realized that monitoring would only produce meaningful answers if they could recruit large numbers of people to do the tagging.[44]

In 1952, Fred published an article—written by Norah without acknowledgment—in *Natural History*, a magazine for New York City's American Museum of Natural History, voicing the first request for volunteers. Norah stated that potential overwintering areas for monarchs have been known in California and Florida, but a migration to those sites remained unproven without firmer evidence. The Urquharts then appealed for help and offered to mail tags free of charge to any request. "If enough people cooperate," Norah announced, "we may some day be able to tell the complete story of this mysterious traveler."[45] A dozen people responded. Norah repeated the request in stories for Canadian and American newspapers. Three years after the initial ask, the Urquharts recruited nearly two hundred monarch taggers spread across five provinces and thirty-nine states. A transnational network of lay participants, thanks to Norah's work, was coming together.[46]

Rather than limit research to trained scientists, the Urquharts tried to share

Norah and Fred Urquhart during a graduation ceremony at Scarborough College, University of Toronto, circa 1970s. Courtesy of University of Toronto Scarborough Library, Special Collections.

the process of knowledge production. In 1962, Fred was asked to leave the Royal Ontario Museum because angry curators caused a ruckus over his promotion of long-serving staff without college degrees. Undeterred by resignation, Fred and Norah established the Insect Migration Association to organize their group of collaborators, whom they dubbed "research associates," or what we today would recognize as citizen scientists. The Urquharts sent out tags to all volunteers and published an annual newsletter to share news about tag recoveries. Research associates came from many occupations—teachers, carpenters, mechanics, physicians, and homemakers—but were united by a shared interest in monarch butterflies. Longtime Canadian tagger Don Davis, who held a job as a social worker, remarked, "You didn't have to be a rocket scientist or a Ph.D. to make a contribution to a significant scientific project. People felt good about that."[47] Even in 1965, by which time the Urquharts had transitioned to Scarborough College, a satellite campus of the University of Toronto where Fred was a zoology professor and Norah a lab assistant, they still congratulated the larger collective after results were published.[48]

Tagging monarchs turned arbitrary movements into a coherent migration. Between 1953 and 1964, volunteers with the Insect Migration Association marked 70,800 butterflies, of which six hundred tags were returned. While most recoveries documented short-length journeys, the Urquharts reported a few long-distance flights spanning thousands of miles every year. Audrey Wilson of Grafton, Ontario, tagged one monarch that was recaptured about a month later in Baton Rouge, Louisiana. David Bridge of Kent Point, Maryland, marked another butterfly that was recovered near Marietta, Georgia. John Peplinski of Erie, Pennsylvania, labeled still another that was recaptured near Tallahassee, Florida. Norah Urquhart maintained a large map of the United States and southern Canada in their home, charting the beginning and end points of returned tags to visualize what they called "release-recovery lines." These lines revealed that the winged creatures traveled in a south-to-southwest direction every autumn, and as noted in Fred Urquhart's book *The Monarch Butterfly*, their best guess by the early 1960s was that the eastern population overwintered along the US Gulf Coast.[49]

But there were too many anomalies to be certain of an overwintering location. In 1967, for example, Mrs. C. R. Orr tagged a monarch butterfly in Houston, Texas, that was recovered in Mexquitic de Carmona, San Luis Potosí, Mexico. The new data prompted the Fred and Norah to reevaluate their prior claims, telling their volunteers in a newsletter, "One of the BIG mysteries of the monarch migrations is, where do they go from Mexico?"[50] The Urquharts suggested that *Danaus plexippus*, "D-plex" butterflies, might head west toward Baja California to join the western population of hibernating monarchs along the Pacific Coast, or they might continue to fly south toward Central America because all other lepidoptera in the *Danainae* family were found in the equatorial tropics.[51]

The Urquharts modified their study to speak to these inconsistencies in flight path. Because less than 1 percent of all tagged monarchs were later recovered, they worked to grow the Insect Migration Association's membership to increase tagging—and to therefore boost the number of tag recoveries. The Urquharts had recruited six hundred research associates by the early 1970s, and the organization received nearly a hundred letters every week. Fortunately, they obtained a $70,000 grant from the National Geographic Society to pay Norah Urquhart for administering the tagging program on a full-time basis. During a sabbatical year in 1969–70, Fred and Norah drove through southern Texas and northern Coahuila and Nuevo Leon searching for roosting sites, covering fourteen thousand miles. They met two volunteers, Charles Lipscomb of San Antonio and Kerry Yeager of Pearsall, who helped them realize that "monarchs do not remain in Texas but continue through Mexico." The Urquharts responded by contacting

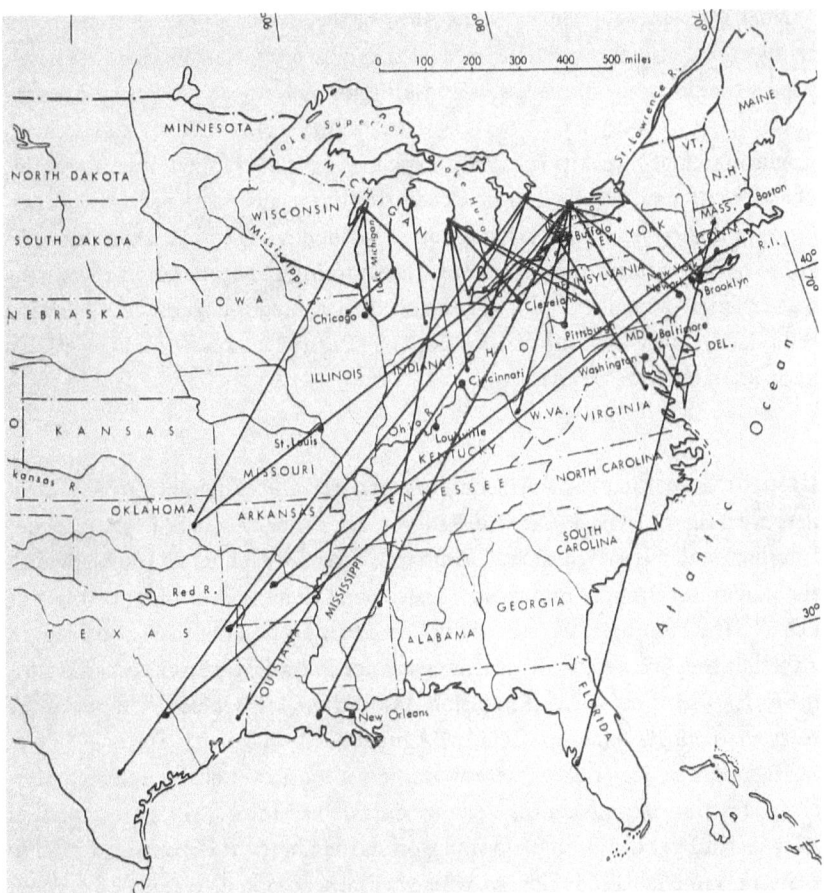

A map, maintained by Norah Urquhart, showing "release-recovery lines" of tagged monarch butterflies in the eastern United States and southern Canada. Fred Urquhart proposed that monarchs overwintered along the coastline of Gulf of Mexico. From F. A. Urquhart, *Proceedings of the Entomological Society of Ontario* (1965). Courtesy of Biodiversity Heritage Library.

entomologists at Mexican universities to ask about potential sightings and by focusing more of their organization's effort in marking monarchs throughout the state of Texas.[52]

The adjustments paid off. In 1971, the Urquharts received tagged butterflies from two locations northwest of Mexico City. Syble Mayberry of Eagle Pass, Texas, marked one monarch that was recovered near Huichapan, Hidalgo. Franz Pogge of Westover, West Virginia, labeled another that was recaptured near San

Miguel Tenochtitlán, México state. Most of the evidence pointed to overwintering areas located somewhere in Mexico. Norah contacted Mexico City newspapers the next year to place stories asking for assistance in locating monarch butterflies. On February 25, 1973, Norah Urquhart published an update in an English-language weekly *The News*, explaining that "to follow the migrations of a particular animal, be it a bird or a butterfly, it is necessary to have the assistance of a great many individuals working cooperatively." The Urquharts, she added, "are at present most interested in obtaining the cooperation of as many naturalists as possible."[53] Norah concluded that anyone curious about joining should write the University of Toronto. Tracking flight necessitated a trinational network of citizen scientists.

KENNETH BRUGGER, an American expatriate and self-taught garment engineer working in Mexico City for Rinbros, a subsidiary of textile giant Jockey International, responded to the Urquharts. "I read with interest your article on the monarch," Brugger told them. "It occurred to me that I might be of some help."[54] The Urquharts were receiving some help from more than just Brugger, although they were unaware of this assistance initially. Inclusion became a defining feature of the Insect Migration Association, but someone left out of the story was Brugger's partner, Catalina Aguado.[55]

Aguado was one of the most important, yet most overlooked, research associates. Born in 1949, Catalina grew up outside of Morelia, the state capital of Michoacán. As a child, Aguado snuck off to the library at Universidad Michoacana de San Nicolás de Hidalgo, where her father worked, to read books about natural history. "I was the girl that played with insects," Catalina recalled. She would lie on her stomach in the mud for hours to watch Mexican bluewings, swallowtails, and other butterflies. The free-spirited Aguado was convinced she would not marry young like her four sisters. At age seventeen, she was living in Mexico City, working in household sales for Philips Company. Aguado spent her free time traveling internationally, roaming from Guatemala and El Salvador to the United States and Canada. In 1971, a Canadian friend introduced Aguado, who was taking English-immersion classes in the Mexican resort town of Acapulco, to a charming gringo named Ken Brugger. He found Aguado captivating. "He followed me around," Catalina recalled. When she departed for another trip, fifty-three-year-old Brugger "wrote [her] letters on the back of a tortilla." Although more than thirty years separated their ages, Brugger and Aguado began a relationship that included looking for butterflies.[56]

Beginning in 1973, Catalina and Ken spent most weekends searching for monarchs. "I made Ken a net so we could chase butterflies," Aguado reminisced. "I taught him about them." Later in November, while driving toward the state of Michoacán, the couple saw a stream of monarchs being pelted out of the sky by a hailstorm. This event led them to the region. They developed a routine: driving a Winnebago motorhome to the mountains around Zitácuaro, located ninety-five miles (150 kilometers) west of Mexico City, then motorcycling or hiking the logging roads for reconnaissance. A year of adventure together shifted Catalina's outlook on marriage. In May 1974, Brugger and Aguado wedded in Texas. They spent their honeymoon in Toronto, where they finally met the Urquharts. "They told us more about their research and how the monarchs disappeared into a 'black hole' after they left Texas," Aguado remarked. Fred and Norah outfitted them with photographs of monarchs at the stages of development—caterpillar, chrysalid, and adult—and paid them a stipend of $1,500 for their labor. When Brugger and Aguado returned to Mexico, they dedicated nearly every day to searching for the winged wanderer.[57]

Aguado's familiarity with culture and countryside proved decisive in locating the butterflies' whereabouts. Local ejidatarios viewed Catalina and Ken with suspicion during expeditions. It was perhaps because Brugger was a white foreigner, or maybe it stemmed from middle-class Aguado approaching working-class campesinos, or it could have been that native residents wanted to keep the biological wonder a secret from outsiders or locals worried about what might happen to their ejidal rights. Whatever the source of uneasiness, Catalina was able to calm apprehensions. "Some said that we were looking for Zapata's treasure or that we wanted to take their minerals," Aguado remembered. "But I'd tell them, oh, no, we're looking for monarchs." Unfamiliar with scientific names for the butterflies, Catalina showed them pictures. "'Why?' they'd ask, and I would say because we wanted to educate people about how they live and travel. They were okay with that." Aguado persuaded one local dubbed Agapito to serve as their guide.[58]

In December 1974, Aguado and Brugger spotted some monarchs flying closer than usual to the forest canopy but were unable to see them land anywhere. The Urquharts had told them that monarchs required a hibernating area of thermal stability—too cold and they'd freeze to death, too hot and they'd perish from burning up all their fat reserves. The two research associates obtained topographical maps of the region to locate places where the altitude was most likely to hold temperatures between 35 to 45 degrees Fahrenheit. Parking their motorcycle at high-elevation trailheads, Catalina and Ken usually hiked eleven miles a day covering the forested terrain of ejidos.[59]

On January 2, 1975, the couple left their Winnebago at 4 a.m. to meet Agapito, who had brought his uncle's horse to carry their backpacks of water, food, maps, notebooks, and cameras, to a trail near Ejido Nicolás Romero. They could hardly see a few hundred feet in front of them because of low-hanging clouds. They finally pierced through the fog at a dramatic summit called Cerro Pelón, which funneled into an *arroyo*, a streambed just trickling from the recent showers. "That's when we saw them," recalled Aguado. Monarchs upon monarchs, most of them clinging to tree branches and trunks, made the evergreen forest look many shades of orange.[60]

Two days later, Catalina and Ken encountered another butterfly colony at Sierra Chincua, thirty miles away from the first cluster, with another important finding. Among the menagerie of monarchs, they found two white-colored tags from the Insect Migration Association: One had the serial number S2-294 glued to its wing, labeled by Mrs. C. Emery of Nevada, Missouri, and another tag was homemade with the number 84 stuck to it, marked by John McClusky of Fredericksburg, Texas. Aguado recalled that she couldn't fathom how butterflies from all over North America had come gather to less than two hundred kilometers from her birthplace. Catalina and Ken raced down the mountains to find a telephone. On January 9, Brugger called the Urquharts. "We have located the colony!" he informed Fred and Norah, voice trembling with excitement. "We have found them—millions of monarchs—in evergreens beside a mountain clearing."[61]

It took an entire year for the Urquharts, now in their sixties and dealing with health issues, to arrange for travel to Mexico. In January 1976, Fred and Norah finally met up with Catalina and Ken, as well as a photographer, to visit the Cerro Pelón colony. On one of their trips to the overwintering site, Fred recovered a tagged monarch bearing PS-397, which traced back to Chaska, Minnesota—two thousand miles away. Besides enjoying their major scientific finding, the Urquharts dedicated much of their three-week stay to tagging ten thousand butterflies with fuchsia-pink labels. They had pieced together the puzzle of the southward journey during fall, now it was time to understand the northward remigration come spring.[62]

In August 1976, Fred announced the discovery to an international audience through an article in *National Geographic* magazine. The lepidopterist was purposely vague in not offering an exact location of the overwintering area. Fred recounted how over thirty years of organizing three thousand lay volunteers to place three hundred thousand tags on monarchs had led him and Norah to a biological spectacle far beyond their wildest dreams. Democratizing who could participate in science was responsible for authenticating the migration, but it was not so gratifying for those in Mexico. Despite adorning the cover photo, research associate Catalina Aguado Brugger was reduced to a footnote; she was mentioned only once in the fourteen-page spread: "Ken Brugger doubled his field capacity by marrying a bright and delightful Mexican, Cathy."[63] Aguado was a crucial member of the Insect Migration Association for helping locate the roosting sites in Mexico, but she was hardly recognized at the time (or thereafter) for her contributions.

Aguado's sidelining was indicative of what scholars of science and technology studies have called "boundary-work," a concept used to discuss how scientists uphold their elite status in society by creating hierarchies of knowledge production between professional and amateur. Participatory monarch research challenged distinctions of expert and layperson while upholding boundaries of race and nationality. Citizen science was therefore Janus-faced: Inclusion and exclusion became dueling sides during the discovery of the trinational migration.[64]

INTERNATIONAL RECOGNITION of the monarch migration heated up smoldering tensions between Fred Urquhart and Lincoln Brower. They held different national loyalties. They came from different intellectual traditions. They carried different conceptions of how a protected area would function. The Urquhart-Brower quarrel set the stage for top-down conservation.

Mistrust developed early due to funding disparities across nation-states. In January 1973, Brower contacted Urquhart to inquire if he knew about any clusters of monarchs within Mexico for studying cardiac glycosides. Fred responded that if any concentrations of butterflies were discovered, he would forward some milkweed samples. When Fred mailed *Asclepias* specimens that Catalina and Ken had collected during their explorations, Lincoln replied thanking him and inquiring if they had made any progress on finding hibernating sites. The letter came on January 8, 1975, only six days after the Bruggers had located monarchs at Cerro Pelón. Urquhart ignored the letter, miffed that Brower, an American lepidopterist, had been awarded a large grant from the US National Science Foundation. Urquhart had applied for the same grant, "since 90% of our Research Associates reside in the United States," but the Canadian lepidopterist didn't get a penny. Fred and Norah complained about operating their tagging project on a "shoe-string budget." Lincoln reached out again in September, this time Fred replied that he was finishing up a piece for the National Geographic Society, which had given him another round of financial support. Once the article hit magazine stands, Fred promised to be forthcoming with the precise locale.[65]

Except that never happened. Fred and Norah met with the National Geographic Society's editorial board, with whom they agreed not to reveal the exact overwintering location until the Mexican government took steps to protect the monarch butterflies. Otherwise, they explained to Brower, "it was anticipated that many people, collectors, film makers, etc. would wish to visit and, as happened in other similar situations, destroy it."[66] The Urquharts had grown increasingly possessive, even paranoid, about who had access to information about this biological site of rising international significance. Brower would have to wait.

Urquhart and Brower also cast doubt about each other's scientific legitimacy by contrasting laboratory and field research. Contrarian, Fred did not think Lincoln's work on the anti-predator strategies of monarchs was worthwhile. At a meeting for American Association for the Advancement of Science, Urquhart questioned their cardenolide-based toxicity by stating that he had once eaten a monarch, and it did not taste bitter but neutral, like "dry toast."[67] In another jab at Brower's approach, Urquhart told volunteers that biologists should "experiment with this so-called distasteful aspect of the monarch butterfly—in nature, not in a cage."[68] Fred's critique, in other words, was that laboratory experiments could never authentically duplicate the real world. Alternatively, Lincoln judged the Urquharts' study to be merely gathering data that any naturalist could do, stating their methodology was "an amateurish, self-serving approach to biology that isn't science."[69] Or put another way, tagging wasn't science. Without manipulating variables in a lab, Brower thought, field observations could only support loose correlations and not solid causation. Lincoln also held the position that advancing biological science required the open sharing of published information so that findings could be replicated or refuted. When Urquhart refused to do so, it pissed Brower off.

Because Fred would not divulge the location, Lincoln conducted some detective work of his own. William Calvert, a postdoc entomologist at the University of Massachusetts, agreed to be his feet-on-the-ground because Brower was tied down with college teaching. Calvert was a Texas native who often explored northern Mexico and was, as Brower characterized him, "a real easy-going sort of hippie. He really did not give a damn whether he had a job or not, as long as he was managing to survive. But he really liked doing biology."[70] Brower and Calvert narrowed their search based on two clues from the Urquharts: The *National Geographic* article had mentioned that the overwintering site was at ten thousand feet in elevation; and in a Lepidopterists' Society publication, they stated that the butterfly colony was in the northern portion of the state of

Michoacán. On a topographical map, Brower and Calvert circled every high-altitude spot within the state to investigate. In early December 1976, Calvert left for Mexico.[71]

On New Year's Eve, Brower received a telephone call with some exciting news: Calvert had already located the hibernating area at Sierra Chincua. He told Brower that locals in the town of Angangueo had pointed Calvert to a resident who was to the nephew of municipal president Manuel Arriaga Nava and who helped him find the roosting grounds. Calvert returned home for schoolwork, so Brower set out in January 1977 for Mexico's Trans-Mexican Volcanic Belt. Coincidentally, Lincoln encountered Fred, Norah, Ken, and Catalina on the mountain tagging butterflies. The Urquharts were wholly baffled by his arrival and treated the rival lepidopterist with contempt. "It was not a pleasant meeting, to use an understatement," Brower remarked.[72] The researchers ignored one another for the remainder of the trip. In the magazine *Natural History*, the same publication that Norah had asked for volunteers, Brower declared that his search team had independently discovered the overwintering home, which he renamed "Site Alpha" in an apparent swipe at the Urquharts.[73]

Egos flared. In a newsletter to research associates, the Urquharts accused Brower's party of several abuses. *Bad*: They had followed the Bruggers to find the unspecified hibernating area. *Worse*: They had purposely set a fire under the trees so that rising smoke would disturb the monarchs for a photo opportunity. *Worst*: They were now widely publicizing its exact whereabouts, and visitors were flooding the site before formal protections could be established. Brower defended himself: Calvert had been lucky to meet someone who already knew the location, his article shared no more information than Fred and Norah had, and their local guide had started the small blaze so that cows grazing at the meadow could feed on smoked-out butterflies. Plus, researching how mice and birds ate the butterflies, Brower argued, yielded new insights "in spite of old uncle Fred." Squabbling aside, the Urquhart-Brower feud was two experts, one Canadian and one American, arguing over who should manage Mexican lands.[74]

The saga underpinned federal-level conservation to come. Both lepidopterists asserted that uncovering a trinational migration compelled international action for habitat protection. "Having made the discovery and realizing what damage might be done to the colonies," Fred and Norah Urquhart told their volunteers, "we have been working with an International Conservation Organization in cooperation with Mexican authorities to take steps to prevent further risks."[75] Brower proposed that "a coordinated international effort should be mounted a soon as possible to protect this extraordinary site from what could be inevitable destruction in the not too distant future."[76] But given the asymmetrical power among the three nation-states, calls for collective action verged on calls for foreign intervention.

THERE'S ANOTHER LAYER to the discovery story that rarely gets told, a bottom-up view that has been more important to Mexicans who live near the monarch butterflies. Jesús Ávila Montes de Oca recalled first seeing a tagged butterfly sometime before 1975, before Aguado and Brugger did, but he declined to inform the Urquharts. Descended from a family of powerful estate owners called *hacendados*, Ávila has been described as big, light-skinned, used to commanding the labor of others, and given to wearing a pith helmet while doing so. Ávila lived in Donato Guerra, México state, and he organized deer hunting parties for friends, bringing them up to Cerro Pelón through Macheros, where he stopped to hire local men as guides. It was on one of these outings that Ávila first saw a marked monarch. Whether or not he recovered a tag, Ávila was fondly remembered in Macheros for another reason.[77]

Ávila attended an audience with the governor in Toluca when "two Canadians," probably Fred and Norah Urquhart, asked municipal presidents for help in locating monarchs. Governor Jorge Jiménez Cantú (who addressed Ávila by the nickname "Bear" since he was fat) asked him: "Hey, Bear, you know where the butterflies are?" Ávila responded that he would find out, leaving for Macheros to summon his hunting guides Valentín Velázquez, Leonel Moreno Espinoza, and Elidió Moreno de Jesús. Valentín and Leonel went looking around Llano de Tres Gobernadores on behalf of Ávila, but torrential rain interrupted their search. The next day, Valentín and Elidió returned and found a massive cluster of monarchs, bringing live butterflies in bags back to Ávila as proof of the colony's existence. A Mexican broadcasting station named Televisa interviewed Ávila and the Urquharts about the discovery. The documentary program drew national attention to the plight of monarchs, as evidenced by the president of Televisa receiving hundreds of letters on the topic from across Mexico.[78]

Once the program aired, Ávila worried about what the rush of sightseers would do to the overwintering site. Around 1977, perhaps earlier, Ávila used his political connections in México state to secure work for the local men from Macheros as *guardabosques*, or forest rangers, through an agency that became the State Commission of Natural Parks and Wildlife (Comisión Estatal de Parques Naturales y de la Fauna, CEPANAF). Ávila approached Leonel, who was serving as ejido commissioner at the time, and told him, "I'm here because I

Left: Close-up photograph of *hacendado* descendant Jesús Ávila with a monarch resting on his commanding pith helmet. Courtesy of Carlos F. Gottfried.

Below: CEPANAF rangers Genaro Reyna, Elidió Moreno, and Valentín Velázquez pose with *armas de retrocargas* (muzzle-loaders) next to Valentín's son Miguel at Cerro Pelón, circa 1977. According to interviews, Ávila sent local men from Macheros to search for the monarch butterfly colonies and Velázquez discovered one of them. Courtesy of Carlos F. Gottfried.

need three people. You are one, and who else do you want us to take?"[79] Leonel chose Elidió and Valentín to travel with Ávila to Mexico City and register all their names with the Ministry of Agriculture and Water Resources. The three men of Macheros now watched over Cerro Pelón to conserve the butterflies, even fulfilling their duties without pay for the first few years since "Don Jesús" had commanded them to do so. As Elidió later recollected, Ávila instructed them all to "'get up there and do what you're told' because, yes, he was the one in charge."[80]

After the CEPANAF positions formalized, Leonel, Elidió, and Valentín became some of the few people in the community of El Capulín with economic security from forest protection. They enjoyed stable employment with paid vacation, health care, and government pensions. The rangers spent most of their time working to prevent logging wherever the butterflies roosted, but they also performed other tasks like maintaining trails, building fire breaks, and regulating tourist use. They managed to protect monarchs and their habitat at Cerro Pelón despite several constraints. Working for a México state–based agency, the CEPANAF crew saw their jurisdiction as ending where the state line did along the crest of the mountain, even though the butterfly colony alternated sides of this border every year. No matter where they patrolled, they also did not have authority to fine or arrest locals for tree cutting. All that Leonel, Elidió, and Valentín could do was talk to the loggers they encountered and make reports to their superiors. A conservation job took care of them, so they took care of the forest.[81]

Whom do we credit with discovering the monarch migration? Fred and Norah Urquhart? Catalina Aguado? Lincoln Brower and Bill Calvert? Jesús Ávila? Elidió Moreno and Valentín Velázquez? From the perspective of citizen science, a transnational network of collaboration was responsible—these relationships set the basis for ecological internationalism. However, in the world of lepidoptery, only lead researchers, usually white men such as Fred and Lincoln, received attention. With discovery came recognition, with recognition came power, and with power came authority over conserving the biological marvel. Despite an international fanfare attracting influential figures from Canada, Mexico, and United States, it was telling that CEPANAF's presence went unrecognized by scientific elites.[82]

Saving the Butterflies

I piled into a van with other foreign tourists to travel from Cerro Pelón to Sierra Chincua, two butterfly colonies separated by a thirty-mile drive. Despite the short distance, it took us a couple hours to navigate the terrain with a speed bump, a fruit stand, and a Virgen de Guadalupe shrine at every hamlet along the way. We passed by ejido after ejido with local Mexicans working on avocado orchards, at poinsettia farms, and in a mine outside the town of Angangueo. Murals of monarchs, painted on the sides of white adobe houses, pointed the way. We had barely left behind the concentration of buildings, punctuated by a Gothic-style cathedral, when I finally saw a large roadside sign for "Sanctuario Mariposa Monarca."

The van shuttled us for most of the elevation gain, so unlike the steep trail to Cerro Pelón, which needed to be navigated on horseback, I decided that the gentle path at Sierra Chincua could be easily walked on foot. Our guide Anayeli Moreno Rojas, one of the few women I saw escorting tourists, led the way to the butterflies. Growing up near Cerro Pelón, Ana recalled that big tour companies from Mexico City and Morelia used to bring buses full of people to the sanctuary. Other than stopping to dine at her mother's eatery, Restaurante de Doña Rosa, very few pesos stayed in her local community. Moreno, as a first-generation college student, chose to major in tourism development to learn ways to benefit from the hundreds of thousands of visitors who come each year.[83]

On our short hike to the roosting site, I asked Ana what I knew was a politically fraught matter: Did she know who "discovered" the monarchs? Moreno skirted the question, sharing that her family members had known about the butterflies long before foreign scientists got involved and that local communities had reasons not to share with outsiders. "Ya sabes," Ana stated, "el silencio es una forma de conservación." (You know, silence is a form of conservation.) Moreno added that other guides would probably give different answers.

Moreno met Aguado once to talk about the historic moment. Ana shared that Catalina didn't say much, perhaps hurt by her erasure from the story. In 1998, her husband, Ken Brugger, died, and many memorialized his passing. Orley "Chip" Taylor, the director of Monarch Watch, was one of them. Taylor had visited Cerro Pelón that winter to help shoot a documentary film. In Macheros, Taylor met an elderly man named Benito Juárez who claimed he had guided Catalina and Ken up to the colony. The ninety-six-year-old recounted how he had discovered the location himself while hunting as a teenager, so Juárez knew where to go looking and "led the Bruggers directly to the site." This version of the story reached public audiences when their exchange aired on television and when Taylor wrote about it in a Monarch Watch newsletter.[84] But this version was false: Catalina and Ken possessed maps showing they traveled through the Michoacán side, not the México state side, to get to Cerro Pelón. Taylor retracted the story after much insistence from Aguado that the details did not add up, but the damage was already done. It sadly revealed a pattern of exclusion in top-down conservation.

As a border-crossing figure in science, US lepidopterist Robert Michael Pyle was cognizant of the contradictions. His path to protecting the winged wanderer arrived along a path of irrigation water. Born in 1947, outside of Denver, Colorado, Pyle grew up chasing butterflies along the High Line Canal. With his mother's support, he joined the Lepidopterists' Society at twelve years old, becoming its youngest card-carrying member. Coincidentally, Pyle met one of the founders, Charles Remington, during the next summer on a family vacation in Crested Butte. While fishing with his father, Pyle encountered a group of scientists with butterfly nets in tow, including Remington, who were all working out of the Rocky Mountain Biological Laboratory nearby. They caught a monarch or two.[85]

Pyle crossed paths with Remington again when he turned twenty-five years old, now a master's student at the University of Washington, writing an illustrated field guide about the state's butterflies. On a Fulbright scholarship to Britain studying at Monks Wood Experimental Station, Pyle learned about Xerces blue, the first known butterfly in North America to go extinct from habitat loss in the 1940s. Pyle realized that wildlife protections mostly focused on large vertebrates, those "furries and fuzzies with backbones," but little was done for invertebrates like butterflies. Pyle understood that the Lepidopterists' Society was committed to collecting, not conservation, so he decided in 1971 to create Xerces Society, the first organization dedicated to protecting lepidoptera. Pyle mailed postcards to all the scientists and collectors he knew asking them to join; Remington replied with an invitation to the annual gathering of the Lepidopterists' Society.[86]

At the 1972 meeting in San Antonio, Texas, Remington held a special session called "Endangered and Extinct Lepidoptera," with Pyle giving a talk to launch the Xerces Society. Remington was impressed and afterward asked Pyle to come to Yale, where he could work on a PhD in butterfly ecology and conservation. Remington also introduced Pyle to one of his former students, Lincoln Brower—Linc and Bob immediately hit it off. Pyle resolved to go to New Haven. There, Pyle grew Xerces membership by organizing annual butterfly censuses called Fourth of July Counts, as well as by throwing expertise behind land-use decisions affecting them, all the while writing a dissertation that placed his work in its historic, scientific, and political context. In 1976, Pyle finished his doctorate and followed his wife, Sally Hughes, a scientific illustrator and Xerces supporter, back to her native Britain.[87]

At Cambridge, Pyle was volunteering for the International Union for the Conservation of Nature (IUCN) when an opportunity came his way. The IUCN, founded in Europe during the late 1940s, was created as a political body for coordinating wildlife protection on a global stage. To raise money for the undertaking, in 1961, a group for British nature enthusiasts, including Sir Peter Scott, son of Antarctica explorer Robert Falcon Scott, launched the World Wildlife Fund. A household name as founder of the Wildfowl Trust and host of BBC's nature series *Look*, Scott headed the IUCN's Survival Service Commission (later renamed Species Survival Commission). The commission contracted inventories of extinction risk, what became known as *Red Data Books*, and appointed specialist groups to prioritize worldwide conservation measures. In March 1976, Scott asked Pyle, whom he had met through the British Butterfly Society, if he would chair a Lepidoptera Specialist Group. The British wildlife painter, who had designed the World Wildlife Fund's famous panda logo, surprised the American lepidopterist with the request. "I was a long-haired, long-bearded hippy," Pyle reminisced. "He was a clean-cut conservative from an aristocratic family." Pyle still accepted the offer.[88]

In August 1976, Pyle convened the first meeting of the Lepidoptera Specialist Group at the Fifteenth International Congress of Entomology in Washington, DC. The task force, initially consisting of three scientists from the United Kingdom, one from Peru, and three from the United States, was charged with drafting an agenda for protecting the world's threatened invertebrates. Although not an official member until later, Lincoln Brower joined the gathering. Thomas Lovejoy, another Yale graduate, who worked as a biologist for the World Wildlife Fund, hosted them all at its Washington headquarters. Pyle and the other specialists debated the threats to many terrestrial arthropods and ranked them accordingly, but one issue rose unanimously to the top. Shocked by recent news of Fred Urquhart's *National Geographic* article, they decided that protecting the monarch's hibernating grounds in Mexico was "the number one priority in butterfly conservation."[89]

Pyle recognized that putting the priority into practice was going to be complicated given uneven power dynamics during third world decolonization movements. In 1977, the government of Papua New Guinea hired Pyle and Sally Hughes to conduct a conservation feasibility study of Queen Alexandra's birdwing butterfly, a rare insect that was distinguished as the largest species of lepidoptera in the world. Likened in size to a small bird with its foot-long wingspan, its name evokes British royalty because the butterflies' habitat was once a British colony, then an Australian protectorate. Even though Papua New Guinea had gained independence two years before their arrival, Pyle and Hughes witnessed the continued presence of neocolonial control: British-owned oil palm plantations were destroying birdwing territory just as black-market collectors were sending bird wing specimens to Europe. Besides meeting foreigners like him, Pyle encountered Indigenous Melanesians while pursuing butterflies in the rainforest, recognizing that local peoples were "too ensconced in the habitat" to advocate for establishing a US-style national park. Pyle and Hughes thus recommended that Papua New Guinea's Wildlife Branch should focus their efforts on curbing the illegal insect trade, and on rearing valuable species, to protect the endangered butterfly.[90]

A year later, Pyle wrote an essay for Xerces Society titled "International Problems in Insect Conservation," reflecting on what lessons he had learned from working abroad and then applying them to monarchs in Mexico. Pyle worried about reinforcing what he termed "conservation imperialism" when first world scientists pressured third world nations for wildlife protections. He commented, "A major problem confronting international conservation efforts is that of external desires versus internal needs. When expatriate biologists perceive preservation needs which differ from national development goals, severe conflicts may arise between sovereignty and interference." Pyle believed that decisions over natural resources should be determined by national governments and their citizenry, not by foreign experts. "Such an approach," Pyle added, "will be essential to the Mexican monarch situation since certain ancient human uses are too well established on the land in question to simply displace them for a total preserve."[91] That is to say, Pyle felt the three countries could work together in protecting the overwintering grounds as the foundation of a transnational migration, but neither Canadian nor American models should be forced on Mexico.

Both internal and external forces pushed Mexico's federal government to do something about its unique monarch habitat. After reading the *National Geographic* story, a Mexico City lawyer named Rodolfo Ogarrio decided to take a trip to the hibernation areas for their proximity to his home. In January 1977, Ogarrio traveled to Sierra Chincua, where, much to his surprise and delight, he met rival lepidopterists Fred Urquhart and Lincoln Brower, who both (separately, of course) showed him around. Ogarrio was inspired by the experience, so much so that he wanted to help preserve the place before the inevitable tide of tourists came to see the spectacle. Ogarrio and industrial engineer Fernando Ortíz Monasterio, also of Mexico City, founded the organization Pro-Mariposa Monarca, A.C. (Asociación Civil, or nonprofit) for such a purpose. Pro-Monarca wanted, in the words of Ogarrio, "to bring together members of various institutions that share a common interest . . . whether they belong to the government, scientific, or private sector, and whether they be Mexicans or foreigners."[92] With the help of scientists at National Autonomous University in Mexico City, Pro-Monarca lobbied federal officials in the Ministry of Agriculture and Water Resources to establish a protected area around Angangueo, Michoacán.

While Pro-Monarca got more hands-on, Pyle flexed the International Union for the Conservation of Nature's political muscles. In March 1979, the US lepidopterist attended a meeting of the IUCN's Survival Service Commission in Costa Rica. Pyle convinced the committee to pass a resolution asking President José López Portillo to take concrete steps ensuring the perpetuation of the monarchs' overwintering areas. Peter Scott conveyed a special message about the Costa Rica resolution to Mexico's president, which was followed up by similar declarations from other organizations such as the National Geographic Society, Smithsonian Institution, British Butterfly Society, and Society of European Lepidopterists. Foreign conservationists, at Pyle's request, were walking a fine line between international concern and foreign intrusion.[93]

A year later, in March 1980, the Portillo administration issued a decree providing vague conservation measures for "areas in which the butterfly known as the Monarch hibernates and reproduces," despite ejido communities having communal ownership over these lands stretching back to the post-Revolution reforms of the 1930s. It is difficult to judge what had the most influence over Portillo's decision to protect the monarch migration, but Pyle felt the decree was "probably the result of Sir Peter's letter to the President."[94] If true, the US lepidopterist had violated his own rule: External desires outweighed internal needs.

BUT PYLE COULD DO BETTER living his ideals and reaching out to his Mexican counterparts, for there were inadequacies with the 1980 edict. The federal government gave protections to monarchs in situ instead of the forest where they overwintered, which meant that logging still threatened their habitat when the butterflies were not present. Brower, now at the University of Florida, called Pyle about Portillo's decree, sharing his concern that "roosts may be gone when the Monarchs return or attempt to shift in response to local conditions." Pyle replied that a joint conference between the Lepidopterists' Society and Sociedad Mexicana de Lepidopterología might offer an opportunity to advance a proposal for a new national park or preserve dedicated to the monarch butterfly. In anticipation of the multinational event, Brower promised to "bury the hatchet" with Fred Urquhart for the sake of conservation.[95]

Visiting one of the butterfly colonies during winter 1980–81 was pivotal on numerous fronts. Brower organized for Pyle and Lovejoy to join him at Sierra Chincua where they toured the overwintering site with Ogarrio and Monasterio of Pro-Monarca, the latter of whom Pyle called Zorro for his dashing looks, as well as Mexican entomologists Leonila Vázquez García and Héctor Pérez of National Autonomous University. Vázquez, a prolific insect researcher who would describe thirty-nine species new to science, also served on the IUCN's

Left to right: Rodolfo Ogarrio, president of Pro-Monarca; Leonila Vázquez García, entomologist and member of the IUCN's Lepidoptera Specialist Group; Vázquez's colleague Héctor Pérez; and monarch researcher Lincoln Brower at El Llano de las Papas, Angangueo, Michoacán, 1980. Brower explains the "string of pearls" concept to colleagues for protecting the monarchs' overwintering habitat. Courtesy of Robert M. Pyle.

Lepidoptera Specialist Group. During the trip, Brower and Vázquez developed a conservation strategy, modeled after UNESCO's Man and the Biosphere program, dubbed the "string of pearls." Like prized gems, the roosting areas were defined as "core zones," where extractive activities would be forbidden, and these nodes were strung together by a chain of "buffer zones," where natural resource use could still happen with government oversight. Brower and Vázquez thought a biosphere reserve could reconcile the tension between biological conservation and sustainable use, especially given ejidos' preexisting claims to the forest.[96]

An extraordinary storm also underscored the desire for new wildlife protections. Starting on January 10, 1981, three consecutive days of rain and snow pelted the Trans-Mexican Volcanic Belt, followed by a straight week of nightly subzero temperatures. A mass die-off of 2.5 million butterflies resulted, 20 percent of the colony. Brower's research team, led by Bill Calvert, detailed the striking impact of thinned forest on monarch survival. A heavy logging operation around Sierra Chincua during 1978–79 had altered forest composition from

400 trees per hectare to 150, causing temperatures to be 3.5 degrees Celsius lower in lumbered areas. In the 1981 event, butterflies, wet from the downpour, froze to death. Thinning the forest was like removing a protective blanket over the monarchs, exposing them to greater weather extremes and thus heightened mortality rates. As Brower recognized, conserving a "closed canopy" of intact forest was crucial to the persistence of overwintering monarchs.[97]

With a conservation proposal in mind, Brower and Pyle received help from the state government to identify core zones. In January 1981, the pair of American lepidopterists traveled to Morelia, the capital of Michoacán, for a meeting with Governor Cuauhtémoc Cárdenas, son of renowned former president Lázaro Cárdenas. Although the exact words of the exchange were not recorded, Cárdenas expressed his support for the endeavor if tourism and other industries could replace local logging income. Pyle agreed, sharing the success of butterfly farming in Papua New Guinea as an example of what alternatives could be developed. "Through protection," Cárdenas later stated, "we are going to have economic development in these regions so that the inhabitants can receive the same yields and productivities that they could get from the rational and intensive use of their forest resources."[98] Cárdenas also authorized Brower to use state-owned helicopters for conducting surveys along the Michoacán and México state border to locate any previously unidentified butterfly colonies. Brower then forwarded the aerial reconnaissance to Vázquez for designing the parameters of a biosphere reserve.

Trinational collaboration culminated with the first-ever Symposium on the Biology and Conservation of the Monarch Butterfly, which began on July 31, 1981, and later became known as MonCon I. Lepidopterists, mostly Mexicans, Canadians, and Americans, gathered just south of Mexico City in Cocoyoc, Morelos, with special government attendees to deliberate the future monarch protections. Brower, as president-elect of the Lepidopterists' Society, started things off by symbolically presenting a silver gavel to Urquhart, though absent from the event, to ease past animosity between the two scientists.[99]

With geographer Carlos Melo Gallegos, Vázquez and Pérez presented "A Preliminary Proposal to Create a National Park in the Chincua, Rancho Grande, and Campanario Mountain Ranges of the States of Mexico and Michoacán" based on Brower's string-of-pearls design. Juan José Reyes Rodriguez of the Ministry of Agriculture and Water Resources agreed to circulate the "Monarca Manifesto" among federal bureaucrats. With Lovejoy securing World Wildlife Fund grants for Pro-Monarca, Ogarrio took responsibility for devising tourism-based job opportunities for ejido communities. "Each of these countries shares in the legacy of the migratory Monarch, the zenith of North American insect evolution," Pyle concluded at the conference's end. "Yet each nation also shares the responsibility for preserving that legacy; no one of them can do it alone." Pyle was expressing the desire for ecological internationalism. MonCon I organized major players of different nationalities behind conservation ideas for political action.[100]

A window of opportunity opened for these demands when President Miguel de la Madrid stepped into federal office in 1982. The new administration, from the Institutional Revolutionary Party like its center-right predecessor, established the first cabinet-level environmental agency in Mexico—the Ministry of Urban Development and Ecology. In January 1983, Pyle, Brower, and Lovejoy made another trip to Mexico City to meet with Vice-Minister of Ecology Alicia Bárcena Ibarra on creating a monarch reserve. The American scientists raised their concerns that Madrid's neoliberal economic policies, opening Mexico to outside capital because of its debt crisis, would come to undermine ecological commitments. The financial restructuring was already underway with a foreign-owned paperboard and plywood factory newly opened at San Felipe, Michoacán, adding regional demand for timber extraction around the butterfly colonies. For the US experts, it was time for the Mexican government to act.[101]

Pyle ramped up international pressure on the Madrid administration. In February 1983, the United Kingdom's Queen Elizabeth and Prince Philip visited Mexico. Pyle asked Peter Scott to forward a message to the Duke of Edinburgh, who acted as honorary president of the World Wildlife Fund. Consequently, Prince Philip discussed his personal interest in monarch butterflies with President de la Madrid, telling the head of state that the larger conservation community would support him when stricter wildlife protections were enacted. Because the monarchy got behind the monarchs, the World Wildlife Fund followed up with increased monetary support for Pro-Monarca for developing tourist regulations at the overwintering areas. Rodolfo Ogarrio helped to set up ticketed access at the butterfly sanctuaries, charging entrance fees to be collected by local guides as another source of income.[102]

Amid the activism, Scott charged Pyle with compiling a detailed list of the world's imperiled invertebrates for the International Union for the Conservation of Nature. With over a million organisms known to science at the time, Pyle recruited two British entomologists to divide the intellectual labor: Susan Wells worked on mollusks and Mark Collins on spiders while Pyle focused on butterflies and moths. Pyle felt that monarchs should be included, but how? The species was not in danger of going extinct, even though Pyle and Brower had independently declared that their unique migratory behavior was threatened by logging, wildfires, and tourism at the overwintering sites. Conventional distinctions such as "endangered species" or "threatened species" didn't work to explain the circumstances. "Since invertebrates often fail to fit into systems developed for vertebrates," Pyle noted, "we are innovating with new categories." In 1983, the *IUCN Invertebrate Red Data Book* was finally published, the first inventory of its kind for judging extinction risk in this taxon. Pyle listed the monarch butterfly migration, a three-thousand-mile-long journey across North America, as an "endangered phenomenon."[103]

Less than a decade after the transnational spectacle was discovered, this particular aspect of monarch biology was teetering on the edge of collapse. "These animals live on an ecological knife's edge," Brower often told Pyle during their meetings, and unless that edge, that habitat, could be maintained, "they are going to fall to one side or the other."[104]

THE SIERRA CHINCUA ENTRANCE was lined with storefronts selling all sorts of food and handicrafts, from tamales and Coca-Cola to scarves and pottery. Foreign sightseers like me intermingled with middle-class Mexicans who drove from the big cities for a weekend getaway in the mountains. Kids ran around the lines of visitors who shuffled through the ejido-operated shops. I bought a bag embroidered with butterflies, sewn by a local woman who said it took her over two hours to weave together the orange, black, and white tapestry. She was proud of her stitchwork, which set me back a hundred pesos, or about five US dollars. I imagined that our exchange was the type of thing Pro-Monarca and other organizations had in mind when they tried to develop economic alternatives to logging. But knowing that half of rural Mexicans lived below the poverty line, I had a hard time believing these tourism-related industries made up the income discrepancy for local communities.

At the entrance gate, the flags of Canada, Mexico, and the United States flapped high up in the breeze. Although the three national emblems were flown at equal height, the stars and stripes and maple leaf seemed to cast a shadow over the eagle sitting on a cactus. The flags projected cooperation, but they also masked unevenness. The same could be said for the history of migrating monarchs. Asymmetrical power between the countries hindered moments of cross-border solidarity, of ecological internationalism.

During the process of listing the phenomenon of monarch butterfly migration as endangered, Bob Pyle came to believe that Xerces Society should establish a separate entity called the Monarch Project. He thought that only a species-focused conservation campaign on a trinational scale would get a biosphere reserve in Mexico across the finish line. However, he started running into resistance. From Pyle's point of view, Ogarrio, Vázquez, and his other Mexican colleagues sent "implicit signals" that he was hindering their work. "I was afraid it was starting to look like the roots of conservation imperialism," Pyle remembered. Xerces Society launched the Monarch Project in 1984, but its attention would be limited to monarchs that overwintered in California because they largely migrated within the United States and therefore were more of a national issue. Pyle withdrew from Mexican conservation.[105]

One less scientist crossed the US-Mexico border, but many more goods would flow in-between nation-states during the coming decades. Oyamel firs were converted into cardboard boxes that held consumer electronics made at tariff-free border factories and destined for the United States. Genetically modified corn, grown across the US Midwest with heavy applications of herbicide, went to Mexico. The forest of winter monarchs and the milkweeds of summer monarchs were both disappearing from the landscape. During an era of neoliberal globalization, migratory butterflies became both symbol and victim of this economy.[106]

CHAPTER FOUR

Stepping Stones

I couldn't remember exactly where on my drive the cornfields started, but on back roads from the large railyards at North Platte, Nebraska, to the large campus at Lawrence, Kansas, all I saw was cultivated maize for eight straight hours. Headed south of state Highway 47, I zipped past a Bayer Crop Science Learning Center surrounded by experimental farms. Bayer Corporation, the German multinational aspirin company, purchased this facility from the Monsanto Company as part of a larger 2018 acquisition valued at over $60 billion.[1] The biotechnology products of Monsanto-turned-Bayer were not only valuable but also responsible for mile after mile of the "clean fields" I motored by. No weeds were in sight across the agricultural landscape of herbicide-tolerant corn.

I arrived at the University of Kansas to visit Monarch Waystation #1, a pollinator garden created by the conservation organization Monarch Watch in 2005, which seemed more like a butterfly oasis amid a desert of maize. After walking under a vine-covered trellis arch, I entered the garden featuring host plants for every kind of butterfly: pawpaw trees with their banana-like fruit for zebra swallowtails, wild senna with its gold-colored flowers matching the golden wings of cloudless sulphurs, and purple plantain for spotted-wing common buckeyes. A patch of common milkweed (*Asclepias syriaca*), usually waist high in the wild, grew taller than I could reach my hand up into the sky. "Enthusiastic" was the word that Monarch Watch founder Chip Taylor kept repeating to describe the gargantuan plants over the hum of chirping crickets. Taylor made the analogy that just as I'd taken several pit stops on my road trip there to eat food, take a leak, or fill up my car's tank, monarch butterflies needed their own waystations along their migratory path. If milkweed and nectar resources were located farther and farther apart—imagine no restaurants, bathrooms, or gas stations along a highway—completing the annual journey becomes more difficult.

Taylor, a University of Kansas insect ecologist who could easily be mistaken for Santa Claus with his snowy beard and cherry nose, explained to me why Monarch Waystation #1 was important. When the Insect Migration Association disbanded in the mid-1990s, Fred and Norah Urquhart left unanswered questions about the timing, pace, and survival of migrating monarchs. Taylor wanted to pivot away from his honeybee research, so he cofounded Monarch Watch to launch a monarch education and tagging program. The response of volunteers, especially from teachers and students, was overwhelming: Since 1992, two million butterflies have been tagged and nineteen thousand recovered. But participants from the US Midwest noticed the monarchs were becoming harder to find, a perception backed up by declining numbers recorded in Mexico. Taylor knew the culprit was habitat loss, estimating from land-use surveys that at least a million acres of grasslands—home to milkweeds and wildflowers—disappeared every year due to corporate agribusiness, plus another million annually from urban sprawl. Monarch Waystation #1 marked the first effort to recover a "milkweed corridor" through the Corn Belt, now a sizable network of fifty-two thousand registered habitats and counting.[2]

Although thousands of new waystations have been added every year, Monarch Watch must run to stay in place. "It brings to mind the Red Queen of *Alice in Wonderland*," Taylor remarked, "who tells Alice: 'Now, here we must run as fast as we can, just to stay in place. And if you wish to go anywhere you must run twice as fast as that.'"[3] Conserving the migration demands the restoration of a million acres every year just to stay in place, and far more than that to get ahead. It's like stepping stones crossing a river: Monarch Watch must frantically add rocks to the water even as floods of herbicides continue to push them downstream.

This tension is further revealed when considering both anti-GMO activists and NAFTA proponents adopted the monarch butterfly as the symbol of their cause. Mexico entered the General Agreement on Tariffs and Trade in 1986, the same year its federal government created a formal protected area for overwintering monarchs. Negotiations between Canada, Mexico, and the United States to expand GATT provisions were formalized under the North American Free Trade Agreement, which went into effect in 1994. Two years later, Monsanto introduced Roundup Ready crops to US markets. Corporate agribusiness ensured not only that glyphosate-resistant soybeans and corn were exported to Mexico but also that scientific concerns about spraying milkweeds were met with neoliberal fixes instead of transnational solutions. Created as a side accord to NAFTA, the Commission for Environmental Cooperation (CEC) opened diplomatic channels to settle trinational issues like conserving the monarch migration. In these forums, regional elites voiced desires for solidarity across borders while centralized governments blamed the inaction of neighboring countries as an excuse to limit their own investments. Local people, whether they were small Midwestern farmers or poor Mexican ejidos, struggled with how to protect milkweed for summer monarchs and forest for roosting butterflies. By the year 2000, conservationists agreed on restoring habitat corridors through the Corn Belt and redrawing the core boundaries of the biosphere reserve.

Stitching together the unraveling migration revealed that neoliberal conservation was more about bailing out big business than beautiful butterflies. Amid these top-down measures, there were brief moments of ecological internationalism, of rooted relationships built across international borders. Centralized states held the purse strings to conserve the larger landscape, but local communities interacted with the trees or milkweeds. Both resources and relationships were key to achieving transnational solidarity.

Campesinos and Conservation

Although scientists Leonila Vázquez García and Lincoln Brower first proposed creating a biosphere reserve for monarchs in 1981, a protected area had yet to be realized by Mexico's federal government. Pro-Monarca and Grupo de los Cien, two groups representing the nascent Mexican environmental movement, finally secured conservation measures from political elites. They argued that the development of ecotourism and other industries could replace lost income for local communities whose rights to the forest would be circumscribed by a new park. But the continuation, even acceleration, of logging inside the Monarch Butterfly Biosphere Reserve showed that economic alternatives were not going to be enough.

The main challenge in protecting overwintering monarchs was minimizing deforestation across the region. Two types of timber harvesting poked holes in the blanket-like forest: "heavy logging" (*tala inmoderada*), well-organized and well-equipped cutting destined for regional sawmills, and "ant logging" (*tala horminga*), small-scale tree felling for domestic purposes like cooking or heating homes. During 1982, the Ministry of Agriculture and Water Resources permitted sixteen thousand cubic meters of pine trees and two thousand cubic meters of fir for heavy logging. Ejidatarios usually received forest-use payments on their land titles and distributed the job of chopping down trees to community members. Ant logging took a toll as rural communities expanded their village footprints. For example, Ejido Cerro Prieto grew from seven families who first settled near the Sierra Chincua colony in 1928 to over five hundred people by 1980. The introduction of agrochemicals during the 1960s allowed villages to replace oyamel forests with maize fields on steep slopes with shallow soils. Both ant logging and heavy logging degraded the forest where monarchs roosted.[4]

Poverty compounded the problem. The Portillo administration, which had issued the 1980 decree for conserving monarchs, tied government spending to petroleum development through Mexico's state-owned company Pemex (Petróleos Mexicanos). When world oil prices fell in mid-1981 because of a global recession and a glut in crude oil from the Organization of the Petroleum Exporting Countries (OPEC), President José López Portillo responded by devaluing the peso to encourage foreign lending. In August, the federal government announced Mexico would not pay its $80 billion in debts—40 percent of which belonged to US banks. Paul Volcker, chair of the US Federal Reserve under the Reagan administration, offered Mexico a rescue package in exchange for promises to privatize and deregulate its economy. In 1982, Miguel de la Madrid, the newly elected president, inherited the debt crisis and decided to adopt the neoliberal policies outlined in the US bailout. Between 1982 and 1988, per-capita government spending on social programs declined by 38 percent just as poverty rates rose by nearly the same percentage. Three-quarters of the Mexican population, especially those living in rural areas, faced severe economic deprivation.[5]

Ejido communities were no exception. On a 1982 visit to the El Rosario colony for the television program *60 Minutos*, Mexico City reporter Jaime Maussan confronted an ejidatario named Macario with axe in hand. "Aren't these butterflies in danger?" Maussan asked, gesturing at the monarch-laden fir tree. "Why are you chopping this tree down?" Macario shrugged and then pointed to the village at the edge of the forest. "For my children," he responded. Artemio Martínez, who also resided near El Rosario with his wife and kids in a one-room wood plank house, suggested that poverty was contributing to population growth. "A family should have eight children," Martínez stated, "because children die so easily [from malnutrition], you need at least eight." Fernando Ortíz Monasterio of Pro-Monarca observed that "ejidatarios know they are extracting more wood than is growing back. But they must eat."[6] The federal government, now committed to austerity budgets under de la Madrid, was not going to commit funds to alleviating poverty or administering a federal reserve.

Pro-Monarca secured foreign capital for the purpose of developing new jobs at the overwintering sites. Carlos Federico Gottfried Joy, an industrial engineer from Mexico City who became president of the environmental nonprofit, received grants from the World Wildlife Fund, MacArthur Foundation, and The Nature Conservancy—all charitable organizations based in the United States. Pro-Monarca members reasoned that because the migration was a trinational phenomenon, it was appropriate for monetary support to come from outside of Mexico. Rodolfo Ogarrio argued that "the monarchs constitute a global resource and a global conservation issue." In 1984, at Pro-Monarca's urging, the Ministry of Urban Development and Ecology launched the Monarch Butterfly Protection Trust, an endowment made up of private donations plus financial commitments from the state governments of Michoacán and México to bankroll development programs for ejido communities.[7]

Another environmental organization, Grupo de los Cien (Group of 100), turned up the political heat on the de la Madrid administration for monarch protections. Grupo de los Cien came together in 1985 when famed Mexican poet Homero Aridjis wrote a declaration, signed by a hundred prominent artists, writers, and intellectuals, denouncing air pollution in Mexico City. They demanded the government reduce automobile smog by directing Pemex to remove lead from gasoline, by requiring the installation of catalytic converters on buses and trucks, and by mandating cars be taken off the streets for one day every week by license plate number. After pushing for better air quality in the federal district, Aridjis turned Grupo de los Cien's attention to other environmental issues, including campaigning for butterflies near his childhood home. In April 1986, Aridjis convinced Manuel Camacho Solís, minister of urban development and ecology, that the federal government should create a biosphere reserve, including at Cerro Altamirano, where he grew up. Cultural elites secured a commitment for conservation.[8]

A strategy was laid out at the Second International Conference on the Monarch Butterfly, or MonCon II, which began on September 2, 1986, in Los Angeles, California. Bringing together scientists and conservationists from every corner of the continent for discussion, Pro-Monarca outlined its programs, in the words of Carlos Gottfried, to "recognize and respect the basic needs of the local inhabitants that share forest resources with the monarch butterfly." To achieve this goal, Rodolfo Ogarrio believed "the most effective means to preserve nature was through the socioeconomic improvement of those communities." María Elena Camas de Castro and Susana Rojas González de Castilla shared that Pro-Monarca had organized ejidatarios to charge visitors entrance fees, constructed facilities to sell local handicrafts, and contacted urban travel agencies to bring more people to the overwintering sites. With fifty thousand visitors during the 1985–86 season, the assumption was that developing an ecotourism industry run by locals could make up the economic difference from lost logging. It was about to be put to the test.[9]

In October 1986, President de la Madrid issued a decree establishing a 150-square-kilometer federal protected area named Reserva Especial de la Biosfera Mariposa Monarca (Monarch Butterfly Special Biosphere Reserve). The presidential edict outlined formal boundaries and regulations along the Michoacán and México state border for five of the thirteen overwintering areas: Cerro Pelón, Cerro Chivatí-Huacal, Sierra del Campanario (also known as El Rosario), Sierra Chincua, and Cerro Altamirano. These sites made up the "core zones," where timber extraction was outlawed, surrounded by "buffer zones," where natural resource use required a government permit. Communal land titles remained in the hands of ejidatarios, but a new ecological reserve for monarchs meant their rights to the land were restricted.[10]

At the dedication ceremony outside of Sierra Chincua, dignitaries told an audience of ejidatarios not to worry because transnational cooperation would lead to economic development. On the one hand, Curtis Freese, representative for the World Wildlife Fund, jubilantly remarked, "The declaration of overwintering sites as an ecological reserve is without a doubt the most significant act for the conservation efforts of this international phenomenon." Ogarrio proclaimed, "The effort, like the flight of the Monarch butterfly, knows no borders and erases the differences between countries, emphasizing that we identify as human beings and nothing more." On the other hand, federal official Manuel Camacho Solís informed the crowd, "Here we have the campesinos of the core zones, with a collective promise to contribute to the protection of those sanctuaries without harming your subsistence." The daughter of Pro-Monarca founder Fernando Monasterio, Valentina, expressed hope that "the woodcutters who live here, although they know they cut trees to make a living, [would] be given some other kind of work, for example, as forest rangers or guides since they already know the place well."[11] However, voices of local communities were conspicuously absent during the celebration.

Near Sierra Chincua, the Mondragón sawmill closed its doors permanently after the federal government suspended its timber leases around the butterfly colony. With financial assistance from the Canadian International Development Agency, Pro-Monarca purchased the facility in 1987 to convert it into a tree nursery for oyamel seedlings, employing former loggers with temporary reforestation projects, as well as establishing a Christmas tree farm for longer-term work. "People told us we were making a mistake," Carlos Gottfried recalled hearing when explaining Pro-Monarca's plan, "because at least General Mondragón was protecting [trees] with his machine guns." Employed by Pro-Monarca, local women raised 350,000 saplings over the next two years and local men planted 150,000 of them inside the federal reserve. Although these opportunities were limited, Pro-Monarca demonstrated that jobs and justice were just as effective conservation tools as guns and guards.[12]

Left: Carlos Gottfried, who became president of Pro-Monarca in 1989, covered in monarch butterflies. Courtesy of Carlos F. Gottfried.

Below: Mexican women, who lived in local ejido communities, plant oyamel seeds at Pro-Monarca's tree nursery. The oyamel saplings would be transferred to the Monarch Butterfly Biosphere Reserve. Courtesy of Carlos F. Gottfried.

After the 1986 decree, administering the biosphere reserve became the responsibility of Pro-Monarca. The thirty-six-year-old Gottfried, who remembered singing "both national anthems" in his Mexican American family, traveled to the World Wildlife Fund headquarters in Washington, DC, to ask for continued monetary support. "They could not believe the number of employees we had, the number of projects, and the number of successes that derived from $20,000 a year," Gottfried stated. "For us in pesos, devalued at 300 percent, it was a huge amount."[13] Because World Wildlife Fund's sponsorship financed most of Pro-Monarca's programs, the environmental nonprofit hosted Prince Philip, honorary WWF president, again in 1987 at the butterfly sanctuary. Gottfried and other environmentalists wanted to show monarchs to the monarchy and secure more capital for conservation.

Some ejido communities received tangible benefits from the new federal reserve. During the 1990–91 season, about ninety thousand tourists visited El Rosario, paying entrance fees, local guides, eateries, and souvenir shops in what amounted to $1.4 million MXN ($350,000 USD) in total revenue. When television reporter Jaime Maussan returned to the butterfly sanctuary, he met Macario again. The ejidatario was not cutting down trees anymore but was escorting people to the overwintering site. "Why are you working here as a guide?" Maussan asked. As in their first exchange, Macario simply replied: "For my children." The tourism income was distributed across about 280 families who lived around the roosting site. Pro-Monarca also developed educational materials about the migration phenomena for teaching in primary schools, which generated interest in improving local classrooms. Transnational solidarity took hold where overwintering monarchs led to rural development.[14]

Other ejido communities, however, resented the federal reserve because they neither consented to its creation nor obtained job opportunities through ecotourism. Silverio Tapia Torres of Ejido Jesús Nazareno near Sierra Chincua explained, "In 1986, when the government created the five monarch butterfly reserves, we thought we'd get some benefits. But that never happened. Finally, in 1990, because our families were starving, some of us decided to log in the area without a permit."[15] The Ministry of Urban Development and Ecology had limited public visitation to El Rosario until the mid-1990s, so only those surrounding villages profited from sightseers. In July 1995, fourteen ejidos formed the Alianza de Ejidos y Comunidades de la Reserva Mariposa Monarca to organize a political defense of their rights to the forest. Dimas Salazaar, who owned three hectares (eight acres) of Ejido Francisco Serrato in Michoacán, was one spokesperson. "To the scientists, if you can speak the language of the monarch butterfly, please thank them," Salazaar stated. "Those little animals give us the opportunity to express ourselves as campesinos."[16] Logging intensified inside the federal reserve due to anxiety that lands might be expropriated from ejidatarios.

Lincoln Brower teamed up with scientists at National Autonomous University of Mexico to calculate forest degradation on the Michoacán side of the reserve. Using aerial photographs of Sierra Chincua, El Rosario, and Cerro Chivatí-Huacal taken in the years 1971, 1984, and 1999, they determined that 44 percent of mixed oyamel-pine forest was thinned out by logging over a quarter century. Most shockingly, the rate of deforestation was 1.7 percent annually from 1971 to 1984 and 2.4 percent from 1984 to 1999, suggesting that timber harvesting had increased after the biosphere reserve was established. Brower rang the alarm bells, but Mexican environmentalists had to tread lightly in their treatment of ejidos. "We had to be careful not to be too associated with Brower, who was like a bull in a china shop," Gottfried observed. "He would say things like 'Why don't you just put these people in prison if they're cutting illegally?'"[17] As a toned-down response, Brower and Homero Aridjis proposed in 1996 to "lease the forests from the peasant communities," making annual payments to ejidatarios for conserved habitat using the interest of a $5 million fund from the US-based Packard Foundation. This solution was deemed more sensible to halt deforestation where monarchs overwintered.[18]

Unlike what Brower documented on the Michoacán side, the forest remained more intact on the México state side because of the presence of CEPANAF rangers. In 1982, Melquiades Moreno de Jesús joined the crew to make for five total ejidatarios who patrolled the Cerro Pelón colony under the oversight of *hacendado* descendant "Don Chucho" Jesús Ávila. Pro-Monarca had minimal presence at this overwintering area because Ávila did not want to give up authority as CEPANAF supervisor. Moreno recollected "walking in the mountains and we ran into loggers occasionally. It was difficult because we had to get them to understand that it was not right and, well, sometimes they paid attention to us and left."[19] While local people still cut down trees on Cerro Pelón, especially those from the neighboring Ejido Nicolás Romero in Michoacán, the state border became a defining line in deforestation. In 1993, Aridjis noted the spatial variance in the newspaper *La Jornada*: "It's absurd to subject the same ecosystem to two different sets of forest regulations, which forbid logging in the State of Mexico but allow it on the Michoacán side."[20] CEPANAF rangers made a difference in monarch conservation.

While CEPANAF managed Cerro Pelón, Pro-Monarca lost its grip on the four other overwintering sites. A power struggle was set in motion during the

CEPANAF ranger Genaro Reyna poses with an *arma de retrocarga* (muzzleloader), provided by his state supervisor Jesús Ávila, in a cloud of monarch butterflies, circa 1985. Photo by Frans Lanting. © Frans Lanting / Lanting.com.

1988 presidential elections when Carlos Salinas de Gortari, handpicked to uphold de la Madrid's neoliberal agenda for the Institutional Revolutionary Party, was challenged by leftist candidate Cuauhtémoc Cárdenas, former governor of Michoacán. Once the ballot tabulating system crashed, Salinas preemptively declared victory before all votes were counted. Mexican citizens, including Cárdenas, assumed fraud but did not contest the results. Government officials later admitted to rigging the election. President Salinas responded to weakness in the electoral process by consolidating his authority over federal agencies, including the management of protected areas.[21]

The Federal Environmental Protection Agency (Procuraduría Federal de Protección al Ambiente, or PROFEPA), created under the Salinas administration in 1992, became known as the "forest police" since the bureau enforced the laws governing the biosphere reserve. The forest police fined ejidatarios for cutting down trees within the core zones. CEPANAF ranger Elidió Moreno remembered arguing with PROFEPA officials as his crew was getting ready to build a firebreak around the butterfly colony. Elidió recalled carrying tools when "those aggressive PROFEPA guys came with their notebooks, and I got scared. I thought those bastards were going to denounce us. They were writing things down and asking, 'Why are you doing that?' [I told them] because we don't want the forest to burn down, and if the fire reaches here, this part will burn. What are we going to put it out with? With a hat? With a branch?"[22] CEPANAF maintained some autonomy at the sanctuary, but Pro-Monarca did not. According to Gottfried, President Salinas met with Kathryn Fuller, CEO of the World Wildlife Fund, and told her that Pro-Monarca had played a minor role in monarch conservation and federal officials were taking over the biosphere reserve. In 1994, the World Wildlife Fund terminated its patronage of Pro-Monarca. "That was the end of our active participation," Gottfried remarked. "The new government wanted to coerce us into their programs."[23]

The North American Free Trade Agreement went into effect that same year. Salinas and Bill Clinton, who won the US presidency by a plurality after third-party candidate Ross Perot split the vote, both pushed NAFTA to compensate for their political vulnerability with the support of multinational corporations. Monsanto Company sought to benefit financially from the trade deal by exporting transgenic crops. Monsanto CEO Robert Shapiro became a close friend of and donor to President Clinton. In 1994, Monsanto filed an authorization request to sell Roundup Ready soybeans. Herbicide-tolerant corn, cotton, and canola hit the market soon thereafter. These products, branded as the new frontier in "environmental sustainability," were called into question when they were found to harm monarch larvae and their host plants, *Asclepias*. The dramatic reduction of milkweeds across the US Midwest put the reproductive capacity of monarchs at risk. Habitat destruction on both sides of the border jeopardized different phases of the migratory cycle.[24]

Asclepias and Agribusiness

According to Monsanto, the company was in transition—from producing chemicals to making biotechnology. For executives, the shift offered to throw off its traditional past as a "scavenger capitalist," generating profits by recycling the waste of other businesses. In 1901, druggist John Francis Queeny had founded Monsanto in St. Louis on the manufacturing of saccharin, an artificial sweetener, from coal tar. By the 1970s, 80 percent of its products were derived from fossil fuels. When the Arab-dominated OPEC imposed an embargo against the United States in retaliation for its military aid to Israel during the 1973 Yom Kippur War, Monsanto's petrochemical feedstocks diminished and its production stalled. Genetic engineering presented a way out from natural resource dependence.

In 1997, Monsanto executive Bob Shapiro commented, "If we grow by using more stuff, I'm afraid we better start looking for a new planet."[25] For industry scientists, biology was supplanting chemistry, a discipline tainted by the environmentalist critique of insecticides like DDT poisoning the food web. They believed wholeheartedly that genetically altered plants would unleash a green revolution in agriculture by lessening reliance on chemical pesticides. Monarch butterflies proved otherwise, challenging Monsanto's claim that its biotech makeover freed farmers from the material bonds between people and plants.[26]

In 1981, a year after the US Supreme Court had ruled that living organisms could be patented, Monsanto's Agricultural Division opened a new lab at Building U to research how genetic engineering might be profitable. Longtime employee and biochemist Ernest Jaworski, who spent his early career developing herbicides, directed the operation: "I concluded that a time would come when you couldn't solve all problems with chemicals."[27] Jaworski recruited a team of molecular biologists to find answers to a question: Was it possible to insert new DNA into a plant cell to cultivate new traits in a plant? They conducted experiments using tissue culture, or raising plants from petri dishes, to see if the soil bacterium *Agrobacterium tumefaciens* could be introduced into petunia or tobacco cells, the lab rats of botany. If feasible, the *Agrobacterium* "pack mule" would bridge gene transfer between single-celled and multicelled organisms. In 1983, Monsanto scientist Rob Horsch announced they had successfully inserted a gene for the antibiotic kanamycin A into a petunia leaf. A transgenic plant was possible; a lab technique was patented.[28]

Monsanto, however, did not have a marketable crop as petunias were easier to manipulate than potatoes. Richard Mahoney, company president from 1984 to 1995, instructed Jaworski's team, "We are not in the business of the pursuit of knowledge, we are in the business of the pursuit of products." Mahoney reminded Jaworski that Monsanto's exclusive rights to Roundup, the trade name for their best-selling weed killer, were set to expire in the year 2000, so "Roundup Ready" crops sold as a seed-and-spray package would undercut its generic competitors. The goal was to find a gene that would immunize plant cells against the herbicide. If achievable, Roundup would kill all plants except the genetically modified crop. In 1987, Monsanto scientists thought to take samples from waste-treatment ponds at their Luling Plant, a factory situated five hundred miles south of St. Louis that manufactured glyphosate, Roundup's key ingredient. Jaworski's lab found some bacteria outside of the factory that had developed a valuable attribute—glyphosate tolerance. The remaining challenge was to locate the gene responsible for this desired trait, then transfer it to foodstuffs.[29]

Other biotech firms faced similar obstacles; one made a breakthrough. Jozef Schell and Marc van Montagu, two Belgian scientists who had consulted for Monsanto, started their own company called Plant Genetic Systems. In 1985, they transferred the first "useful" gene into a tobacco plant from *Bacillus thuringiensis*, or Bt. A Japanese bacteriologist had identified the soil bacterium almost a century ago when the substance with insecticide-like properties decimated the nation's silkworms; now organic farmers applied the Bt toxin for targeted pest management in their fields. But biotechnology created something new under the sun. Inserting the gene for *Bacillus thuringiensis* meant that the natural poison was released permanently throughout the plant, generating a threat to all insect populations, from the harmful like tobacco hornworms to the useful like ladybug beetles. In 1987, the scientists at Plant Genetic Systems published their results and began licensing the technique to other companies.[30]

While Plant Genetic Systems developed Bt tobacco, the US government formulated policies about the use of genetically altered plants. Michael Taylor, an attorney who worked as legal counsel for Monsanto, also served under Presidents Ronald Reagan and George H. W. Bush in the revolving-door ties between industry and government. In 1986, the Reagan administration issued the Coordinated Framework for the Regulation of Biotechnology. Although the judicial branch declared that genetically modified organisms were new innovations, and therefore subject to issuing patents, the executive branch maintained they were not, allowing them to be approved without new regulations. Biotechnology was simply a new way to produce old things, neoliberal politicians argued, so federal agencies like the FDA for foods, EPA for pesticides, and USDA for plants would "regulate" them under existing laws. With Taylor selected as deputy commissioner of the FDA, the Bush administration released a 1992 guideline titled "Foods Derived from New Plant Varieties." The policy memo stated that if a genetically modified plant was similar to a traditional food proven safe for human consumption, then the foodstuff required minimal testing under the doctrine of "substantial equivalence."[31]

In practice, substantial equivalence sped up government approval of genetically altered crops. Dan Glickman, who joined the Clinton administration in 1995 as secretary of agriculture, remembered feeling intense pressure from the biotech industry. "Most of the regulatory climate was basically focused on approvals, approvals of the crops, facilitating the transfer of the technology into agriculture in this country and pushing the export regime for these," Glickman remarked. "I found that there was a general feeling in agribusiness and inside our government in the US that if you weren't marching lockstep forward in favor of rapid approvals of GMO crops, then somehow you were anti-science and anti-progress."[32] Consider the rapid turnaround for N4640-BtII, a Bt maize resistant to agricultural pests like European corn borers. In July 1995, Mycogen Seeds (later called Novartis) submitted an authorization request with prior field trials for approving this Bt corn variety. Six months later, in January 1996, the US Department of Agriculture certified it for commercial use.[33]

However, a simple laboratory experiment set off a bombshell about the federal testing regimen. John Losey, an entomologist at Cornell University, and his team had been dusting milkweed plants with the pollen of Bt corn, then feeding them to monarch caterpillars. While no casualties were reported from conventional maize, 44 percent the monarchs died within four days after eating Bt-coated leaves. Losey and his colleagues acknowledged the uncertainty about whether caterpillars encountered similar toxic levels of Bt pollen on milkweeds outdoors as they did under artificial lab conditions. Nonetheless, the study suggested that genetically altered crops might be killing the winged wanderer and other non-target species. "These results have potentially profound implications for the conservation of monarch butterflies," Losey wrote. On May 20, 1999, the Cornell researchers published their findings in the high-profile journal *Nature* under the headline "Transgenic Pollen Harms Monarch Larvae."[34]

Major news outlets around the world picked up this story. The *Washington Post* framed the potential ecological hazards of genetically engineered crops as "Biotech" versus "Bambi of the Insect World." A previous study, using stable isotopes to trace origins of flight, showed that *50 percent* of all monarch butterflies in Mexico grew up as caterpillars in Corn Belt states like Nebraska, Iowa, Ohio, and Indiana. In 1998, nine million acres (10 percent of the total US corn crop) had already been planted with Bt maize. "Why is it that this study was not done before the approval of Bt corn?" asked a director of the Union of Concerned Scientists. "This should serve as a warning that there are more unpleasant surprises ahead."[35] The *Times* of London reported that John Beringer of Bristol University, who acted as chairman of the biotechnology committee overseeing transgenic crops in Britain, was also concerned. "This is a warning bell," Beringer stated. "It would be remiss now to approve large-scale use of these crops in the U.S., if they have a significant effect on these insects."[36] The press questioned the passive regulatory framework for authorizing genetically modified plants.

Biotech firms and supporters mounted a defense of their products. Willy De Greef of the Switzerland-based corporation Novartis, which produced the

Bt variety that had been used by Losey and his colleagues, asserted the Cornell study amounted to "force-feeding" caterpillars. Monarch larvae, De Greef insisted, would only encounter trace amounts of Bt toxins in fields. Novartis and Monsanto scientists requested private meetings with Losey, suggesting that the entomology professor might hurt his academic career with overreaching claims. *Wall Street Journal* columnist Michael Fumento, writing for the Hudson Institute, a conservative think tank, stated the "hysteria" over "Frankenfoods" would dissipate because Bt crops lessened the need for pesticides. Fumento concluded, "We ought not allow such progress to be brought to a screeching halt by a beautiful little insect."[37]

The Bt corn controversy worried the Clinton administration in terms of how it might harm big business, not butterflies. In June, the European Union's Environment Council decided that transgenic crops must be proven to have no environmental or health risks before they could be released on the market. A month after the Cornell study was published, the European Union Commission announced a five-year moratorium on importing all Bt varieties. "The EU is proving increasingly problematic," an internal memo to President Clinton detailed, "in part due to ... preliminary research indicating that some biotech corn pollen may harm Monarch butterflies."[38] The US Department of Agriculture estimated that American agribusiness could lose $200 million annually in corn exports alone from the European ban. Agriculture Secretary Glickman reported to Clinton that "the findings, which USDA scientists are still reviewing, could add fuel to the increasing controversy over the spread of genetically modified organisms."[39] To guarantee foreign buyers of transgenic crops, the federal government and their corporate allies needed to calm the storm about monarch butterflies.

Monsanto, Novartis, and Pioneer Hi-Bred (of DuPont) recruited a consortium of university scientists, called the Agricultural Biotechnology Stewardship Working Group, and funded their research during the summer of 1999 to address issues raised by the Cornell findings. Their results were presented on November 2 at a symposium in Chicago attended by media outlets, industry representatives, and government officials of the US Department of Agriculture and Environmental Protection Agency. The purpose of the conference was to discuss the studies and determine what scientific information was inconclusive or missing.[40]

Lincoln Brower attended the conference as a monarch expert to make a fair and critical evaluation of the scientific presentations on Bt corn. He observed that "some results indicated possible major impact, others suggested minor impact, and most agreed that the current research base could not resolve the problem."[41] But that's not what the public heard. Carol Yoon, a *New York Times* science journalist, received a press release during the meeting with the reassuring headline "Scientific Symposium to Show No Harm to Monarch Butterfly." Yoon asked if other participants had been notified of the industry-written article, to which they unanimously replied no. Two days later, on November 4, Yoon reported the fiasco with the headline "No Consensus on Effect of Genetically Altered Corn on Butterflies," but most newspapers blindly reprinted what the companies had provided.[42]

Two studies put the Bt corn controversy to rest. A group of entomologists at Iowa State University confirmed that monarch larvae died in the field as the Cornell researchers had observed in the lab. For their experiment, the Iowa State researchers collected *Asclepias syriaca* growing near Bt corn farms and placed caterpillars on them. Twenty percent of monarch larvae perished. The 2000 study revealed that wind-borne pollen from transgenic crops could hurt the relationship between monarchs and milkweeds, but there were caveats.

A year later, scientists with the Agricultural Biotechnology Stewardship Working Group finally published a comprehensive risk assessment for Bt corn and monarch butterflies. They found that different Bt maize cultivars had varying levels of toxin, some dangerously high, others reasonably low. Losey and his colleagues had used a Bt variety with insecticide levels forty to fifty times higher than other cultivars, and which Novartis was now phasing out of production. Moreover, Bt pollen on milkweeds only seldom reached concentrations that were toxic to caterpillars. Taking these factors into consideration, they concluded the overall threat of Bt corn to monarchs was "negligible."[43]

The end of the Bt saga was just the beginning for biotechnology and butterflies. As the ecological effects of transgenic crops were questioned, Monsanto figured out how to insert a gene for herbicide tolerance into soybeans and corn. Roundup Ready crops posed no danger to monarch caterpillars, but Roundup weed killer did affect the host plant *Asclepias*. Cutting trees hurt winter monarchs at the southern end of their border-crossing migration; now spraying herbicides in the US Midwest and southern Canada imperiled summer monarchs on the northern end. While early protections focused on Mexico, monarch conservation became an increasingly trinational issue.[44]

ANOTHER LAB TECHNIQUE opened the floodgates for Roundup Ready crops. The challenge for scientists was that plant cells mostly rejected the genes they wanted to introduce. In 1983, plant geneticist John Sanford approached his colleague Edward Wolf of Cornell's Submicron Facility about using lasers to cut cell walls and insert DNA. It was impossible, Wolf responded, but they decided that a more feasible idea might be soaking microscopic particles of metal with DNA and shooting them into plant cells. For an experiment, Sanford and Wolf fired tiny tungsten beads at onion cells. The laboratory's smell, Sanford recollected, was a mixture between a McDonald's restaurant and a shooting range. The "gene gun" worked—high-velocity projectiles delivered the DNA to living cells. In 1987, the same year Monsanto identified bacteria resistant to glyphosate, Sanford and his colleagues published about the gene gun. Monsanto was a buyer, and company scientists could shoot Roundup-tolerant genes into plants.[45]

Although Monsanto had rebranded itself as a "life-sciences company" by the mid-1990s, it was no seed business. And getting into agriculture, selling biotechnology to farmers, required seeds. Robert Fraley believed that Monsanto would license its genes to seed companies just as Microsoft had licensed its Windows operating system to computer manufacturers. If Monsanto owned the software, then Pioneer Hi-Bred held the hardware in corn and soybeans. A deal was struck. In 1992, Monsanto agreed to a one-time payment of $500,000 for the unlimited use of its glyphosate-resistance gene in Pioneer's soybean varieties. In exchange, Pioneer Hi-Bred would print "Roundup Ready" on all its seed bags and require farmers who purchased the new product to sign a contract agreeing to not save seeds during harvest. Roundup Ready crops were more expensive with a "technology fee," but farmers would save money because Monsanto's popular weed killer was cheaper than other herbicides. For the company that claimed to be moving away from chemicals, the Roundup gene became a tool for selling more Roundup.[46]

When new seeds finally hit the US market, most growers eagerly bought them. In 1996, Roundup Ready soybeans covered one million acres of farmland; two years later, they covered twenty-five million acres—one-third of the total crop planted. "Roundup Ready soybeans made bad farmers into good farmers," remembered Mac Ehrhardt, a small seed dealer from Albert Lea, Minnesota. "I don't know if I'll see anything like it again. Farmers were just crazy to get Roundup Ready soybeans. They bought every bag."[47] Ehrhardt was skeptical of genetically altered crops on many levels. As a "God-fearing Protestant," he believed that tinkering with nature upset the moral order of the universe. As a liberal Democrat, he distrusted giant corporations like Monsanto. And as an environmentalist, he thought that new technologies always brought unintended consequences to the land. Despite all these misgivings, Ehrhardt worried that his family-run business wouldn't survive without adopting GMOs. Ehrhardt's

Albert Lea Seed House signed a sales contract. For every acre planted, Monsanto earned fifteen dollars—one-third from the seeds, two-thirds from the chemicals. The biotech firm was raking in revenues in the hundreds of millions of dollars.

Monsanto brought Roundup Ready corn to farmers by buying out Pioneer Hi-Bred's rival, Dekalb Seeds. In 1998, about thirteen thousand growers planted 850,000 acres with the transgenic variety. Many were taken with the "greening" of agricultural practices through biotechnology. Farmers typically plowed with tractors to prep their fields before planting, but Roundup Ready crops facilitated "conservation tillage." Spraying herbicides basically replaced turning over the topsoil, thus slowing down soil erosion. "With Roundup Ready corn, I can get the benefits of Roundup without taking a yield reduction," stated Tim Seifert of Auburn, Illinois. "I'm saving money on every acre, and I'm helping to protect the environment."[48] In addition to promoting "con-till" methods, farmers appreciated that Roundup was a broad-spectrum herbicide, killing off everything except the genetically altered crop and thereby requiring fewer applications than conventional herbicides. Wayne Stouder, who had converted most of his Iowa farm to Roundup Ready corn, agreed. "With other corn," Stouder noted, "I have to use three different chemicals to control weeds, and if one doesn't work, I have to come back with another."[49] While individual farmers used a third less herbicide overall, Monsanto sold more Roundup than ever before.

Spraying glyphosate on agricultural lands destroyed over a hundred different weed species, including *Asclepias*, the only food source for monarch caterpillars. Roundup works by entering milkweeds through their leaves, from which it is carried through vascular tissue to the roots. There, the chemical inhibits an enzyme needed for amino acid production, which leads to the decrease in chlorophyll, the green pigment responsible for photosynthesis. This causes necrosis of the plant cells, leading to death of the plants. During growing seasons, farmers sprayed glyphosate on Roundup Ready corn in early May, then again in early June. For Roundup Ready soybeans, the first spraying came in late May, the second in late June. The massive application of Roundup weed killer on fields across the US Midwest just so happened to coincide with the arrival of monarch butterflies on their migration.[50]

In contrast to Norah and Fred Urquhart, who had used tags to follow monarchs on their fall journey southward to Mexico, Lincoln Brower utilized "cardenolide fingerprints" to track their spring movements northward to the United States and Canada. Brower already knew that different milkweeds had distinct levels of cardenolides found in discrete regions. For example, *Asclepias viridis* (popularly known as spider milkweed or green antelope horn) yielded high cardenolides and was restricted to the south-central United States (Texas and Oklahoma), whereas *Asclepias syriaca* (common milkweed) possessed low cardenolides and grew on the northern range of the United States and Canada. After collecting butterfly specimens over the late 1980s, Brower found a pattern across two latitudinal transects: Monarchs captured during late March and early April in the US South, which were laying eggs on emergent milkweeds, held low cardenolide levels and chemical fingerprints identical to those in Mexico. Conversely, monarchs found during May and June across the US Midwest held high cardenolide levels and chemical fingerprints typical of southern milkweeds. Brower delivered evidence to support the stepping-stone hypothesis, first proposed by Anna and John Comstock almost a century earlier, which posited that the butterflies completed the northward migration over multiple generations. If the migratory cycle was akin to a relay race, it also indicated that the second leg of monarch offspring had been given the baton when Monsanto stuck out a foot to trip them up.[51]

University of Minnesota ecologist Karen Oberhauser was interested in understanding the handoff between generations. In 1984, the Wisconsin native left her job as a high school biology teacher to begin a PhD program at Minnesota studying monarch reproduction. Adult butterflies copulate in a coercive mating ritual in which males chase females through the air, pin them down on the ground, and attach themselves with abdominal claspers for a postnuptial flight lasting up to sixteen hours. Oberhauser explained that the sperm packet (or spermatophore, in technical jargon) that male butterflies deliver contains a nutritional protein gift, increasing the eggs that female butterflies lay on milkweeds. Her dissertation study, completed in 1989, involved detailing parental investment in offspring. Oberhauser stayed at Minnesota as a professor and looked for ways to continue the research. "The challenge was there are way more monarchs than monarch scientists," Oberhauser recalled. In 1996, the same year that Monsanto began selling Roundup Ready soybeans, she and graduate student Michelle Prysby launched a citizen-science effort called the Monarch Larva Monitoring Project.[52]

The project asked volunteers to go out once a week to the same patch of milkweeds—found in roadsides, ditches, home gardens, and restored prairies—and count all the monarch eggs and caterpillars they saw on host plants. They recruited lay participants through schools, nature centers, and online mailing lists like Monarch Watch's D-Plex listserv. Oberhauser and Prysby trained hundreds of people in data-collecting protocols to minimize errors when surveying populations over the summer breeding period from late May to September. Three

University of Minnesota ecologist Karen Oberhauser (*far left*) examines a milkweed plant for signs of monarch caterpillars with high school and college students near Minneapolis, 1998. Two years prior, Oberhauser and graduate student Michelle Prysby had launched the Monarch Larva Monitoring Project. Photo courtesy of Michelle Solensky.

years of citizen-science information ended up useful as a background control during the Bt corn controversy.[53]

In 2000, the US Department of Agriculture recruited Oberhauser to the industry-funded Agricultural Biotechnology Stewardship Working Group to evaluate the likelihood that monarch larvae might be exposed to Bt pollen. For the study, Oberhauser, Prysby, and entomologists from Iowa State University compared the per-plant density of monarchs at locations that the Monarch Larva Monitoring Project was censusing with those found on maize farms in Iowa, Minnesota, and Wisconsin. While the Bt pollen shed was insignificant, they discovered that breeding monarchs laid eggs on milkweeds more than previously assumed: Per-plant concentrations were as high or higher at agricultural sites than nonagricultural sites. Biotech executives were pleased to hear the first part but were anxious about the second. "Regardless of risks imposed by transgenic corn," Oberhauser and her associates concluded, "changes in agricultural practices such as weed control or the use of foliar insecticides could have large impacts on monarchs by affecting milkweed density and condition."[54] Any threat to milkweed was a threat to monarch larvae; any threat to monarch larvae was a threat to the migratory cycle.

John Pleasants was one of the Iowa State University collaborators who monitored *Asclepias* on farms. Chip Taylor of Monarch Watch suggested that more research needed to be done on the agricultural landscape: "The use of corn and soybeans genetically modified to resist applications of herbicides could lead to significant reductions of milkweeds within and adjacent to these row crops, thereby reducing the milkweed base upon which the Monarch population depends."[55] Pleasants continued examining plots in fields near Ames, Iowa, for the next eight years. In 2000, he counted thousands of milkweed stems; by 2008, there were practically none. Given that six out of every ten acres of maize were planted with Roundup Ready corn and nine out of every ten acres in Roundup Ready soybeans, Pleasants estimated that crop fields infested with milkweeds had dropped from 51 percent to 8 percent.[56]

Signs of trouble appeared early on. In spring 2004, Chip Taylor received an email from David Wurdeman, a grower from Leigh, Nebraska, expressing his worry about how the eradication of milkweed on his farm harmed monarchs. "Sirs," Wurdeman wrote,

> I am concerned that the recent large use of Roundup ready crops (which I use), and the subsequent widespread use of Roundup herbicide (which I also use), had led to the virtual elimination of milkweed in fields and crops. As someone who has raised crops, I can personally attest to scarcity of the milkweed plant today compared to, for example, 20 years ago....
>
> Although I have always considered milkweed a rather troublesome "weed," and have appreciated how modern herbicides have controlled it, I am concerned about the effect this might have on the monarch butterfly population. I consider the monarch a very beautiful insect, and have noticed that they appear to be scarcer than in years past.
>
> Is there some advice I can get, as a grain farmer with economic considerations, to balance both my needs and that of the monarch?[57]

The message stuck with Taylor, who had developed robust tagging and education programs at Monarch Watch but had little to offer Wurdeman in terms of habitat conservation.

As a result, Monarch Watch initiated a new program a year later called the Monarch Waystation. Taylor explained that "in the 1800s, waystations were typically places where steam driven trains stopped to take on water and coal." Swap trains with butterflies, water and coal with milkweed and flowers, and the concept of a monarch waystation was born as a migration stopover site. Taylor asked his network of citizen scientists, who were engaged in rearing and tagging

Kathy Davis's second-grade class from Hillcrest Elementary in Lawrence, Kansas, watches in awe as Monarch Watch director Chip Taylor demonstrates how to catch, tag, and release a monarch. Photo by Early Richardson. Courtesy of Monarch Watch.

monarchs, to become gardeners. They should plant *Asclepias* at their homes, schools, parks, zoos, nature centers, and roadsides, Taylor insisted, to counteract "the rapid adoption of herbicide resistant crops in the last 5 years that appear to have eliminated 80–100 million acres of monarch habitat." To become a certified waystation, participants submitted paperwork showing they had planted a patch of at least ten milkweed plants among annual and perennial flowers and had committed to watering regularly and keeping the site free of all pesticides. By the decade's end, the Monarch Waystation program registered over 3,500 pollinator habitats. Taylor and other Monarch Watch staff witnessed how the public participation surged from six hundred waystations in the first year to, now, an average of three thousand new waystations per year.[58]

Taylor liked to quote a famous line from the 1989 film *Field of Dreams*: "If you build it, he will come." In the film, an Iowa corn farmer played by Kevin Costner hears a voice instructing him to build a baseball diamond in his fields. After he does, Shoeless Joe Jackson and other famous baseball players emerge from the cornfields to play ball. Taylor reinterpreted the quote for butterflies: "If you build the habitat, the monarchs will come." He judged Monarch Watch's approach of building milkweed corridors among the maize to be "pragmatic" because they helped butterflies to adapt but did not address the underlying policies and practices of corporate agribusiness. In fact, Taylor and Oberhauser found in one multiauthored scientific review that taking marginal lands out of agricultural production would be the most effective strategy for restoring the billion *Asclepias* plants lost to herbicide-tolerant crops because the land on the fringes was the most important, historically for monarch reproduction.[59]

In 1996, a Republican-controlled US Congress passed an omnibus farm bill that permitted growers to plant on marginal lands. Under Newt Gingrich's Contract with America, which promised to rein in federal spending, the new law gradually phased out price subsidies for foodstuffs while repealing acreage limitations to maximize crop production for export. Farmers still received generous government aid through emergency legislation—$32 billion in the year 2000 alone, for example—but farmlands set aside as natural areas under the Conservation Reserve Program declined substantially. Enrollments dropped by 4.7 million acres between 1996 and 1999. The US government traditionally rented marginal lands to prevent soil erosion and restore wildlife habitat, such as restored prairie grasslands with milkweed and nectar plants for monarchs.[60]

Conserving the migratory cycle now required major changes, not minor accommodations, something that neoliberal governments in all three countries were reluctant to do. Federal officials talked eloquently about trinational solidarity, but they were quick to point fingers north or south toward the real cause of monarch decline. Ecological internationalism was founded on relationships at the bottom, but it needed resources from the top.

The Neoliberal Order

Critics of the North American Free Trade Agreement said that while the treaty would do a lot to enhance the movement of capital and goods among Canada, Mexico, and the United States, it would do little to address transborder issues relating to labor and the environment. How would NAFTA's trade liberalization produce worker displacement when cheaper wages could be found across the international line? How would foreign investment from the two rich countries lead to "pollution havens" in the one poor nation? To placate these critiques from labor unions, environmentalists, and those on the political left, President Bill Clinton proposed that two side accords be ratified in conjunction with the trade deal. Clinton called his centrist stance "NAFTA plus."[61]

On January 1, 1994, the North American Agreement on Environmental Cooperation went into force alongside NAFTA, its chief objective to "increase cooperation between the Parties to better conserve, protect, and enhance the environment, including wild flora and fauna."[62] The treaty established the Commission for Environmental Cooperation, headquartered in Montreal, Canada, to facilitate state-to-state collaboration on transboundary problems. Since the monarch butterfly migrated across all three countries, Homero Aridjis and Grupo de los Cien suggested the insect be chosen as the official symbol for NAFTA's environmental accord.[63]

In the spirit of collaboration, Prime Minister Jean Chrétien of Canada and President Ernesto Zedillo of Mexico signed a joint declaration in October 1995 creating an International Network of Monarch Butterfly Reserves. The Canadian government immediately designated three areas in southern Ontario as monarch reserves under the agreement: Point Pelee National Park, Long Point National Wildlife Area, and Prince Edward Point National Wildlife Area. All were formal protected areas before the declaration, but Canadian officials promised that more would be done to enhance milkweed and nectar sources at those sites along the Great Lakes. It was the launch of "sister parks" for monarch conservation. Scientists with the Canadian Wildlife Service commented, "Without ensuring the continued existence of suitable breeding habitat within its own borders, Canada is in no position to offer aid or advice with regard to the current crisis at the Mexican overwintering sites."[64]

The crisis was a winter storm. On December 30, 1995, twelve inches of snow fell on the Trans-Mexican Volcanic Belt during an unusual cold spell. Tens of millions of monarchs froze to death at the biosphere reserve. Homero Aridjis and Lincoln Brower estimated that 30 percent of the overwintering population had perished, later revising their calculation to 10 percent. For Brower, the severe weather event punctuated two facts: Intact forest not only acted like a blanket, regulating temperature extremes for butterflies, but it also served as an umbrella, shielding them from the elements. Wind dislodged monarchs from clinging to tree branches and trunks; cold water on the insects translated to higher mortality rates as fatality occurred at −0.5 degrees Celsius (whereas the lethal threshold was −8 degrees Celsius under dry conditions).[65] Surviving butterflies moved around and formed colonies at new locations to adapt to the erratic weather. For Aridjis, it signaled that the Commission for Environmental Cooperation should step in. "That Mexico, the United States, and Canada, the NAFTA trio," Aridjis wrote, "are unable to defend the physical integrity of the oyamel forests, and together guarantee the survival of the monarch's migratory phenomenon is proof that the so-called environmental accords are merely cosmetic."[66]

The news got monarch scientists talking about how the Commission for

Environmental Cooperation might be mobilized. Two months after the winter storm in February 1996, Karen Oberhauser traveled to Angangueo, Michoacán, where she joined Mexican companions like Eduardo Rendón-Salinas of National Autonomous University. Oberhauser and Rendón-Salinas talked about having an international platform to share the most up-to-date research on monarch biology in both English and Spanish. In June, Chip Taylor and Bill Calvert met Rodolfo Ogarrio and Hector Luis Ruiz Barranco at a Houston hotel to discuss monarch conservation. Ruiz, a federal official with the Instituto Nacional de Ecología, presented a sustainable development plan to the group, and they agreed about the "establishment of a trilateral commission to advise government agencies on the state of the Monarch butterfly." In August 1996, the CEC's Council passed resolution #90-04 supporting the trilateral study of monarchs and the ecosystems on which they depended, as well as recommending additional protected areas or management protocols to ensure monarch habitat across the three countries. The CEC released institutional funds for organizing an international conference.[67]

On November 10, 1997, the North American Conference on the Monarch Butterfly (or MonCon III) began in Morelia, Michoacán. Sponsored by the Commission for Environmental Cooperation, the four days of meetings featured the usual participants from academic scientists and government officials to journalists and NGOs. However, conference organizers like Karen Oberhauser made certain to involve a previously neglected group of stakeholders: the ejidatarios of the Monarch Butterfly Biosphere Reserve. Even though the Mexican ambassador to Canada, Jürgen Hoth, stated that the purpose was to build "bridges between often adversarial groups," US writer Sue Halpern of *Audubon* magazine observed that attendees acted like they were at a wedding ceremony: Conservationists sat on the groom's side while campesinos sat on the bride's. Keynote speaker Lincoln Brower told the three-hundred-person audience that the two biggest threats to the monarch migration were the extensive use of agricultural herbicides in the breeding areas of Canada and the United States and illegal logging in the overwintering areas of Mexico.[68]

During roundtable discussions, one issue was amending the Monarch Butterfly Biosphere Reserve to satisfy human needs without depleting natural resources. Local representatives from Alianza de Ejidos y Comunidades de la Reserva Mariposa Monarca laid out collective demands that all butterfly sanctuaries be opened to tourism, rural infrastructure be improved, and the 1986 decree be scrapped and "replaced with the goal of maintaining and improving the forest through sustainable resource management programs." Rosendo Caro Gómez, a Mexican forester with Ministry of the Environment, Natural Resources, and Fisheries, remarked, "The problem is that legislation and regulations are the formal expressions of a set of political interests where the campesino organizations have little political power." Alianza director Pascual Sigala Páez responded that as a bare minimum, any revisions to the biosphere reserve "should include the active participation of the area's agrarian groups" and "consider compensation for the campesinos" who had suffered diminishing income. Diverse interests debated the most effective strategies to conserve monarchs alongside their Mexican neighbors.[69]

Another issue was building transnational solidarity in monarch conservation, broadening the scope beyond the overwintering phase in Mexico. Jesús Manuel de Jesús of the San Felipe los Alzati Indigenous community expressed appreciation for the interest of Americans and Canadians in the biosphere reserve, but he asked that "northern friends take real steps for the conservation of monarchs in their own countries, too." Manuel suggested launching an "adopt-a-sanctuary" program in which ejidos of the biosphere reserve were paired up with communities in Canada and the United States to hold each other accountable. Brooks Yeager, deputy secretary of the US Department of the Interior, noted, "The trinational migration of the monarch butterfly does not fit easily under existing U.S. domestic regimes for conservation." Interior Secretary Bruce Babbitt, who attended at the urging of Brower, commented that the US Fish and Wildlife Service's Wildlife Without Borders program should award grants for monarch-related projects. Ejidatarios believed any money should go directly to them, not filtered through middlemen like government bureaucrats or NGOs. Mexican campesinos wanted ecological internationalism, not empty promises.[70]

The conference's outcome was solidifying a proposal to modify Mexico's Monarch Butterfly Biosphere Reserve. On November 10, 2000, President Ernesto Zedillo issued a decree expanding the protected area from 16,000 to 56,300 hectares (from 60 to 217 square miles). The threefold enlargement, according to Mexican scientist Leticia Merino Pérez, "cover[ed] possible fluctuations in the areas where colonies [were] located."[71] In other words, extending the core and buffer zones for one contiguous park allowed the butterflies to move and roost where the mixed oyamel–pine tree canopy was most intact, adapting to a thinned forest. To discourage logging, the Fondo Monarca (Monarch Fund) would compensate all ejidatarios who lost access to forest resources from an endowment created by the Packard Foundation and other international NGOs. Every year, they received $360 MXN ($18 USD) per cubic meter of forfeited timber because of the new core zones, plus another $250 MXN ($12 USD) per hectare of conserved forest. The 2000 edict was a compromise between conservationists and campesinos. The former wanted the expansion of the biosphere reserve; in exchange, the latter demanded annual payments for preserved ecosystem services.[72]

Thirty-one communities impacted by the new reserve boundaries (80 percent of the total) chose to participate in the Fondo Monarca. Government bureaucrats with the Ministry of the Environment, Natural Resources, and Fisheries conducted field visits to assess deforestation rates; World Wildlife Fund officials administered the trust. If less than 2 percent of the forest was degraded, ejidatarios would receive the payment. If less than 3 percent, they were warned but still received money. If greater than 3 percent (above what Brower and his colleagues figured was the deforestation rate between 1984 and 1999), they were denied the money. In 2003, Fondo Monarca disbursed almost $500,000 USD across thirty communities; all but one ejido received compensation. Scientists have estimated that the financial incentives slowed down logging by seven hundred hectares (1,700 acres) in core zones.[73]

But Fondo Monarca's effectiveness at conserving forests through economic stabilization was limited by the patriarchal and undemocratic structure of landownership. Although the ejido system improved the lives of landless campesinos who once toiled on haciendas, the 1930s land reforms were still based on an unequal distribution of resources. The title of "ejidatario" became an inherited position handed down from father to youngest son, which allowed them to participate in local governance and to receive trust payments. Seventy to eighty-five percent of local people had no formal rights in their ejido. Some ejidatarios distributed funds to the wider community; others kept money for themselves. Economic instability meant an outmigration of young men and women; half of those who grew up around the biosphere reserve left in search of jobs in Mexico City or the United States. Cutting trees was always an option for those who stayed.[74]

And those who did were criminalized. At MonCon III, the Ministry of the Environment, Natural Resources, and Fisheries promoted the formation of vigilance committees to help the federal government patrol the biosphere reserve. In 2003, for example, Vicente Guzmán Reyes of Donaciano Ojeda petitioned the Vicente Fox administration for money to pay small salaries for his forty-member crew. Guzmán wrote, "In order for our community members to give more attention to the forest, who instead leave for the cities in search of work, we ask you to support us in our productive activities."[75] Federal officials responded by purchasing radios through the World Wildlife Fund to coordinate the watch and turn their accosted neighbors over more easily to PROFEPA. The forest police responded to forty-two complaints of illegal logging in the year 2003 alone, assessing fines when people were caught in the act.[76]

The Mexican federal government expanded the protected area, and yet its administrative response was haphazardly implemented or excessively punitive. The Vicente Fox administration (2000–2006) enacted a zero tolerance policy, increasing law enforcement through the Ministry of National Defense to crack down on illegal logging at the sanctuaries. Some ejido communities wanted state intervention but believed government officials turned a blind eye, colluding with sawmill owners; others believed their birthrights as ejidatarios were being persecuted unfairly with arrests and massive fines for tree felling; still others felt the money should go to financial assistance for rural development projects rather than paramilitary. Instead of addressing rural poverty, federal officials criminalized poor people who had rights inside the park.[77]

Take the example of Emilio Velázquez Moreno from Ejido El Capulín. According to family lore, Emilio's father, Valentín, found a colony of monarch butterflies on Cerro Pelón, and Jesús Ávila rewarded him with a position as a CEPANAF forest ranger in 1977. Valentín died in a brawl four years later, and the job transferred to his oldest son, Miguel. But Miguel was fired from the state agency for insubordination and needed a way to support his six younger siblings, including Emilio, so he entered the logging business. Soon all the brothers joined the clandestine timber trade. In an underground market, "you already have your contacts," Emilio learned. "When I decide I'm going to cut a tree, I know where I'm going to sell it and what price I'll get for it." Buyers, who usually paid half the market price for the lumber, were rarely penalized; it was poor people hauling wood out of the biosphere reserve who ran the risk of running into the authorities. Velázquez recalled an unhappy day when PROFEPA officials caught one of his crew members with a downed tree and fined him $80,000 MXN ($3,500 USD) for the park infraction. He remembered how long this financial catastrophe set his family back: "We took out loans against our land titles; we sold what we had."[78]

Other job options involved migrating across borders, like the butterfly. Because illegal logging involved risks, Velázquez took a different job as an apple picker in Washington state. "When you go somewhere else for work, you eat badly, you sleep badly," Velázquez remarked. "All to try to save money so you can buy food for your family."[79] Velázquez took his wife and child with him to the orchards one season so that they did not have to be apart for the four long months. However, US Immigration and Customs Enforcement arrested him and his wife for lacking work visas, and their daughter spent six months in foster care. When the family was finally reunited, Velázquez was further traumatized because his child no longer spoke Spanish. Velázquez, whose plight represented the economic insecurity of many ejidatarios in a post-NAFTA world, was treated like a criminal no matter where he went.

Policing criminality at the biosphere reserve has become more complex after 2006, when the Mexican drug war began. Drug cartels have utilized the absence of people in protected areas as cover to smuggle cocaine and other illicit substances from Colombia to the United States. Moreover, the cartel-dominated *aguacate* industry in Michoacán has a vested interest in illegal logging around the monarch sanctuaries because deforestation opened land for new avocado orchards. To prevent tree cutting and thus preserve their payments for ecosystem services, ejidos and Indigenous communities organized local patrols (*rondas del buen orden*). The practice of *rondas* dated to historic traditions for regulating communal forest use, but some communities now faced cartel-led violence during these duties.[80]

The "organized crime" narrative has also allowed federal administrators to adopt a position of nonintervention at the biosphere reserve. At Cerro Pelón, CEPANAF rangers know that loggers are not dangerous, just desperate—and young. "Most [are] twelve- or fourteen-year-old kids who are no longer going to school," Patricio Moreno stated. "They are not [dangerous] because we have sometimes had experiences of finding some and, no, they are simply scared. When you get there, they are scared and we're wearing our uniform, but it is not a police uniform—it's a uniform of park guards—but people are scared. We know that they are unarmed."[81] A stable job, more than a yearly payment, was the difference between pilfering or protecting the butterfly forest.

One promising approach to monarch conservation employs local communities. In 2019, Emilio Velázquez finally found economic security through an arborist position with a grassroots conservation nonprofit called Butterflies and Their People. "Now I have a secure job and I will not cut down trees again because it's very dangerous," Velázquez stated.[82] He performs a myriad of tasks like trail maintenance and forest patrolling, and he enjoys returning home to his family every night. Velázquez offers a glimpse of what a more equitable future might look like, but too few people in too few communities have long-term employment. Instead of payments for ecosystem services going solely to ejidatarios like Velázquez, the money might be distributed to everyone in the ejidos as a form of universal basic income.[83] Environmental justice for the monarchs' Mexican neighbors means broadening conservation approaches beyond market-based strategies such as ecotourism, which comes and goes like the butterflies.

BY THE FIRST DECADE of the new millennium, monarch conservation needed a metamorphosis, but it was unclear what type of response was going to emerge from the cocoon. Monarch numbers in Mexico dipped to their lowest ever recorded. Administrators of the biosphere reserve calculated population size since the mid-1990s by tallying forest coverage. Between 1994 and 2003, the butterflies covered nine hectares (twenty-two acres) on average; between 2003 and 2012, they overwintered on only five hectares (twelve acres). During the 2013–14 season, they had shockingly dwindled to a mere 0.67 hectares (1.5 acres). At the general rule of fifty million butterflies per hectare, monarchs had plummeted from their highest recording of one billion in 1996–97 to an all-time low of thirty-three million in 2013–14, which translated to an 87 percent decline over twenty years. Any species that has dropped by 90 percent in population size the International Union for the Conservation of Nature considers to be "critically endangered." Monarch butterflies came to represent the dire circumstances of many insect species.[84]

Creating pollinator corridors was one conservation policy. In 2013, John Pleasants and Karen Oberhauser uncovered a correlation between the decline of overwintering monarchs in Mexico and the disappearance of breeding habitat in the US Midwest. The annual spraying of seventy million kilograms of glyphosate on Roundup Ready crops was equivalent to emptying 185 water towers of herbicides onto the Corn Belt every year. Given that egg and larva density was highest on milkweeds in agricultural fields, Pleasants and Oberhauser calculated an 81 percent decrease in monarch reproduction between 1999 and 2010, when most farmers adopted herbicide-tolerant soybeans and corn. Population size fell during the breeding phase as it did during the overwintering phase. "The decline in monarch population since 1994 coincides with the NAFTA years," Homero Aridjis wrote for a newspaper editorial. "If our three countries cannot prevent the extraordinary monarch butterfly migratory phenomenon from disappearing, then what's the point of this agreement?"[85]

On February 14, 2014, Aridjis and Lincoln Brower penned an open letter, cosigned by 175 renowned scientists, environmentalists, and nature writers, to present to the heads of state—Barack Obama, Enrique Peña Nieto, and Stephen Harper—during a "Three Amigos" summit in Toluca, México state. Aridjis and

Brower asked the national leaders to collaborate on making a transborder conservation pathway for the monarchs. "A milkweed corridor stretching along the entire migratory route through our three countries must be established," asserted the Mexican environmentalist and US lepidopterist. "This will show the political will of our governments to save the living symbol of the North American Free Trade Agreement."[86] During talks about oil pipelines and immigration enforcement, Obama, Peña Nieto, and Harper made vague promises to create a trilateral task force on conserving the migration. Monsanto jumped into the conversation a week later to greenwash its image: "We're eager to help rebuild monarch habitat along the migration path."[87] To spur US involvement, President Obama issued a memorandum in June creating the "National Strategy to Promote the Health of Honey Bees and Other Pollinators," including a blueprint for milkweed restoration.[88]

Instead of facing corporate agribusinesses, President Obama accommodated them in 2015 by announcing plans for the Monarch Highway. The initiative set out to establish a migration corridor that followed Interstate 35 from Laredo, Texas, to Duluth, Minnesota, aligning with the central flyway of the eastern population of the monarch butterfly. The idea was that strips of flowers and milkweeds would be planted along the right-of-way for the 1,500-mile-long route that cuts through the Corn Belt. These pollinator islands would expand over time as annual late-season mowing dispersed the seeds. Monsanto pledged a $3.6 million donation over three years to the US Fish and Wildlife Service to support the endeavor as part of a public-private partnership, reflecting the cozy relationship between industry and government for neoliberal conservation. Chip Taylor of Monarch Watch, who consulted with Monsanto executives on the topic, estimated the project would cost somewhere between $5 million and $10 million total, depending on whether plugs (starter plants) or seeds were used. Butterflies died from vehicle collisions in the hundreds of thousands every year, but Monarch Highway proponents believed that additional roadside habitat would boost numbers overall by the millions. The transportation departments of Texas, Oklahoma, Missouri, Iowa, Kansas, and Minnesota agreed to participate within the first year of the presidential announcement.[89]

The Monarch Highway proposal arrived on the heels of another strategy, a petition to review the monarch butterfly's status under the Endangered Species Act. In August 2014, the Center for Biological Diversity, Xerces Society, and Center for Food Safety, along with Lincoln Brower, submitted a request that the US Fish and Wildlife Service list *Danaus plexippus* as a threatened species. While status review was normally a ninety-day process, federal officials received

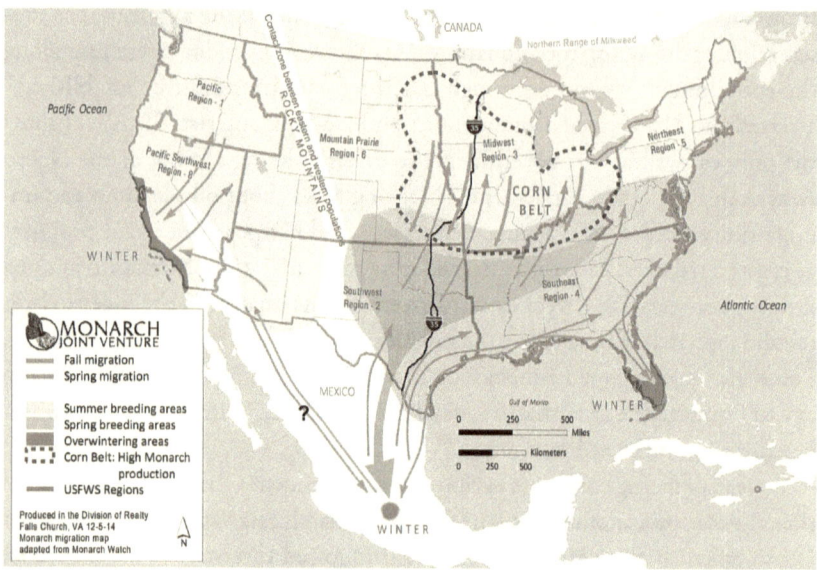

Map of the Interstate 35, "Monarch Highway," a roadside pollinator corridor stretching from the US-Mexico border to the US-Canada border. First proposed in 2015, the Monarch Highway seeks to restore milkweed and nectar sources for the migration. Courtesy of US Fish and Wildlife Service.

an extension to gather information because population size alone was not a sufficient criterion for listing. The migratory phenomenon, not the species itself, would be judged for risk of collapse. Monsanto publicly opposed listing monarchs, with policy director Eric Sachs stating it would not "do anything to help solve the problem." Worrying about the loss of company sales in Roundup herbicide and its corporate image as a leader in sustainable agriculture, Monsanto strove to frame itself as a monarch ally. After almost seven hundred published posts on its *Beyond the Rows* blog, the first time that Monsanto mentioned its support for monarch conservation was the year of the Endangered Species Act petition. Monsanto then donated money to the same agency that would be reviewing a case with ramifications for herbicide-tolerant crops.[90]

To deflect criticism of the corporation, Monsanto stirred public fears or pointed to other factors. The company argued that designating milkweeds as critical habitat under the federal law would result in government control of farming. One midwestern grower remarked, "If this petition looks like it is going to pass, Monsanto will have a banner year as every farmer I know will be spraying every stalk of milkweed we can see for miles around." The firm also emphasized that monarch decline was multicausal, blaming the illegal logging in Mexico. Instead of transnational solidarity, the scapegoating method pitted Americans and Mexicans against each other. Monsanto delayed potential regulation by pushing the idea that Washington, DC, should only act if Mexico City did, too. Chemical capitalists had a vested interest in thwarting ecological internationalism.[91]

IN CONTRAST WITH the Three Amigos, consider the bottom-up responses of another trinational relationship. Joel Moreno Rojas grew up in Macheros at the entry of Cerro Pelón, where his father, Melquides, worked for CEPANAF beginning the year he was born. "When my dad got the job as a forest ranger, it changed our lives," Joel recollected. It afforded his family a degree of economic security that too few others enjoyed. It also meant that Joel spent a great deal of time at the biosphere reserve, learning to appreciate the monarch's beauty. In 2000, the eighteen-year-old Moreno migrated to the United States, following the butterflies and his brothers, who worked at a landscaping business in New York City. There, Joel learned to speak English and honed his customer service skills at a Korean American–owned nail salon as he saved money to buy a house back home. He returned to Macheros ten years later, purchasing an avocado orchard and working as a butterfly guide at the biosphere reserve. In late 2011, Joel gave a tour to Ellen Sharp.[92]

Sharp was on a break from her PhD studies in cultural anthropology. Hypothetically, the South Carolina native could have met Joel earlier when she worked for an immigration services nonprofit in New York City, but their paths did not cross in the densely populated metropolis. Ellen learned Spanish by teaching English classes to nonnative speakers and by spending any time off traveling across Latin America. In 2008, she enrolled at the University of California, Los Angeles, to research vigilante justice in the Indigenous village of Todos Santos Cuchumatán, Guatemala. Ellen and Joel were opposites: She was an only child; he came from a close-knit family of ten. She was completing a doctorate; he had never read an entire book. She was a middle-class American; he was a working-class Mexican. Despite their differences, the two fell in love watching monarchs.[93]

Joel had noticed that big tour operators took visitors to Cerro Pelón, but none of the money stayed in his community, which lacked basic services like mail delivery. Ellen observed that the best hotels were run by international couples, "a local with land and connections and a foreigner attuned to tourist tastes." In 2012, Joel and Ellen opened JM Butterfly B&B (later renamed Cerro Pelon Butterfly B&B). Joel converted his house into a four-room hostel with a shared bathroom; Ellen set up a website for reservations. As more people—mostly Canadians and Americans—came to see the butterflies, they reinvested the profits into their business and community. Joel and Ellen came to operate a fourteen-room hotel with a dozen local employees, and the growing tourist traffic kept guides busy taking people up the mountain to the butterfly sanctuary.[94]

One of the first patrons was Darlene Burgess. The Canadian woman lived in Leamington, Ontario, on five acres of pastureland abutting Point Pelee National Park. "My backyard is basically a straight shot to the park," Darlene remarked, pointing to the protected area where Fred and Norah Urquhart once tagged. Monarch butterflies search for the easiest routes to migrate across the Great Lakes on their southern journey, and flying from the Pelee peninsula is the shortest way over Lake Erie. Darlene began rearing caterpillars on her property after 2013, when the overwintering population hit an all-time low, taking monarch larvae that she found on *Asclepias* to a sheltered place she calls "the mon cave" for ensuring metamorphosis. In 2015, she certified her tended lot of milkweeds, goldenrod, and aster as Monarch Waystation #10275. Darlene remembered missing the butterflies when they left by October, so she visited JM Butterfly B&B to see them in February.[95]

When Darlene made the southern pilgrimage during the 2014–15 overwintering season, Cerro Pelón featured a new generation of CEPANAF rangers. Joel's younger brother Patricio "Pato" Moreno Rojas inherited one of the positions. Pato commented that having the same conservation job as his father was "an honor because since I was little, I remember that he always took me with him."[96] But the State Commission on Natural Parks and Wildlife had different plans for the transition. In 2017, the state agency reassigned Pato and the two other CEPANAF rangers to a different, more profitable location. Illegal logging at the biosphere reserve skyrocketed in their absence, and Ellen organized a successful petition to bring them back to Cerro Pelón.[97] The temporary void exposed just how important these jobs were for their community and for forest health. Using the CEPANAF model, Joel, Ellen, and Pato founded a grassroots conservation nonprofit called Butterflies and Their People. And they invited Darlene to join the volunteer board, solidifying a trinational relationship.

Butterflies and Their People secured grants and fundraised to employ three local people who would coordinate forest monitoring with the CEPANAF rangers. Soon they expanded to six full-time workers, three from México state and three from Michoacán. Their constant presence resulted in an 80 percent drop in illegal logging at Cerro Pelón. Osvaldo Esquivel Maya, who lives in the community of Nicolás Romero, noticed that Butterflies and Their People was different from Mexican federal agencies in its commitment to people and place: "Outside organizations come for an inspection, people come to take a look, like those from the biosphere reserve who come to census the population, but they only pass through and go back home or to work."[98] Francisco Moreno Hernández of Ejido El Capulín put the contrast more bluntly, observing, "They just drop by when the butterflies are here, because the butterflies leave and they all disappear."[99] Looking after the monarchs was a four-to-five-month duty, but taking care of the forest was a year-round responsibility, as tree cutting hit hardest when the butterflies migrated and the tourism jobs left with them.

Ecological internationalism challenged the powerful elites behind neoliberal conservation instead of accommodating them. Because the creation of Butterflies and Their People suggested that ongoing strategies addressing the livelihoods of ejido communities were inadequate, Joel and Ellen irked official administrators who interrogated them with "We have these programs already, so why are you doing that?" The excruciating contradiction was that billionaire Carlos Slim had been sponsoring the World Wildlife Fund's programs at the biosphere reserve since 2003, using the profits he amassed after buying Telmex-Telcel from the Mexican government, and yet his private company did not even bring phone and internet services to Macheros at the park entrance.

Joel constructed homemade signal towers himself just to get a spotty connection for making ecotourism bookings. The Telmex-Telcel arrangement with the World Wildlife Fund also translated to a scientific monopoly at the protected area, withholding research permits to local organizations like Butterflies and Their People. Joel Moreno hoped the transnational network they fostered could be used as political leverage against an unjust system of conservation.[100]

It was conceivable that a monarch butterfly mated and then left Cerro Pelón, flying northward all the way to Texas to lay breadcrumb-sized eggs on a patch of spider milkweed. The offspring hatched and fed on their host plant, but most larvae did not survive—one choked to death on the milky latex-like substance oozing from a leaf; another was eaten by a yellow jacket; still another perished from an internal protozoan parasite called OE. Two caterpillars matured and formed jade-green chrysalides on the underside of a leaf, transforming into adult butterflies. The first went farther north and was harassed by a hungry robin while searching for milkweeds in a clean field of Roundup Ready corn. The second luckily found common milkweed in Kansas at Monarch Watch's Monarch

Waystation #1 to deposit its eggs. A grandchild caterpillar grew up and metaphorized into a butterfly, traveling across Lake Erie to Darlene Burgess's Monarch Waystation #10275 in Ontario. Darlene reared a great-grandchild on butterfly weed, which as an adult then glided southward on a three-thousand-mile journey from Point Pelee to Cerro Pelón, back to the forest under watch by Joel and Ellen's Butterflies and Their People guardians.[101]

Multiply this single migratory cycle by millions to get full appreciation for the stepping stones of large landscape conservation. A healthy population of butterflies requires sustaining trees, milkweeds, and flowers across all three countries; any weak link along this chain of habitat can break the migration. In a post-NAFTA world, the monarch butterfly naturalized economic integration for national governments in Ottawa, Mexico City, and Washington, DC, just as it politicized shared ecologies for everyday Canadians, Mexicans, and Americans. Relationships formed the basis of a transnational solidarity so that civil society could hold their countries responsible for resources.

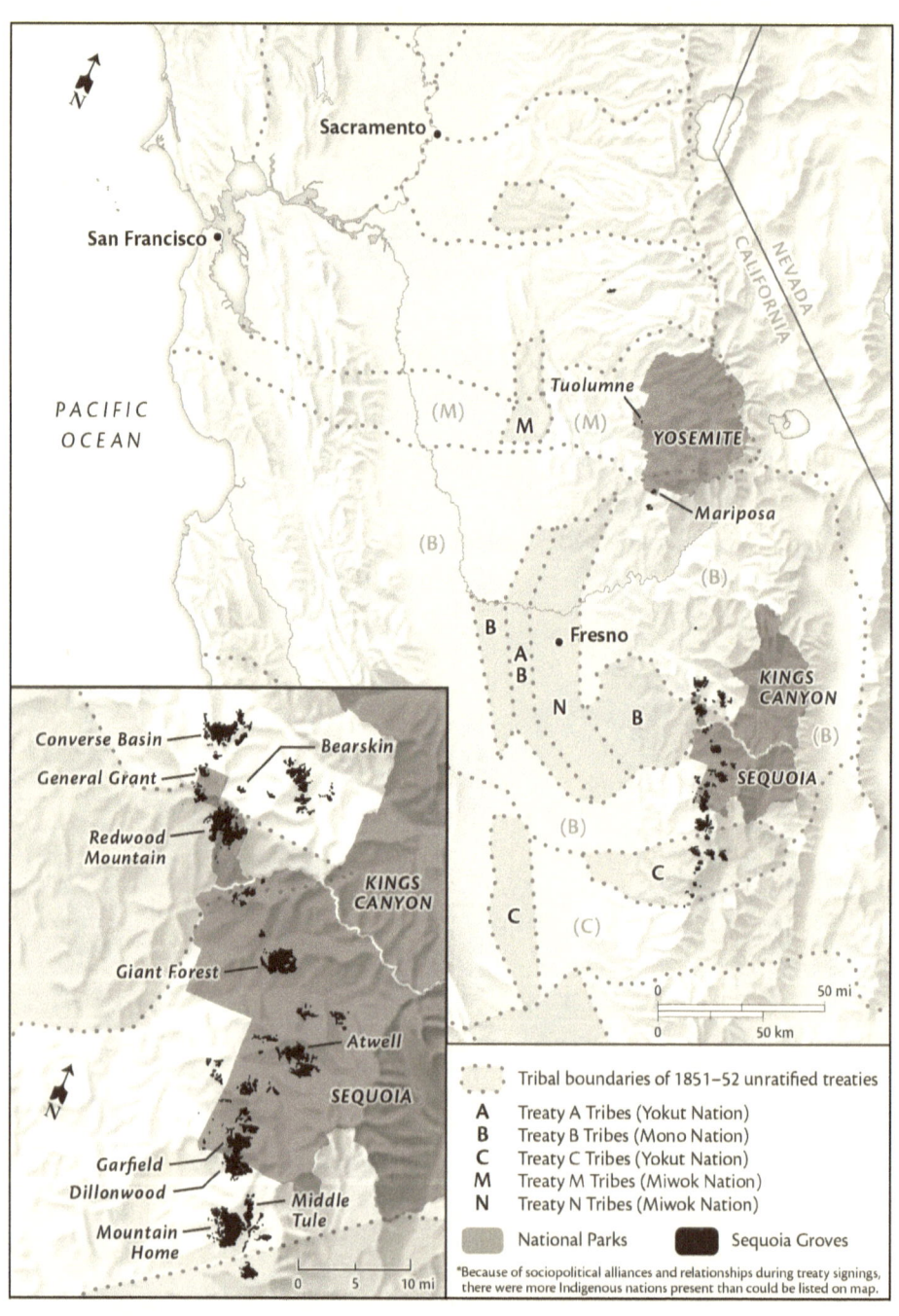

Giant sequoia groves and Indigenous treaties of the Sierra Nevada. Map by Erin Greb Cartography.

PART III

Giant Sequoias
(Sequoiadendron giganteum)

CHAPTER FIVE

Creative Destruction

One summer, I trekked along a loop path at Big Stump Basin to see if this place on the western slopes of California's Sierra Nevada lived up to its name. After descending from the trailhead surrounded by thick stands of conifers, I reached a gap in the forest where a meadow filled the opening. The scene resembled a cemetery because, from a distance, all I could make out was what appeared to be headstones protruding upward from the emerald grasses. Moving closer, however, I realized that the stone-colored markers were the mutilated stumps of giant sequoia trees. Most of these remains stood far taller than I and were much wider than my outstretched arms. Over a century earlier, in 1883, the Comstock Lumber Company had purchased this tract of old-growth forest to begin commercial harvesting.

Standing among the fallen giants, I envisioned how owner Smith Comstock lit his usual cigar, pursed between his cancer-ridden lips, when overseeing the lumbering operations at his newly established mill. A towering figure, standing at six feet four inches, Comstock himself was impressed by the tall trees for the lengths of board feet they produced. Ponderosa pine and white fir grew to heights of multistory buildings, but none of them could surpass giant sequoias. The Big Trees, as they were more often called, stood fifty feet higher than San Francisco's first skyscraper, the 218-feet-tall Chronicle Building, upon reaching full maturity. Comstock saw a wide array of goods in those trees from which to make money: One bulky sequoia could produce enough fence posts to surround an eight-thousand-acre ranch, plus enough roof shingles to cover eighty houses. Redwood products were soon integrated into local, national, and global networks of timber trading. Most of these Big Trees met the axe and saw in the process of transforming raw nature into marketable commodities.[1]

Mounds of what looked like fine crimson sand were piled on one area of the meadow. Bending over to pick up a handful, I grasped that the substance was redwood sawdust from all those years ago. I imagined how an earthy musk of woodchips must have punctuated Comstock's tobacco smoke while laborers toiled in the Sierra Nevada. Comstock hired Anglo Americans as mill supervisors, Italian immigrants as woodsmen, and Chinese "coolies" as laundrymen and cooks in a tiered wage-labor system, a common practice for capitalist proprietors during the nineteenth century. Ethnic tensions masked the pittance wages they received, which amounted to hundreds of dollars for a whole season, whereas Comstock made handsome profits in the tens of thousands. Giant sequoia trees fell to the ground at such a rapid pace that the timber company spent only six years cutting over the area, leaving behind desecrated terrain. Big Stump Basin, I decided, was a fitting name.[2]

The irony of this whole experience was that I stood within a protected area now called Kings Canyon National Park. Just a few miles away from this resource destruction, the famous Grant Grove was enchanting park visitors. Back in 1880, Theodore Wagner, the federal surveyor general for California, unilaterally instructed the US General Land Office in Visalia to withdraw from sale a four-mile section of land around this magnificent stand of sequoias. The space was closed off, Wagner stated, because the Big Trees made "a remarkable and rare curiosity which should be preserved."[3] Congress followed up ten years later with more formal protections by establishing the nation's second and fourth national parks, named Sequoia and General Grant. Park borders, however, were not fixed upon their founding. They represented shifting regimes of territoriality around the Big Trees, going from Indigenous nations to socialist communes to timber capitalists to park conservationists. The final group insisted that the federal government enlarge these protected areas over time, even renaming one of them, to include more groves of sequoias. Hence why I stood where I did.[4]

For a while, General Grant National Park remained an island preserve among a larger sea of timber extraction. Bigger business took hold to the north and south. Comstock's teams of oxen moved the Sierra redwoods out of the forest, but they were soon replaced by flumes and railroads. Money poured into the remote mountain region to finance the construction of new transportation systems and larger mills. By 1920, logging operations chopped down at least one-quarter of the original sequoia forest. Timber barons had incorporated human labor and giant sequoias, shipped near and far, into a world system of commerce. The gridded landscape of giant sequoias turned out to be an outgrowth of global capital and an underpinning for global science. A geographer and an astronomer eventually took interest in these cutover regions of the Sierra Nevada. Both scientists wanted to understand how the Big Trees, because they lived for so long, might record a chronology of past climates. From their redwood studies, mounting evidence of supranational meteorological forces came to undermine the park's administrative lines.[5]

Bringing the Big Trees to Market

The general onslaught on the Sierra Nevada forests came in twos. Two mining booms—the California Gold Rush of 1848–49 and the Victoria (Australia) Gold Rush of 1851—initiated waves of newcomers, placing market demands on lumber for mine shafts, railroad ties, and other wooden structures. But timber felling did not include sequoias, at least not at first. The Big Trees were too remote and costly for logging. These two barriers were lifted within a generation after the United States incorporated California as the winnings of war with Mexico. By 1880, two railroad corporations merged, operating branch lines across the Central Valley—the Central Pacific pushed southward from its transcontinental terminus near Sacramento and the Southern Pacific moved northward from Los Angeles. Two political measures, in state surveys and federal legislation, erased Indigenous land tenure by imposing the modern grid system on the mountains and opening the way for privatizing these parcels. And a decade later, two San Francisco–based capitalists sought to further enrich themselves by reinvesting wealth in the redwoods.[6]

In 1851–52, the US government signed treaties with the Indigenous nations to placate violence between newcomers and Natives during the California Gold Rush. Johann Sutter, a Swiss immigrant of Mexican and US citizenship who had founded Sutter's Fort (Sacramento), called for a "war of extermination" against all Indians standing in the way of mineral wealth. Endorsed by regional elites, armed militias like the Mariposa Battalion killed Miwoks, Yokuts, and other Indigenous peoples with reckless abandon. State-sponsored massacres resulted in the loss of tens of thousands of lives, amounting to what one scholar has argued can only be defined as genocide. At the request of John C. Frémont, a US senator for California who had organized the Mariposa Battalion, the federal government sent three commissioners to negotiate treaties with Indigenous nations. They demanded the cession of gold-rich lands in exchange for the payment of annuities and the establishment of reservations.[7]

The treaties, however inadequate because of intercultural barriers and military presence during their negotiation, delineated the territories of Indigenous nations along the Sierra Nevada. In March 1851, representatives for "Treaty M Tribes," comprising various groups of the Miwok Nation, agreed to "the country between the Mercede [Merced] and Tuolumne rivers, extending above said described district to the Sierra Nevada mountains."[8] In May, leaders of "Treaty B Tribes," including the Wuksachi and Potwisha bands of the Mono Nation, accepted a territory stretching from the Kings River as the northern boundary to Kaweah River as the southern edge. "Treaty C Tribes" of the Yokut Nation acceded to lands around the Tule River watershed a month later. In most cases, reservation boundaries surrounded sequoia groves, and the US Bureau of American Ethnology later mapped out tribal lands.[9]

As legal agreements between one sovereign nation and another, these treaties had to be ratified by the US Senate. When debate came to the Capitol, elected officials raised alarms that the negotiated reservations were too generous in size, thereby squeezing out Anglo American immigrants. They were also uncertain whether Mexico had recognized the territorial claims of Native peoples in California before acquisition by the United States. For these reasons, the US Senate rejected all eighteen treaties. To Indigenous nations, the unratified agreements meant they were living on unceded territories, so all customary uses of traditional lands where the sequoia–mixed conifer forests grew were never extinguished. To US settlers, Native communities were frequently described as "landless" because they did not adhere to Western notions of bounded private property.[10]

Small-scale logging outfits arrived after midcentury when wagon roads provided access to the sequoia belt of the southern Sierra. In 1873, Hyde's Mill began cutting Big Trees on Mono lands between the Kings and Kaweah Rivers at Redwood Mountain. Lumbermen harvested a few hundred trees over the next five years, ignoring the largest ones because they were too difficult to get out of the forest. Loggers also set up mills at the Dillonwood and Mountain Home Groves on Yokut lands. The Tule River region opened to Anglo American settlers with the dispossession of Indigenous inhabitants. A genocidal campaign led by the US Army against the Mono and Yokut nations during the Tulare Expedition of 1856 overlapped with a series of outbreaks in diseases of Eurasian origin—smallpox in 1833, cholera in 1849, and measles in 1887. One harrowing tale from the conflict involved a white man named Orson Kirk "O. K." Smith who led a volunteer brigade to hunt down Natives for apparently setting fire to his mill site. Limited logging of the Big Trees ensued after combat and contagions.[11]

Government actions spurred on land expropriation and timber extraction. Beginning in 1863, Josiah Whitney of the California Geological Survey charged his cadre of trained geologists to chart land and resources in the Sierra Nevada. They recorded locations of some Big Tree groves, climbed peaks (naming Mount

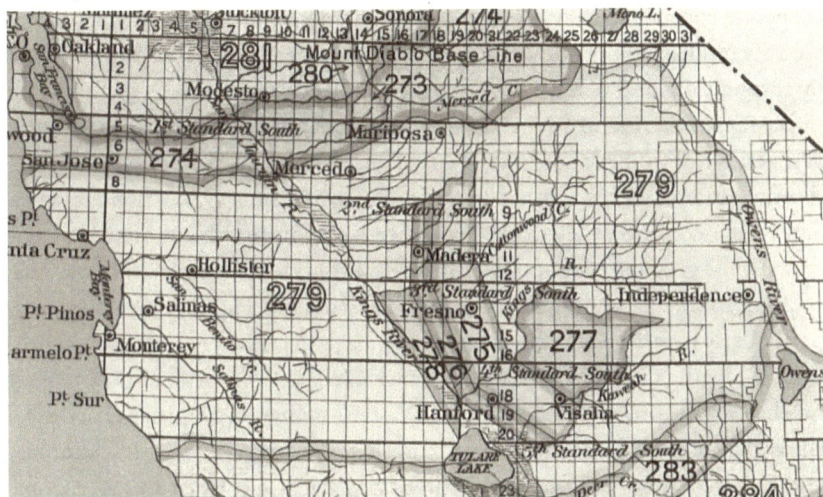

Map of Indigenous territories in California as delineated by the 1851–52 unratified treaties. For example, "Treaty B Tribes"—comprising various bands of the Mono Nation—signed for boundaries corresponding to 277, 278, and 279, which included sequoia–mixed conifer forest of the Sierra Nevada. Note the gridded section lines, representing US notions of property, over the top. From Charles C. Royce and Cyrus Thomas, *Indian Land Cessions in the United States* (1899). Courtesy of Library of Congress.

Whitney from afar), and made contour line maps. The US General Land Office proceeded with formal cadastral surveys, applying the Cartesian logic of township and range sections to the mountains. Invisible lines on maps made the landscape legible to the settler state and, as followed, sellable. US Congress passed the Timber and Stone Act of 1878 to enable the swift transfer of forested lands to private hands. Bounding territory, then, made sequoias free for the taking.[12]

The Timber and Stone Act stipulated that lands in the western public domain that surveyors had deemed "unfit for cultivation" could be opened up to logging or mining. The measure was intended to help small timber producers, just as the Homestead Act of 1862 had helped their agricultural counterparts, by giving them the opportunity to purchase 160-acre tracts for $2.50 an acre. Capitalists, though, soon abused the law. Lumber companies did so by recruiting applicants (with booze and dollars) to fill out entries at the local land offices. Upon receiving the legal title, individuals turned around and sold the land to businesses in prearranged deals. Lawsuits arose to contest these extralegal agreements, but the federal government essentially turned a blind eye. With de facto approval, corporate consolidation of Sierra forests became the norm. One furious Tulare County resident complained that "most of the timber land in the State now held by corporations and rich firms has been fraudulently secured."[13]

Businessmen Austin Moore and Hiram Smith used this tactic to acquire vast acreage of timberlands in the Sierra Nevada. In 1868, thirty-seven-year-old Austin Moore and his brother moved from their Pennsylvania hometown to San Francisco, where they worked together as import merchants to supply residents of the growing metropolis with goods. Profits soared, and within a couple of years, Moore had hundreds of thousands of dollars. He turned to investing, particularly in railroads and lumber, where his path crossed with Hiram Smith. From a logging family in Wisconsin, Smith came to California in 1874 to grab his share of the timber resources. With Moore's money and Smith's experience, the partnership backed ventures in California, Washington, British Columbia, and Durango, Mexico. The pair incorporated the Kings River Lumber Company on April 24, 1888; only then did hordes of dummy claims at the Visalia land office make sense. Government officials sent a special agent to investigate, but alleged payments from Moore and Smith to local congressmen influenced a favorable report. Illegally, Moore and Smith obtained close to forty thousand acres of prime forest in Fresno and Tulare Counties.[14]

And here, the story takes an unexpected left turn. In 1885, fifty-three men rushed into the Visalia land office and filed claims under the Timber and Stone Act on more than eight thousand acres. The land was located within the Kaweah River watershed near an expansive sequoia grove called the Giant Forest. Given the circumstances, one might assume that these individuals were going to pass the property on to a wealthy tycoon or real estate speculator. They were not. All were members of the Co-operative Land Purchase and Colonization Association, a San Francisco utopian commune that sought to replace mainstream capitalist society with a socialist alternative. Led by lawyer and labor organizer Burnette Haskell, a collection of several dozen men and women reorganized as the Kaweah Co-operative Commonwealth with the goal of launching a workers' colony in the Giant Forest.[15]

George Stewart, newspaper editor of the *Visalia Delta*, became alarmed when he saw in a mandatory public notice that claimants had filed under the same address. Stewart contacted the General Land Office to raise suspicions about fraud. The federal agency dispatched an investigator who, unable to resolve the basis of these entries, succeeded in withdrawing the land from sale. However, the Kaweahans believed their claims would be recognized, so they started building a wagon road and sawmill in the Sierra Nevada forest. They named the bulkiest

Kaweah colonists pose in front of the Karl Marx / General Sherman Tree in the Giant Forest. Courtesy of the Yale Collection of Western Americana, Beinecke Rare Book and Manuscript Library, Yale University.

sequoia in the world after Karl Marx (now called General Sherman Tree) and pledged to protect these redwoods. "It would be nothing short of vandalism," Haskell commented, "to indiscriminately destroy these sentinels of past centuries, as has been done in several parts of California, by ruthless ravagers of the Competitive system."[16] As the communards moved to the mountains, Moore and Smith looked to annex more land. The partners joined Smith Comstock, who made his small fortune at Big Stump Basin, to form the Tule River Lumber Company. The three men absorbed an additional three thousand acres into their timber trust with the Mountain Home Grove.[17]

The episode was revealing of the political economy of the United States. Indigenous nations had their land claims go unrecognized by the settler state. While capitalists unlawfully acquired vast swaths of sequoias on Native lands (and Uncle Sam did nothing), socialists legally bought lands with Big Trees only to have their transactions suspended. Conservation of another sort would seal the fate of Native communities and Kaweah colonists.

Moore and Smith's Kings River Lumber Company, capitalized at a million dollars, took control of sixty square miles in the Sierra Nevada. Large-scale logging required industrial infrastructure. In 1889, the company constructed the lower Abbott Mill and the upper Sequoia Mill, as well as a dam at Mill Flat Creek to create a millpond reservoir named Sequoia Lake. In 1890, the firm completed a fifty-four-mile flume through the lower Kings River to connect the mills to Sanger, a small town east of Fresno where the Southern Pacific Railroad transported finished products from the company's factories. And in 1891, Moore and Smith hired Chinese laborers to begin laying down a narrow-gauge track called Sequoia Railroad to the edge of the largest single grove of Big Trees at Converse Basin. That same year, with hundreds of thousands of dollars sunk into these projects, the Kings River Lumber Company commenced full logging operations.[18]

The industrial town of Millwood rapidly sprang up near the upper mill since the company employed some five hundred men, largely immigrants from Sweden, Ireland, Poland, Denmark, Norway, and Germany, many of whom arrived with wives and children. Families lived in crude shacks while single laborers shared dormitories and a cookhouse. The spatial layout of Millwood reflected paternalistic attitudes toward workers: Moore and Smith's headquarters overlooked the townsite featuring a schoolhouse, post office, general store, butcher shop, blacksmith shop, and doctor's residence, as well as two hotels and four saloons. Some women earned incomes through employment, like Irish immigrant Cilicy Kanawyer, who was employed as a local schoolteacher, but most women faced unpaid domestic work. Woodsmen toiled in the surrounding forests for twelve hours a day, six days a week, receiving wages that ranged from $1.50 to $2.50 per day, much of which was funneled back to Moore and Smith through company-owned amenities.[19]

Characteristic of a crisis, economic downturn intensified sequoia extraction. A tight money market during the Panic of 1893 gave banks a reason to collect debts as less currency was in circulation. Moore and Smith's $700,000 in loans were no exception. To make obligatory payments, the pair turned to foreign markets as a substitute for a crippled national economy. For example, the company installed lights at the two mills in early 1894 so that a large Australian order for Sierra redwood could be handled in back-to-back shifts. Moore and Smith also focused on logging sequoias because, more than any other species, the Big Trees had the potential for big money. The firm constructed a third mill in the heart of Converse Basin, where the largest grove of botanical giants in the Sierra grew, and then stationed a mechanical winch at a place called Hoist Ridge to drag fallen trees over the divide.[20]

But both these measures turned out to be insufficient to service debts. Moore and Smith held enough financial sway to reorganize the enterprise with new sponsors as the Sanger Lumber Company before they lost effective control of the business by 1895. Bankers immediately seized the corporation and recouped investments by stripping the land. Over the next decade, the timber harvesting at Converse Basin totaled 191 million board feet; one-third to one-half were sequoias. Or put another way, the amount of redwood coming out of the Sierra during this time could have built, plank by plank, every new home that was under construction in Fresno.[21]

Fresh investors finally stepped in. Ira Bennett, manager for a logging firm in New Mexico, approached wealthy timber baron Thomas Hume of Muskegon, Michigan, to convince him to buy out the Sanger Lumber Company. Since the mid-1890s, when the rapid depletion of Great Lakes forests impelled capitalists to turn elsewhere, Hume had been slowly acquiring old-growth tracts across North America—from Canada to the Carolinas to California. On November 22, 1905, the two incorporated the Hume-Bennett Lumber Company with capital-stock at $400,000. The company decided that the first item of business was to move logging operations eastward to take advantage of unexploited timber stands along the south rim of the Kings River Canyon. Hume-Bennett extended the flume by seventeen miles, relocated one mill from Millwood to a townsite called Hume, and constructed another mill, railroad, and reservoir (eventually named Hume Lake). A second boom period followed during the teen years when the company averaged 17.5 million board feet annually until the new sawmill burned down in 1917.[22]

TWO YEARS BEFORE Hume-Bennett entered sequoia logging, export expert William Wheatley left California for an extended overseas trip. Owners of the state's redwood industry hired Wheatley for the sole purpose, the *Los Angeles Times* reported, of making foreign contacts in order "to skim the cream of the lumber business in all the important markets of the world."[23] Timber capitalists trusted Wheatley's ability to foster trade because of his many years with a San Francisco–based export company. While the coast redwood of northern California was the primary sale, Sierra redwood received secondary benefits because it was sold undifferentiated from its arboreal relative. On a tour with stopovers in England, France, Germany, South Africa, China, Japan, and Australia, Wheatley's competition would be fierce. California redwood entered the consumer fray with Scandinavian pine as its softwood rival while Australian eucalyptus and Indian teak held premier spots among the large hardwoods. To understand how giant sequoias entered this world of commodity flows, we follow the labor systems that brought Big Trees from mountain to market. A stationary plant could become quite mobile.[24]

In the Sierra Nevada, woodsmen began by erecting scaffolding around the sequoia trunk or by drilling holes into it to hold springboards on which to stand. From an elevated platform, ten to twenty feet above the ground, depending on where the tree straightened up, they stripped away the bark on one side. Choppers then swung double-bitted axes at the tree—*thwack, thwack, thwack*—spending up to one week chiseling out a notch as tall and deep as themselves. Once the undercut was finished, they applied a crosscut saw on the opposite side for another two days to make the back cut. For the largest of Big Trees, because standard blade lengths could not handle their girth, lumbermen welded two ten-foot-long saws together to create an almost unwieldy tool. Steel wedges and copious grease kept the weight of the mammoth plant from halting the blade as sawyers moved its razor-sharp teeth back and forth. When a narrow wood strip called the hinge remained, crackling noises gave loggers a reason to take cover.[25]

Boom.

One spectator described how the Big Trees struck the ground with such force that their impact sounded like a thunderclap and felt like a "tremble as from an earthquake shock."[26] After dust and debris settled, loggers sawed or blasted the fallen sequoia into more usable sections for hauling. In smaller operations, teamsters coaxed oxen or mules with words and whips to drag logs over wood-plank skidways or down trench-like chutes. In larger enterprises, timber barons purchased "donkey engines," the nickname for portable steam-powered winches, for workers to use. Patented in 1882 by California lumberman John Dolbeer, this machine drove a steel cable loop around a pulley system with hooks to drag logs to a central point. In Kings River country, a network of railroads also carried timber through the forest. Whether by animal or artifice, redwood arrived at the mills.[27]

A deafening drone of the steam-driven saws drowned out the shuffle of millworkers as they pushed sequoia chunks past humming disks—*buzz* after *buzz* dropping board after board. At the Converse Basin Mill, Sanger Lumber Company supervisors installed a special bandsaw, known as the redwood splitter, with a one-thousand-horsepower engine and a blade circumference of ninety feet. Laborers carried sawed planks into the yard to be stacked into many-story piles for a month of seasoning, open-air drying meant to reduce the moisture content in the wood and to prevent warping. Transport methods at this point varied by location. In the Tule River drainage, oxen remained beasts of burden in hauling timber to the train depot at Porterville. In the Kings River country, the flume operated as a lumber ditch. Workers fastened twenty-foot-long boards together by affixing iron clamps onto each end, and these bundles were ready for liquid transport to the factories of Sanger.[28]

Each morning, laborers opened an outlet at Sequoia Lake, or later at Hume Lake, to start a wave-like *swoosh* of water down the V-shaped flume. The shallow stream carried packages of redwood, pine, and fir out of the mountains at an average rate of five miles per hour, with speeds reaching fifty miles per hour on the steepest grades, for a half-day trip. Herders lived at one of fifteen isolated stations along the route, patrolling the wooden ditch to fix leaks or to break up congested areas with miner-like pickaxes called picaroons. At the end of the day, millworkers nailed an upright board called the joker onto the last bundle to signal that no more lumber would be sent down the flume. While the channel reached its daily capacity of 250,000 board feet at times, the flume conveyed more than just wood. Teenagers looking for amusement stole flume boats and enjoyed thrill rides on occasion. One of the most notable voyages involved an expectant mother who, because of child-bearing complications, rode down the flume with nurse escort Martha Hanify to reach a hospital. The birth of a future wageworker was enabled by the death of trees.[29]

The town of Sanger, the self-proclaimed "Flumeopolis of the West," functioned as the manufacturing base. Company grounds contained a box factory, sash and door factory, planing and shingle mill, storage warehouses, and engine house. Workers unloaded milled lumber from the flume and piled it onto tramways where the loads were carted into their respective buildings to be processed. At the planing and shingle mill, factory workers cut sequoia boards into housing materials from molding to paneling to shakes. Like Chicago's meat-processing plants bragging that they used everything about the hog except the squeal, the Sanger factory was touted as a model of industrial efficiency. "Every scrap is used up," a local newspaper boasted. "No unnecessary ground is covered."[30] The Hume-Bennett Lumber Company sold pine boxes as fruit containers to Central Valley growers with redwood sawdust as cushioning. Whatever the intended use, lumber reentered the grounds as finished wood products for laborers to pack onto railcars. In 1904 alone, 2,200 carloads left the Sanger plant on the Southern Pacific line.[31]

The Big Trees prove difficult to track during resource consumption because they were often lumped together with their more commercial cousin, coast redwoods, and simply sold as the nondescript "California redwood." Botany professor Willis Jepson observed that "Sequoia gigantea lumber is marketed

182 | CHAPTER FIVE

A logger on a ladder holds a welded crosscut saw next to the cut end of a giant sequoia while a woman and man stand on the ground for scale. Courtesy of Museum Collection, Sequoia and Kings Canyon National Parks.

as redwood" within his 1909 survey of California's trees.[32] Little distinction between *Sequoia gigantea* (now designated *Sequoiadendron giganteum*) and *Sequoia sempervirens* was made at the time because these two species held similar wood properties and consumer uses.[33] On account of its natural ability to withstand decay, redwood served as grape stakes and fence posts to farmers and ranchers of the San Joaquin Valley. Due to its durable and decorative traits, homebuilders in Los Angeles and elsewhere routinely used redwood for doors, beams, roof shingles, and window sashes, particularly with a bourgeois appetite for Arts and Crafts architecture. Because of its performance under liquid pressure and resistance to water damage, redwood was employed in storage tanks, stave pipes, and even the flume that transported sequoias out of Converse Basin. Whether the Sierra or coast variety, the range of redwood products appeared boundless.[34]

The same reason that giant sequoias are tricky to follow during domestic consumption also makes them challenging to trace during foreign exporting. In December 1899, for instance, *American Lumberman* vaguely reported that "a cargo of California redwood arrived last month on contracts" to Liverpool merchant Edward Chaloner & Company.[35] Although coast redwoods comprised a much larger share of the timber trade, adequate evidence confirms that the Big Trees were sold in international markets. In 1908, one Fresno County promoter boasted that the Hume-Bennett Lumber Company "shipped to all parts of the world, even as far as Australia."[36] The Big Trees saturated transpacific commerce to the point that an Australian woodworker commented, "The timber is well known here where it is used extensively for many purposes," especially as termite-resistant railroad ties.[37] Trade journals indicate that Big Trees arrived in Europe from transatlantic voyages where Sierra redwood was made into everything from plywood to pencils. What if we assumed that only 5 percent of timber exports labeled "California redwood" were giant sequoias? At this conservative estimate, the figures were still noteworthy: During the twelve years between 1894 and 1906, nearly seventeen million board feet were sent abroad.[38]

Capitalists enclosed territory under the banner of private property within the Sierra Nevada to recruit a (relatively) stationary plant and, through cheap human labor, transform the Big Trees into mobile commodities of wealth. George Sudworth of the US Division of Forestry offered an evaluation of these landscape changes at the beginning of the twentieth century in his magisterial work *Forest Trees of the Pacific Slope* (1908). Although advocacy for public forest reserves colors the account, Sudworth restrained this impulse with empirical knowledge gained from the multiple horseback trips across the mountains where he took hundreds of photographs and extensive field notes. And, as a federal dendrologist since 1895, professional foresters relied on his accurate scientific descriptions for tree identification and distribution.[39]

Of the seventy-five sequoia groves in the Sierra, Sudworth reviewed the legal and ecological status of twenty-six stands, from the well-known Calaveras Grove to the north to the Deer Creek Grove some two hundred miles to the south. Rattled by the fact that over half of the surveyed Big Trees were privately owned, many of which were intensively logged, he reiterated the popular view about how "much concern has been expressed regarding the probable extinction of these trees."[40] In just fifty years, one-third of the sequoia forest in the Sierra Nevada—or, more poignantly, the entire world—was gone. Designating protected areas, Sudworth advised, could halt the unrestricted sale of giant sequoias, preventing their "total destruction for commercial purposes." In nationalistic terms, Sudworth decided that "some of these magnificent forests should be preserved untouched as monuments of American respect and love for nature's noblest legacy."[41] Bounding space as a legal tool would change the management of Big Trees from capitalists to conservationists.

Conserving Giants, Bounding Space

I drove northward from Grant Grove on the Generals Highway. At one curve in the road, I glimpsed out the car window at a white vertical sign nailed to a pine tree stating "US Boundary NPS." I was leaving the national park and heading into the national forest. Back in 1935, the US Forest Service bought some twenty thousand acres from the Hume family for $319,000 in what amounted to corporate welfare: old-growth forest purchased from the government for $2.50 an acre; then clear-cut forest later sold back to the government for $14.93 an acre.[42] The present boundary line essentially separated those primeval Big Trees that had survived from those that had not. Arriving at Princess Campground, just a few miles from the old lumbering headquarters at Hume Lake, I found camping spots interspersed among more sequoia stumps of the Indian Basin Grove.

Near the campground walking trail, I stumbled upon a redwood trunk with its colossal shoot resting on the ground. The mammoth tree, broken into a jagged row of puzzle-like pieces, measured over one hundred paces from cut to crown. Hume-Bennett loggers had chopped down nearly two hundred mature sequoias in this grove between 1905 and 1907, but obviously not all of them made it to the mills.[43] I stood before a felled tree that capitalists would have considered unsellable, and yet here the redwood endured to the present with few hints of aging. Naturalist John Muir reached a similar conclusion when he traveled through these Kings River forests over a century ago. In August 1875, Muir set out from the Yosemite Valley, heading southward with his mule companion named Brownie, to investigate in Humboldtian terms the geographic distribution of giant sequoias that ran along the Sierra. Over the next eight weeks, Muir trekked some six hundred miles of terrain. What the wandering Scot encountered during his journey both amazed and enraged him.[44]

One striking feature about the Big Trees was their ability to withstand the effects of time. Muir came across fallen sequoias, knocked down by lightning strikes or blown over by the wind, but they displayed little indication of decomposing. The naturalist mused, "Nothing hurts the Big Tree. I never saw one that was sick or showed the slightest sign of decay."[45] Giant sequoias deteriorate very slowly, almost imperceptibly to our limited human senses. One shocking aspect was just how brittle Sierra redwood was and, as a result, how much waste resulted from clearing the forests. When Muir reached Redwood Mountain, a series of groves south of Indian Basin, he bumped into Hyde's Mill loggers cutting down Big Trees and squandering a lot of them. "When felled," Muir objected, "the sequoia breaks like glass—from twenty-five to fifty percent unfit for the mill."[46] Muir saw how biophysical characteristics held the potential to either preserve or destroy what he reverently called "the greatest of living things." Upon returning from the mountains, Muir took it upon himself to make the plight of the Big Trees known to a wider audience of Americans.

In popular accounts, Muir receives full credit for the ways in which his poetic voice inspired the nation to defend giant sequoias by creating some of its first national parks. For historians, however, the reasons why the Big Trees were conserved are debatable. One interpretation emphasizes how people like Muir transformed sequoias into cultural icons, personalized with names like General Sherman. In this view, patriotic citizens protected hallowed sequoias because anything less was sacrilege. Another explanation highlights the Southern Pacific Railroad's influence in the founding of Sierra Nevada parks. From this perspective, corporate elites desired to increase rail traffic by enticing the new bourgeois class to take vacation trips to mountain sanctuaries. Still another understanding accentuates the political triumph of agricultural interests over the timber industry. Elected officials came to value sequoias more for their living role in regulating the hydrological cycle during irrigation than in their deceased state as sawed lumber. The last telling highlights the Big Trees' geographic isolation from markets. When compared to the coast redwood, which was both closer to major population centers in San Francisco and Sacramento and more accessible for transport via rail or ship, stands of Sierra redwood were in remote mountain areas.[47]

But these accounts overlook two crucial points: first, how the Big Trees' biophysical properties resisted the economic logic of capitalism, and, second, how park borders operated within a gridded landscape to include the sequoia groves. These factors were more important in planting the seeds of large landscape conservation.[48]

ONE OF MUIR'S OBSERVATIONS during his odyssey was scientific in nature. He noticed that giant sequoias were more sparsely populated along the northern half of the Sierra Nevada and more densely clustered around its southern half. About forty miles separate the Calaveras and Tuolumne Groves today, with as many miles between the Merced and Mariposa Groves, but a more or less continuous stretch of sequoia forest extends across the Kings, Kaweah, and Tule drainages. To explain the spatial difference, Muir employed what we would now call "deep history"—that is to say, he took a long view on landscape change to understand why the Big Trees came to live where they do.[49]

At the annual meeting of the American Association for the Advancement of Science in 1876, Muir spelled out how the Big Trees populated ecological niches that were opened by retreating glaciers at the end of the last Ice Age. "When the Sierra soil-beds were first thrown open to preemption on the melting ice sheet," he asserted, "Sequoia may have established itself along the available portions of the south half of the range." By Darwinian chance, the Sierra redwoods gradually outcompeted other species for habitat in the warm southerly climes. Muir struck a general rule for their post-Pleistocene geography: "*The wider the ancient glacier, the wider the corresponding gap in the Sequoia belt.*"[50]

During the Cretaceous period, about a hundred million years ago when dinosaurs roamed the Earth, the climate was mild and moist. An ancestor to giant sequoias covered most of the Northern Hemisphere. Archaeologists have excavated fossil remains of a sequoia lineage in such disparate places as Siberia, Japan, Germany, and Yellowstone National Park. However, climatic changes steadily began to restrict the range of these trees. By the time of the Miocene epoch, some twenty million years before the present, one offshoot of the sequoia family moved to present-day southern Idaho and western Nevada, where the population largely remained until colder temperatures over the Pleistocene epoch forced it southward toward the Sierra Nevada. As Muir argued, giant sequoias then relocated to the western slopes after glaciation. Contemporary pollen analyses from Log Meadow have detected sequoias in their present location no sooner than 4,500 years ago, meaning that today's oldest Big Trees, remarkably, are only the *second generation* to occupy the area. Other branches evolved over the millennia so that we now have three related species: dawn redwood in north-central China, coast redwood of coastal Oregon and California, and Sierra redwood in eastern California. While the trees share a common progenitor, each plant diverged enough biologically to become a genus of one.[51]

Fight and flight have been two classic survival strategies: Unable to flee natural enemies, the Big Trees were forced to build their defenses. Through the struggle for existence against pests and pathogens, giant sequoias probably came to possess distinct biophysical traits. Tannins, the same molecule that gives wines their dry taste, comprise over 40 percent of the water-soluble organic compounds found within the plant's cell walls. Extractives of this kind inhibit the work of fungi and other decomposers, giving redwood the ability to resist decay that would eventually be valued by capitalists for industrial purposes and scientists for tree-ring data. And the same reason that the sequoia at Indian Basin, despite being felled over a century ago, endured for admirers like me.[52]

Brashness, the high likelihood that wood fractures, is more speculative but possibly arose from a couple selective pressures. The Big Trees might have earned their namesake during past ages when the organism vied with other rival megaflora for sunlight, but the mechanical properties required to stand atop forest canopies also increase momentum when their sheer bulk finally goes down. Climatic influences, especially periods of drought, might have also demanded structural adaptations to wood fibers to assist vascular tissue called the xylem in transporting water upward from the roots to the rest of the plant. Whatever the causes, the resulting brittleness of Sierra redwood would be abhorred by those with money-making motives as biological legacies carried forward into the late nineteenth century.[53]

Muir and other observers witnessed how waste was a common feature for all logging companies due to redwood breakage. Muir's 1875 encounter at Hyde's Mill taught him that "the timber is very brash, by this blasting and careless felling on uneven ground, half or three fourths of the timber was wasted."[54] Giant sequoias, in contrast with coast redwoods, are more brittle as old-growth trees, and accordingly, these mature stands pursued by lumber corporations were the most likely to break apart.[55] "Lumbering of the big tree is destruction to a most unusual degree," one federal forester reported. "Such a tree strikes the ground with a force of many hundreds or even thousands of tons, so that even slight inequalities are sufficient to smash the brittle trunk at its upper extremity into almost useless fragments."[56] Timber capitalists worried that sequoia breakage would lead to financial ruin. In large part, they were right to be concerned: The Hume-Bennett Lumber Company went most years without turning a profit.[57] The Big Trees, in effect, resisted the capitalist logic of being transformed into a commodity.

Logging operations tried to reduce breakage by using special techniques. Wherever possible, teams of choppers and sawyers purposefully felled sequoias toward upward slopes to lessen their blow upon striking the ground. However, when they encountered large sequoias on flatter topography, woodsmen expended weeks building what were known as "feather beds." A dozen or so workers started by digging a trench in the same dimensions as the Big Tree; they proceeded by chopping down small pines, firs, and cedars in the vicinity; and they finished by dragging the branches across the channel to absorb the tree's impact. Despite the best efforts, feather beds were usually no better at safeguarding the brittle timber. A local photographer observed how one redwood possessed a twenty-two-foot circumference and promised 725,000 board feet of lumber. But when it was finally cut down, the upper third shattered into unmarketable debris. Evolutionary accidents in biology proved to be one of the saving graces for conserving giant sequoias.[58] Bounding land under a different legal regime proved to be another.

WALTER FRY WAS a working-class convert to conservation. Born in 1859, Fry heard tall tales of the Big Trees during his Kansas youth. After facing a series of economic challenges as a young adult, Fry moved out west with his wife in search of better opportunities. In 1887, the couple settled in Tulare County, California, and Fry found work at the Southern Pacific Railroad. He later took a job with Smith Comstock in Big Stump Basin, where Fry and four other men were tasked with felling a mature sequoia. They chopped and sawed at the giant's base for five long days, and after the tree struck the ground, Fry began counting the rings to calculate its age. What did he learn from the tally? A plant that was at least three thousand years old had been taken down in less than a week. The episode disturbed his conscience, and consequently, Fry remembered, "I closed my career as a woodsman on the spot."[59]

Fry soon found a vocation to which he would dedicate the rest of his life: protecting the Big Trees. In 1901, he landed a job as a road foreman in Sequoia National Park. This position led to another job as one of the park's civilian rangers, and he moved up the ranks to become chief ranger. By 1914, when military administration of the park officially ended, Fry was hired as Sequoia's first civilian superintendent. After this role, Fry founded the park's naturalist program, publishing dozens of articles on its flora and fauna to assist rangers and guides in educating the public. Fry offers a democratic counterpoint to conventional histories of conservation: No wealth. No elite upbringing. No college degree. Nothing but working with nature. Back in 1890, Fry understood that new political boundaries, or a "fence of laws" in the words of one historian, needed to be drawn around the sequoia groves if these biological resources were to be conserved. When a petition circulated to establish Sequoia and General Grant National Parks, Fry's name was the third signature on the list.[60]

The founding of Sequoia and General Grant was based on local exigencies of capitalist extraction. In 1890, the US General Land Office restored three sections of land in the Sierra Nevada for private entry. One of the sections for sale was township 18 south, range 30 east, which contained the Garfield Grove of Big Trees, located on the north side of Dennison Ridge in the Kaweah drainage. George Stewart, editor of the *Visalia Delta*, who earlier raised fraud allegations against the Kaweah colonists, feared that the loose handling of the Timber and Stone Act would allow this section of sequoias to fall into "the holdings of the big lumber company."[61] Some twenty years earlier, in 1864, President Abraham Lincoln had issued a federal land grant to the State of California for park purposes, which became a model for subsequent Big Tree protections since the measure included the Mariposa Grove. Stewart joined forces with other regional elites—fellow *Visalia Delta* employee Frank Walker, rancher John Touhy, and land office receiver Tipton Lindsey—to campaign for federal designation of a forest reserve or a public park to prevent a corporate takeover of the Garfield Grove. Beginning in July 1890, they held meetings in Fresno, wrote national magazines and newspapers, and elevated the profile of giant sequoias in Washington, DC.[62]

Their argument was simple: If bounding territory under the existing land laws privatized sequoias for lumber companies, then enclosing space under a different legal framework would protect the Big Trees. "The main object of the bill," Touhy explained to one US senator, "is to preserve from speedy destruction and for the people of the UNITED STATES forever the only remaining township left to the government of the SEQUOIA GIGANTEA, the grandest specimen of trees and the only ones of their kind in the world."[63] Their pleas found a sympathetic ear with General William Vandever, US Representative for central California, who introduced H.R. 11570 for a public park in township 18 south, range 30 east. Stewart, Walker, Touhy, and Lindsey had originally envisioned "a large park" embracing the entire west side of the southern Sierra with the Kings, Kaweah, Tule, and Kern drainages. Given the need for swift action, however, the four realized that "it was not an opportune time to ask for so much and we deemed the proper course to be confining our efforts to saving the big trees, then in immediate danger."[64] Lindsey recommended that township 18 south, range 31 east be added to the bill, as well as units 31, 32, 33, and 34 of township 17 south, range 30 east, since they too possessed giant sequoias. The House Committee on

Public Lands amended the bill to include those additions. In short, the spatial instrument for privatization was also used for conservation.[65]

And now, the motives behind park expansion get curious. On September 25, 1890, Congress passed Vandever's bill and President Benjamin Harrison signed it into law, creating what would become known as Sequoia National Park. The legislation withdrew seventy-six square miles of Sierra Nevada forest from entry—the very same sections that Stewart and other conservationists had recommended. Only six days later, on October 1, 1890, another park bill—from an unknown sponsor this time—became law. H.R. 12187 nearly tripled the acreage of Sequoia National Park by adding five new townships for over 250 total square miles, containing two dozen Big Tree groves, and adopted the previously withdrawn four-mile section of sequoias as General Grant National Park. The same act established another national park around the California-owned Yosemite Valley tract.[66]

The mysterious H.R. 12187, scholars agree, most likely came from the Southern Pacific Railroad, though why the corporation got involved is unclear. Historians have suggested that the company wanted to attract eastern tourists on its transcontinental lines through park resorts or to stabilize agricultural commodities on its Central Valley lines through watershed protection. This much is certain: Railroad agent Daniel K. Zumwalt visited Vandever in Washington, DC, during the time when Congress passed the legislation for park enlargement. The earliest known map of Sequoia National Park—dated October 10, 1890, well before the preserve was known to the wider public—was produced by the Southern Pacific's Passenger Department. But by focusing on *who* made the map, historians have overlooked *what* the map represents for the early spatial practices of conservation. As depicted in cartography, the boundaries of Sequoia and General Grant neatly followed the right angles of the township and range sections. For protected areas to work in the Sierra Nevada front country of private property, national parks and their biological resources needed to fall within the legal demarcations of the modern grid system.[67]

There was probably another reason for the Southern Pacific's involvement in national parks: the Kaweah colonists. Four or five years earlier, communard J. J. Martin had visited Charles Crocker, president of the Southern Pacific, to propose a joint venture in which the Kaweah Co-operative Commonwealth would use Crocker's railroad to market their lumber under a different economic model. Unlike the wage labor system, which had enriched only a few individuals, the Kaweahans agreed to pay workers at their sawmill according to the time-check

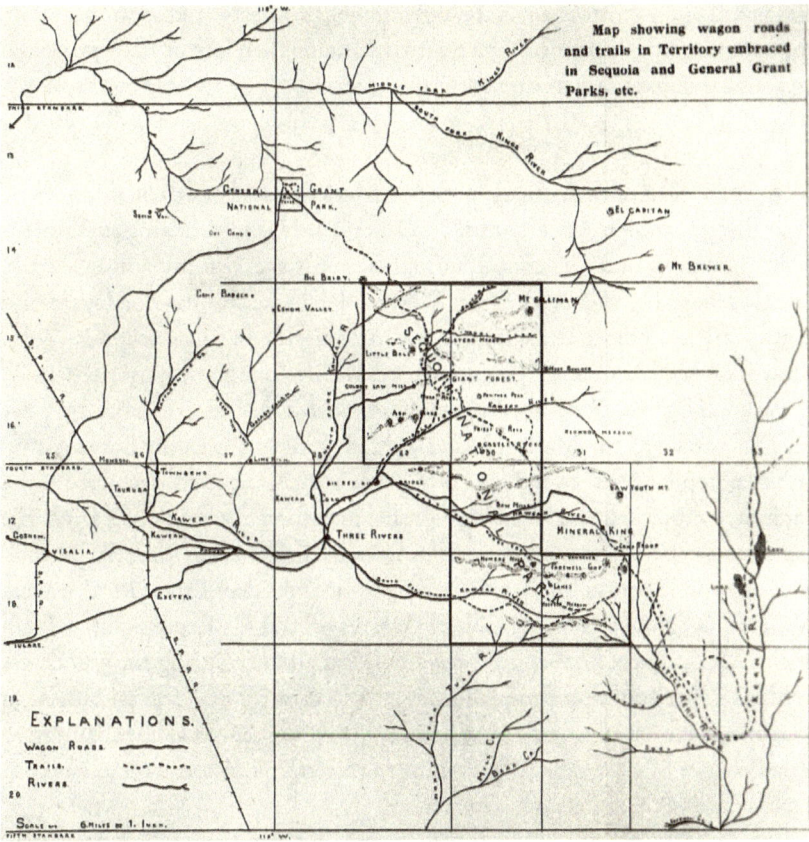

An 1893 map of Sequoia National Park and General Grant National Park as depicted within the township-and-range sections. Courtesy of Museum Collections, Sequoia and Kings Canyon National Parks.

system, where, in words of Martin, "the 'minute' takes the place of the 'cent' as the unit of value."[68] Crocker rejected Martin's offer, but the capitalist presumably felt threatened by the socialist's pending collective landholdings. They challenged the Southern Pacific's monopoly on forest resources in the Sierra Nevada, and rumors even implicated railroad syndicates in Moore and Smith's timber operations.

For these reasons, the second park bill was significant. H.R. 12187 not only expanded the federal reserve to include the Giant Forest and other sequoia groves, but it also jeopardized the Kaweahans' territorial claims since they were now squatting on public lands. On November 30, 1890, US marshals arrested four

Kaweah Colony trustees on charges of "timber trespass"—and soon thereafter, the communards disbanded. The Southern Pacific therefore utilized protected areas as a tool to eliminate unwanted competition.[69]

OVER THE turn of the century, park boundaries turned out to be more elastic than first conceived. As early as 1891, Captain J. H. Dorst, acting superintendent of Sequoia and General Grant, advocated for extending administrative lines eastward to the Sierra's Great Western Divide, as well as northward and southward to embrace the Kings and Kern watersheds. That same year, John Muir endorsed protecting all sequoia groves within "one grand national park."[70]

Park enlargement remained a conservationist desire, as evidenced by what became known as the Mather Mountain Party. In 1915, wealthy borax manufacturer Stephen Mather, in his new position as assistant to the secretary of the Interior, handpicked seventeen high-profile men for a horseback trip across the southern Sierra. The group mostly came from backgrounds in industry, journalism, and politics, including Southern Pacific vice president Ernest McCormick, *National Geographic* director Gilbert Grosvenor, and US Representative Frederick Gillett. Mather hoped to forge a political alliance during the wilderness excursion that could pressure US Congress to create a new federal bureau to administer the national parks. On July 14, the party met at a local restaurant in Visalia where Mather spoke of enlarging the Sierra's protected areas to establish a mega-reserve that he termed Greater Sequoia.[71]

The next day, on July 15, 1915, the Mather Mountain Party traveled by car to Sequoia National Park and set up camp in the Giant Forest. They slept under the redwoods that evening, resting their heads where the Kaweah colonists once congregated. These cultural, economic, and political elites spent the following day touring the Big Trees.

Who was their guide? Not Stephen Mather. Not George Stewart. No, both men were present but only had a shallow understanding of the crimson giants. It was humble Walter Fry—who had gained a deep knowledge of the sequoias not by reading about them but by working among them. The party then slowly made its way toward Mount Whitney via Franklin Pass, starting and concluding each day with meals prepared by Tie Sing, a veteran Chinese American cook for the US Geological Survey. A week into the journey, on July 21, Mather, Grosvenor, and Horace Albright, Mather's private secretary, decided to climb up the Kern Canyon for a look at the vista. "This part of the Sierra was a new world to all three men," Albright recalled later. When they reached the precipice, he said,

Superintendent Walter Fry studying the local flora in Sequoia National Park. Courtesy of Museum Collections, Sequoia and Kings Canyon National Parks.

"they were awestruck by the bold grandeur of the valleys and the mountains. They vowed to include the Kern, Kings, and Whitney regions in an enlarged Sequoia or even an entirely new park."[72]

A few years after the formation of the US National Park Service in 1916, Mather and other conservationists pushed for a "Roosevelt-Sequoia National Park" on the suggestion of the Boone and Crockett Club to honor former president Theodore Roosevelt. The NPS director contacted a congressman from Fresno, Henry Barbour, who in 1921 introduced the first of a series of bills that basically moved to extend park boundaries as the Mather Mountain Party had earlier imagined as Greater Sequoia. Irrigation and hydroelectric companies objected to the expansion immediately based on anxieties that any new designations might block future reservoir sites. Robert Marshall of the US Geological Survey allayed most fears by completing a formal survey of all drainages in the southern Sierra, concluding that park enlargement would not harm water development. Infighting between federal agencies delayed action on Barbour's bill. William Greeley, chief forester of the US Forest Service, wanted guarantees that any land transfer to the National Park Service would hold few timber and minerals—as historian Alfred Runte argued, it had to be "worthless land."[73] Sierra Club members opined that the region, especially the High Sierra, was more appropriate as a national park site because the alpine landscape possessed more recreational than extractive value.[74]

Greater Sequoia received a nudge over the legislative hump by conservationist Susan Thew, who believed that what the proposed enlargement lacked in economic resources made up for in natural beauty. Born in 1878 to a wealthy Ohio family, Thew originally traveled to California with her industrialist father to escape the harsh winters. Thew first encountered the Big Trees in 1918 on a mesmerizing trip to Giant Forest, where she soon met Fry and Stewart. Thew became an advocate for park extension upon learning about Barbour's bill. Beginning in 1923, she devoted summer after summer to traveling through the High Sierra to document the region east and northeast of Sequoia with photographs.[75]

After covering hundreds of miles of trails, in March 1926, Thew published a booklet titled *The Proposed Roosevelt-Sequoia National Park*, the most complete pictorial collection of the region at the time with about 140 views of the rugged alpine scenery. "I am trying to tell the reason for saving this domain," Thew stated, "by showing the character of the country with photographs rather than through the written word."[76] Thew distributed copies of her gazetteer to every member of Congress; Mather called the album a "splendid booklet" and Barbour remarked that it had done a great deal "to create a favorable sentiment of the bill"[77] If people could not go the mountains, Thew wanted to bring mountains to the people.[78]

Politicians invoked Thew's booklet on the debate floor in support of the park, although Congress passed amendments to drop the Kings River area and remove Roosevelt's name. With these changes, on July 3, 1926, President Calvin Coolidge signed Barbour's bill into law. H.R. 9387 took a hybrid approach in demarcating the enlarged protected area: In the frontcountry, the park traced the same private property boundaries as gridded survey lines; but in the backcountry, it followed more ecological boundaries with the Sierra Nevada divide. Overall, Sequoia National Park doubled in size by absorbing national forest lands at the headwaters of the North Fork of the Kern River and the Whitney region.[79]

In standard narratives of large landscape conservation, general recognition that protected areas are "terrestrial islands" usually does not emerge until after 1967 as land managers begin to apply the theory of island biogeography. But the tension between bounded space and biological diversity appeared much earlier.[80] John Roberts White, the longest-serving superintendent of Sequoia, from 1920 to 1947, wrote in a midcentury book, "Our national parks are often but *islands* of rescued scenery and other natural resources in the midst of an exploited region." In the same text, White put it even more bluntly: "What has been saved of the Sequoia groves [is] in the Parks, and what has been irreparably lost on private lands [is] outside the Parks."[81] White essentially summed up my point: Conservationists formalized park boundaries by redistributing lands around the Big Trees and other biological assets where capitalist extraction was permitted and where it was not. The national parks essentially became terrestrial islands through a "fence of laws." At the same time, scientists would break down these legal barriers with climatological findings, raising new concerns about a bifurcated Sierra Nevada landscape.

Studying Tree Rings and Climate

On the campus of the University of Arizona, the Byrant Bannister Tree-Ring Building was designed, architecturally speaking, to resemble a forest. A wide cylindrical pillar served as the lobby and trunk, a series of steel beams supported a canopy of offices, and a veil of aluminum pipes for shade looked a lot like foliage. In the foyer, I strolled up to a polished cross-section of a giant sequoia hoisted on its side with significant dates in world history. Since the pith (core) dated to 212 CE, the tree was twenty-three years old when the Chinese invented a magnetic compass, 359 years when the Prophet Muhammad was born, and 1,564 years when the American War for Independence began. All these events could now be accurately dated, of course, because of the work of A. E. Douglass, founder and first director of the Laboratory of Tree-Ring Research. Back in 1931, Superintendent White had offered the massive slab, taken from a dead sequoia blown over by the wind, as a gift. The Arizona State Museum exhibited the chunk for many years until it made its way back to the Tree-Ring Lab for public viewing.[82]

Near the display, I met senior research specialist Chris Baisan, who offered to take me to the basement to see the Tree-Ring Lab's wood collection. In the underground vault with a footprint wider than the building itself, over two million samples filled the archive from all over the world—cedar from Lebanon, alerce from Argentina, and what most interested me, sequoia from California. We marched past row after row of stacks until Baisan halted at one blue-beige unit. He pulled a triangular piece of redwood from the shelf, handed it over to me, and I gazed at the mosaic of wide and narrow rings. Baisan explained that the light-colored earlywood, laid during the growing season, and the dark-colored latewood, put down over the dormant period, represented one calendar year. Inspecting the section, my thoughts turned to how this tree sample from the Sierra Nevada ended up in Tucson. That journey, I realized, tells us a lot about how the sequoias destabilized administrative boundaries by contributing to scientific understandings about climate change. Supranational forces unsettled national parks.

Andrew Ellicott (A. E.) Douglass was talented at recognizing patterns. Trained as an astronomer, Douglass first came to Arizona in 1894 to establish the Lowell Observatory for gazing at Mars. After choosing a hilly site near Flagstaff for the telescope, he grew mindful of how the atmosphere was integral to any celestial observations. During travels across the Utah-Arizona border, Douglass also noticed what seemed to be a steady relationship in the mountains between altitude, precipitation, and vegetation irrespective of whatever state he was in. At higher elevations, where cooler and rainier weather prevailed, larger and more numerous trees grew; and at lower elevations, where the weather was warmer and drier, trees were smaller and less abundant. While studying at Trinity College in Connecticut and then working at Harvard College in Massachusetts, Douglass developed the habit of looking at the latest scientific papers. He read one study that postulated that sunspot activity might influence solar radiation and thus affect weather patterns. Could tree rings, Douglass wondered, record annual changes in climate that might correlate with variations in sunspot activity?[83]

In 1904, Douglass went to the woodyards of the Arizona Lumber and Timber Company to collect some tree specimens. He obtained twenty-four samples of ponderosa pine by having the mill owner cut thin slices from the ends of logs or from the tops of stumps. Douglass studied these specimens during his spare time while teaching at Northern Arizona Normal School. In 1906, he secured a faculty position in physics and astronomy at the University of Arizona; he received an honorary doctorate from his alma mater two years later for prior work in setting up observatories in Peru and Arizona. He continued with his tree analysis by taking measurements of ring widths and tabulating the results. In June 1909, Douglass published an article in the *Monthly Weather Review*, issued by the American Meteorological Society, suggesting that periodic tree growth possibly revealed an eleven-year sunspot cycle.[84]

The true breakthrough came by developing techniques still used today. First, Douglass established the chronological tool of cross-dating. Identifying the exact year in which a tree ring was produced could be found out by comparing arrangements of wide and narrow rings in multiple samples. For example, a series of three broad rings, followed by two thin rings, then five medium rings in one specimen might overlap with the same pattern in another. Douglass saw compressed rings of 1879–85 in virtually all pine samples, and this commonality allowed him to jump from the known to the unknown. If he knew the date of an outermost ring in one tree, cross-dating allowed Douglass to determine uncertain dates in other trees. Using the method, Douglass could now reasonably stitch together a master timeline.[85]

And second, Douglass recognized that tree-ring width could serve as a proxy for climatic influences. In 1911, he gathered more pine specimens near Prescott, Arizona, because traditional weather records had been kept there since the mid-nineteenth century. After cross-identifying the samples, Douglass created statistical relationships between temperature, rainfall, and ring widths. He validated the approach by checking results against a history of the Southwest by Hubert H. Bancroft, noting how narrow rings corresponded with recorded periods of drought in 1748, 1780, and 1820–23. The Prescott group was also significant for demonstrating larger environmental forces since their growth patterns mirrored those in Flagstaff some seventy miles away.[86]

Douglass obtained a sabbatical for the 1912–13 academic year and used the leave as an opportunity to expand his geographic base. Henry Graves of the US Forest Service contacted forestry departments in Europe on Douglass's behalf to inquire if they would be willing to host the professor and to procure tree samples for him. Traveling abroad for four months, Douglass visited England, Germany, Austria-Hungary, Norway, Sweden, and Denmark. Dr. H. H. Jestrup of Copenhagen, Dr. A. Cieslar of Vienna, and Professor Gunnar Schotte of Stockholm were helpful in taking Douglass to local sawmills. Each stop added to his wood collection, the bulk of which ended up being Scots pine from those countries around the Baltic Sea. As with the ponderosa pine from Arizona, Douglass observed what appeared to be agreement in the tree rings over a greater region.[87]

Compiling tens of thousands of rings from two varieties of pine, Douglass reconstructed a five-hundred-year chronology of past weather conditions stretching back to the 1400s. He would need a much longer time frame, however, to substantiate the climatic effects of sunspot cycles. What Douglass desired was a unique tree species that lived not for centuries but for millennia.

ENTER ELLSWORTH HUNTINGTON. The geographer earned notoriety—and well-deserved criticism—as a eugenicist, but he was known by peers as an unorthodox thinker about how climate influenced human societies. These ideas came during his engagement in the moral and scientific appendages of the US empire. After graduating from Wisconsin's Beloit College in 1897, Huntington received an invitation from the American Board of Commissioners for Foreign Missions to assist Dr. James L. Barton, president of Euphrates College in the Harput region of Anatolia. The school, founded two decades earlier by Protestant missionaries, served a minority population of Armenian Christians that Barton had once described as "the Anglo-Saxons of Eastern Turkey."[88] When not teaching, Huntington used his geography skills to survey the Euphrates River, mapping the watercourse for the British War Office. In 1901, he returned to the United States to join the master's program in geology at Harvard. Huntington studied with one of the foremost experts in physiography, later called geomorphology, Professor William Davis.[89]

A year later, in 1902, geologist Raphael Pumpelly contacted Davis, his former mentor, to see if he would be interested in accompanying Pumpelly on an expedition to Russian Turkestan. Sponsored by the Carnegie Institution, the expedition's goal was to collect geological and archaeological evidence to learn whether the Aral, Caspian, and Black Seas were remnant water bodies of a larger Mediterranean and, relatedly, if greater aridity led to the disappearance of a lost "Aryan" race. Davis accepted the offer and took Huntington along as a research assistant. Departing in April 1903, the three US scientists traveled eastward over the next seven months from Constantinople (Istanbul) to the Tian Shan Plateau with local Kyrgyz attendants. On the trip, Huntington observed what he believed to be physical clues of an interglacial shift in central Asia's climate, such as moraine sediments on highlands and terracing near riverbeds. He thought that Shor Kul, meaning "Salt Lake" in the Kyrgyz language, was "the most important indicator of climatic change." Ancient lacustrine deposits surrounded the marshy alkali flat on the hillsides, which Huntington suspected had once been part of a bigger lake that was now shrinking.[90]

Huntington's ideas about the societal consequences of desiccation crystallized in 1905 when he joined another of Davis's former pupils, Robert LeMoyne Barrett, on a trip over the Himalayas to Chinese Turkestan with support from the Association of American Geographers. From Bombay (Mumbai), the two explorers moved northward through British India toward the Vale of Kashmir, crossing over the divide near Pangong Lake, until they reached the arid Tibetan Plateau. At this point, the pair split up, Barrett heading east to Khotan and Huntington skirting a lake called Lop Nor northward to Turfan. The division of labor during travels assumed imperial overtones, with indentured Islamic servants carrying the supplies in sixty-pound packs while the American "Sahibs," as their porters called them, toured most of the way in tall horse-drawn carriages known as ekkas. Along the route, Huntington saw a series of abandoned villages near withered streambeds and concluded that the lack of water was not due to irrigation diversions or seasonal variations but was "another suggestion of change of climate." The geographer was onto something, for the Tarim Basin deserts of central Asia, particularly the vanished Lop Nor, had been drying after the conclusion of wetter years during the Little Ice Age.[91]

The Pumpelly and Barrett expeditions convinced Huntington of two things: that climatic disruptions, or "pulsations," as he called them, triggered the collapse of past human civilizations and that present climatic conditions were headed toward greater aridity. While a Hooper Fellow at Harvard, Huntington published these thoughts in a 1907 travel narrative called *The Pulse of Asia*. The book was infamous for its climate determinism and its universal application. "Climatic changes have been one of the greatest factors in determining the course of human progress," he asserted. "This conclusion applies primarily to Central Asia, but there is strong reason to believe that it is equally applicable to western Asia, north Africa, and Europe. Apparently, the same is true of America."[92] Just as the United States was transforming into an empire, Huntington held that long-term global weather patterns could be used to track the rise and fall of empires.

Ellsworth Huntington's desire to link Old World and New World climates led him to the region of the US Southwest and the work of A. E. Douglass. In 1910, Huntington traveled to Tucson to join botanist Daniel MacDougal at the Carnegie Desert Lab and to study another arid environment with evidence of collapsed societies. At a dinner party, MacDougal introduced Huntington to the astronomer. Huntington wrote of the occasion to a sibling: "I have met another person here whom I was especially glad to know, namely Prof. Douglass. It happens that he has done a piece of work which gave me exactly the clue for which I had been looking."[93] That clue Huntington and Douglass discussed was tree rings.

Huntington read Douglass's *Monthly Weather Review* article, which left him intrigued by the methodology's potential to bring evidence to his pulsations hypothesis. Huntington needed a tree species that lived as far back as the Asian civilizations he studied. Luckily, Huntington knew a life-form that was exceptionally old. "The connecting link between the past and the present, between the ancient East and the modern West," he supposed, "is found in the big trees of California."[94]

IN MAY 1911, Huntington departed for the Sierra Nevada, driving from Sanger through General Grant National Park to a tract owned by the Hume-Bennett Lumber Company. Huntington and assistants from Yale University, where he was now teaching, scoured a logging area called Camp Six for freshly cut sequoias. Equipped with magnifying glasses and metal rulers, two of them lay on the tops of the stumps and counted off ring widths in sequences of ten while the third recorded the figures in notebooks. "Had the lumbermen seen us," Huntington recalled, "we should have appeared like crazy creatures as we lay by the hour in the sun and rain calling out 'forty-two,' and being answered by the recorder, 'forty-two'; 'sixty-four,' 'sixty-four'; 'seventy-eight,' 'seventy-eight,' and so on, interminably."[95] Field research halted only with the crashes of falling trees.

Huntington returned to the Sierra Nevada the next year. And after two summers of counting, he tallied tree rings from about 450 giant sequoias. His main discovery was that of all the Big Trees sampled, only seventy were older than two thousand years and three were older than three thousand years.[96] Most stumps did not date as far back as once assumed. Undeterred, Huntington charted sequoia growth rates from the oldest specimens over the last three millennia, overlaying the graphed line on another line from climatological data near Palestine.[97] He published an article in a 1912 issue of the Royal Geographic Society's *Geographical Journal* that not only replicated the chart but also claimed to verify worldwide environmental patterns. Huntington asserted that "the main climatic changes of America are synchronous with those of Asia and are of the same kind."[98] Bringing in contributions from other scientists, Douglass included, Huntington repeated his argument in the 1914 compilation, *The Climatic Factor as Illustrated in Arid America*.[99] For Huntington, the Big Trees were crucial to his argument for historical fluctuations in the global climate system.

Douglass was unconvinced; he felt that Huntington allowed his expectations to mold the evidence in any way he saw fit. Huntington never cross-dated the trees. Huntington, in short, was a sloppy scientist. In August 1915, Douglass traveled to the Sierra Nevada, "following Huntington's route," to reexamine many of the sequoia stumps that the geographer had previously counted. Unlike Huntington, Douglass took "radial samples" from cutover groves in the Kings River country so that he could inspect them more carefully later. The astronomer hired two Hume-Bennett woodsmen who used a crosscut saw to make two angled incisions on the top of stumps, taking out what were called V-cuts, and to mark each trunk with the capital letter *D* and a number for cataloguing purposes. The D-5 specimen would be taken an hour after the Big Tree was felled, giving him the exact year of the outermost ring from which to cross-date other sequoias. Douglass recognized the paradox of knowledge creation and resource destruction, later commenting in a radio lecture, "I do not like to think of those giant trees being cut down and I hope no more will be sacrificed, but yet it is the stumps which have supplied most of my material."[100] He nonetheless shipped a total of fifteen triangular pieces of redwood, taken from Camp Six and Indian Basin, back to Tucson.[101]

To prove sunspot cycles, Douglass still wanted a longer timeline. The first trip allowed him to reconstruct a two-thousand-year chronology. In June 1918, he made another trip. For this outing, Douglass had secured a grant from the Carnegie Institution on the recommendation of ecologist Frederic Clements at the University of Minnesota. Clements studied cycles as they related to vegetation communities, and he wanted to know if climatic cycles might be a driving factor in his theory of plant succession.[102] In the Sierra once again, Douglass began searching for useful Big Trees in Redwood Basin. Little turned up of deep chronological worth. Moving west to the Converse Basin near Hoist Ridge, just five miles north of General Grant National Park, Douglass finally hit the arboreal motherlode. Two sequoia stumps, which he labeled D-20 and D-21, reached back farther in time than any others; and the latter, the most senior of them all, dated back to 1305 BCE—the ancient Big Tree turned out to be over three millennia old.[103]

Accurate dates proved easier to obtain from exacting standards and applied technologies. In the Tucson lab, Douglass and a handful of undergraduate assistants started by breaking up the Big Tree samples into standardized periods: The last year of every decade received one pin prick; every fifty years, two pin pricks; every century, three pin pricks; and every millennium, four pin pricks. Using a cathetometer, a microscope that moves over a graduated ruler, they measured ring widths in millimeters and charted them on graph paper to make what was called a skeleton plot. Compressed growth rings, representing years of drought, garnered vertical marks across the sheet—the smaller the ring, the longer the line. Comparing skeleton plots allowed Douglass to better visualize growth patterns from which to cross-identify samples.[104]

Douglass occupied an intellectual border between field and laboratory that allowed him to catch something that Huntington had missed: Not all sequoia specimens provided satisfactory tree rings for studying.[105] After scrutinizing the value of field samples in his lab, Douglass told Huntington that "two very distinct varieties" became apparent when their physical location was taken into

University of Arizona astronomer A. E. Douglass revisits a Big Tree stump with the V-cut notch that he had previously sampled. Courtesy of the Laboratory of Tree-Ring Research, University of Arizona.

consideration. Big Trees that had grown in areas where moisture was consistently available, such as land with poor soil drainage or near streambeds, produced uniform growth patterns that were almost worthless for weather-related information. Douglass called them "complacent" rings. And yet Big Trees that had matured under hydrological duress, like those situated on steep slopes or well-drained ridges, formed highly variable growth rings that were more useful. He named them "sensitive" rings.[106] In the first volume of his magnum opus, *Climatic Cycles and Tree-Growth* (1919), Douglass determined that "complacent trees contribute much less to a knowledge of climatic variations, and some of them have to be discarded because of uncertainty in the dating of their rings."[107] Field conditions thus mattered in determining the efficacy of laboratory results.

In 1920, Douglass wrote an article for the first issue of the journal *Ecology* in which he outlined techniques for a new scientific field, soon to be called "dendro-chronology." He combined the Greek roots *dendron* (tree), *khronos* (time), and *logos* (reason) to arrive at the name.[108] In the paper, Douglass summarized procedures like why it was important to use stressed trees that provided variable rings and to cross-identify multiple samples for establishing reliable dates. Similarities in ring patterns over a wide expanse, however, was most stunning. Douglass remarked how compressed rings in giant sequoias for the year 1851 also appeared in ponderosa pine samples from Arizona, 450 miles away, and in Douglas fir specimens from Colorado, 750 miles away. Douglass was amazed by, in his words, "the marked resemblance found in certain individual rings over a wide extent of the country in which climate is the only common factor."[109] Douglass essentially viewed the Big Trees as part of a larger landscape held together by climatic forces.[110]

Regional parallels soon became planetary corollaries. In 1922, British astronomer E. Walter Maunder wrote to Douglass with a question. Do you have any tree-ring data, Maunder asked, that could provide further evidence to back up one of my scientific findings? He and his wife, Annie Russell Maunder, had detected a correlation over time between the variation in sunspot numbers and the climate of the Earth. From past telescopic records, they noticed a "prolonged minimum," an extended lull, in solar activity from 1645 to 1715, which appeared to overlap with a worldwide cooling trend. (Today, the trend is associated with the Little Ice Age.)[111] Douglass responded with exciting news. After revisiting climatological graphs derived from redwoods, he validated the hypothesis by acknowledging how "the sequoias show strongly the flattening of the curve" over roughly the same period. "It seems to me," Douglass told Maunder, "that you have brought to light a very important corroboration of the relationship between solar activity and terrestrial conditions."[112] The Big Trees, in effect, confirmed astronomical phenomena that would later become known as the Maunder Minimum.[113]

These discoveries were not lost on the US National Park Service. The first two civilian superintendents of Sequoia National Park, Walter Fry and John Roberts White, asked Douglass to forward his publications for studying.[114] They also began conducting experiments related to tree rings, comparing the growth rates of sequoia saplings and mature giants to weather data gathered within the park. In 1930, Fry and White published a book titled *Big Trees*, which at the time was the most synthetic treatment of the charismatic megaflora. The naturalist-writer duo appreciated to an extent how larger environmental forces acted on giant sequoias. "The older trees revealed a fascinating historical record of climatic changes affecting the growth of plant life during past centuries," they commented. "The annual rings show that there were periods, lasting for centuries or more, during which the tree made abnormally large growth, only to recede again to normal or below normal."[115] A year later, Douglass visited the park for the first time, giving a public talk on tree rings and taking core samples from the Karl Marx / General Sherman Tree. From these interactions, park officials started to comprehend that variations in climate could have potentially tremendous effects on sequoia development.[116]

IN THE BANNISTER BUILDING, I sat down in the Tree-Ring Lab office of Chris Baisan. He told me that he first came to dendrochronology by accident—it was a job. In the late 1970s, Baisan spent his summers working as a firefighter in the Gila National Forest of New Mexico, where he met Thomas Swetnam on a helicopter crew. Swetnam eventually pursued graduate studies at the University of Arizona; Baisan stuck around Tucson and procured samples for the Tree-Ring Lab, then located in the football stadium. In 1986, Baisan enrolled in the college forestry program, working in the lab as an undergraduate student assistant and later as a research technician. His chainsaw skills translated to acquiring slabs for tree-ring work through which he learned the discipline. In my mind, Chris Baisan was a latter-day Walter Fry.[117]

As we talked about the Tree-Ring Lab's ongoing research, I became intrigued by the concept of teleconnections. Baisan explained that scientists are coming to understand the causal relationships of meteorological phenomena over long distances. A prime example is the El Niño–Southern Oscillation cycle, which includes El Niño's opposite, La Niña. During La Niña years, trees along the

Pacific coast of South America register intense drought-like conditions, while trees of Australasia, some nine thousand miles across the ocean, record an inverse pattern of monsoonal rains. During El Niño years, the opposite occurs. I asked Baisan where this idea came from. He then moved over to a bookshelf, pulled out two volumes, and plopped them in front of me.[118]

The works were by Douglass and Huntington. I got his subtle answer. For different reasons, both scientists had detected signals of broad weather patterns that tied the Sierra's giant sequoias to a larger global environment. Douglass reiterated this point to paleontologist John Merriam of the Carnegie Institution: "Tree rings have proved a means of measuring time because of certain climatic characters in them that may be recognized over large areas."[119] Douglass later requested funding from Carnegie "for beginning a tree-ring laboratory" at the University of Arizona.[120] In 1937, with a collection of tree samples already in hand, he formally established the Laboratory for Tree-Ring Research. Capitalist destruction therefore led to contradictory ends. It spurred on a conservation practice of bounding nature in national parks to prevent further resource loss. It also assisted in the creation of dendrochronology science, in which climatic forces transcended the boundaries of those protected areas. The Big Trees stood at the center of this tension.

The metaphor that Douglass often used to explain the nascent field to lay audiences was translating a book from another language. "Nature is a book of many pages and each page tells a fascinating story to him who learns her language," Douglass remarked. "The trees composing the forest rejoice and lament with its successes and failures and carry year by year something of its story."[121] In his 1922 presidential address to the American Association for the Advancement of Science's Southwestern Division, Douglass suggested broader applications of dendrochronology, not only in reconfiguring past climates, but also in reconstructing fire histories.

About a half century later, Baisan, Swetnam, and other researchers once again took interest in the book of those "talkative tree rings," this time to understand and interpret giant sequoia's story of past blazes. But if tree rings talked, would resource managers listen? Native communities of the Sierra Nevada already knew the language. They conducted the burns, and awareness that fires operated on a landscape scale cracked open the final strand of ecological internationalism: the collaboration between Indigenous nations and US resource agencies.

CHAPTER SIX

Baptized by Fire

I unfolded a map of Redwood Mountain with a patchwork of colors. Green, brown, pink, blue, yellow—the many different hues represented forested areas that park personnel had set ablaze over the last sixty years—1969, 1974, 1981, 2006, 2012. Tony Caprio, fire ecologist for Sequoia and Kings Canyon National Parks, gave me a copy so that I could find the experimental plots where the first prescribed burns were reintroduced in 1965 within any national park unit of the western United States.[1] Hiking on the Redwood Canyon Trail toward these sites, I saw many signs of fire—some ancient, some recent. Mature sequoias held old, blackened scars that ran a quarter way up their trunks; one such lesion was deep enough to practically swallow me. Other Big Trees exhibited charred trunks as dark as burned toast. Still others displayed only faint traces of past blazes with pepper-like speckles covering the cinnamon-colored bark. A little over a mile into the forest, I finally noticed the plots tucked discreetly off the path.

Redwood Mountain became a formal protected area in 1940 as part of the creation of Kings Canyon National Park. During the 1930s, Secretary of the Interior Harold Ickes renewed legislative calls for the Kings Canyon area to be included in the park system to complete the Greater Sequoia vision. The US Forest Service managed these lands as a "primitive area," but political jockeying among male-dominated conservation organizations like the Sierra Club, Save the Redwoods League, and Wilderness Society hindered the many bills concerning which federal agency should have control. Women's civic groups eventually tipped the scales in favor of the National Park Service with endorsements from Native Daughters of the Golden West and Federation of Church Women, among others. Rosalie Edge, founder of the Emergency Conservation Committee, directed a grassroots campaign to persuade Congress that Redwood Mountain rounded out the southern sequoia belt. On May 4, 1940, President Franklin Roosevelt signed a park bill into law—conferring 460,000 acres of the upper Kings Canyon as a new park unit, as well as absorbing General Grant National

Park. Two months later, under the authority of this park legislation, Roosevelt transferred a large area of Redwood Mountain to Kings Canyon National Park.[2]

Redwood Mountain offered an ideal place for scientific experimentation. Removed from heavy visitor use, ecologists could research the effects of alternative land management practices without garnering too much public attention.[3] Concerns about preserving the Mariposa Grove at the hundredth anniversary of the Yosemite tract inspired a research agenda at the southern park. In 1965, Drs. Richard Hartesveldt and Thomas Harvey of San Jose State College delineated four- to eight-acre plots within the sequoia–mixed conifer forest, half of which were left untreated and half were intentionally burned. Hartesveldt and Harvey returned a year later to measure new sequoia regeneration within the "control" and manipulated plots. They observed a stark contrast between the parcels: The undisturbed ground was choked with a thick layer of litter and duff while scorched earth exposed bare mineral soil for a carpet of saplings to grow. Their studies challenged the dominant NPS philosophy of total fire suppression. After a century-long absence of wildland fire, park managers decided to bring an ecological process back to Sierra Nevada public lands. Looking at the early plots, I observed how the young sequoias grew in clusters from intense burning, sprouting up from the former "hot spot" holes in a manner that reminded me of Swiss cheese.[4]

I reached Redwood Mountain's western ridgeline on what became the Sugar Bowl Trail, until a bizarre sight stopped me in my tracks. Piles of dead branches surrounded the trunks of several giant sequoias, forming what looked like orange-tinted halos around the base of each tree. It appeared that these sequoias had chosen to preferentially shed their older limbs closest to the ground because, when I looked upward, the crimson giants still held a few brittle shoots. And yet the mammoth trees showed no apparent signs of damage or disease. What in the hell, I thought, was going on?

I later discovered the foliage die-back was a reaction to climate change. US Geological Survey ecologist Nathan Stephenson, whom park staff affectionately called "Dr. Sequoia," was stationed at the Western Ecological Research Center near Sequoia's Ash Mountain entrance. Stephenson explained to me that Sierra forests have been subjected to what scientists call a "hotter drought" or "global-change-type drought" over recent years: when higher temperatures worsen the effects of low rainfall. Stephenson earned the title of Dr. Sequoia because very few people have spent more of their personal and professional lives among the Big Trees, and no one detects subtle changes like him. Back in 2014, the most severe drought year in California for at least 120 years, Stephenson also noticed unusual amounts of lifeless sequoia needles. Signs of drought stress have been a worrisome problem for resource managers; during the summer months, one mature sequoia consumes up to eight hundred gallons (three thousand liters) of water per day.[5]

To understand the spatial extent of this issue, Stephenson led an interagency research team to study how physiological responses to the hotter drought of 2012–16 might vary across the sequoia belt. Combining binocular surveys of eight different groves—Redwood Mountain included—with tree-ring data, they found uneven patterns: Big Trees at lower elevations, on grove edges, and on steeper slopes felt the impacts of warmer, drier climatic conditions most intensely. Because parts of the landscape were more vulnerable to drought-induced mortalities than other areas, park officials could then redirect what limited funds and personnel they had to conservation practices where it was most needed. In the context of global warming, prescribed burning took on new meaning for the survival of the Big Trees.[6]

At the intersection of wildland fire, climate change, and giant sequoias, we can see the evolution of thinking about park resources: from objects to ecosystems to landscapes. Each step required broadening the scale and scope of environmental management. At first, park officials focused on preserving the aesthetics of their most venerated Big Tree groves—tourist favorites such as the Giant Forest—by shielding them from every wildfire. This guiding principle shifted after two catastrophic blazes and ecological research compelled park managers to consider the interactions between living organisms and elemental forces, between sequoias and fire. When the 1963 Leopold Report became official NPS policy, park administrators turned to prescribed burns as a management tool for restoring past fire regimes. They increasingly recognized that reestablishing historic processes meant recovering long-term Native burning practices that had sustained these ecologies. Supranational climatic changes then forced the parks to reenvision relationships with subnational communities. In managing for global warming, the US National Park Service had to turn its gaze to processes that went beyond the administrative boundaries of any park. Setting flame to the Sierra forest became a method of building resilience among sequoias and a means of bringing Indigenous nations back to their lands.[7]

Light Burning or Total Suppression?

On August 8, 1928, two vacationers traveled to Bennett Creek, a tributary of the South Fork of the Kaweah River, to do some fishing. One of the anglers made a serious blunder by tossing a cigarette butt into the chaparral, which ignited a brush fire. The small flames quickly turned into a large inferno—spreading to ten thousand acres in just five days. Federal and state authorities dispatched nearly 150 firefighters to try to contain the conflagration. By August 10, however, the wildfire was still advancing northeast toward some Big Tree groves in Sequoia National Park.[8]

The blaze initiated a firestorm of controversy. On August 14, Chief Ranger Lawrence Cook and his fire crew—under the orders of Superintendent John Roberts White—intentionally lit a backfire along twelve miles of the park's boundary. The backfire performed beautifully in that deliberate torching deprived the approaching wildfire of fuel and arrested its progress, or so White thought. A state forester charged that the backfire actually had interfered with suppression efforts and had accounted for a large portion of the $15,000 in overall costs. Newspapers used the public grievance as an opportunity to criticize White's views on "light burning." The local press accused White of withholding park staff from firefighting activities based on his conviction that human-ignited fires kept debris from accumulating on the forest floor and thus prevented large disastrous wildfires. A systematic governmental review of the South Fork Fire found these allegations about White were mostly unfounded, but Assistant Director Horace Albright, a fierce proponent of all-out fire exclusion, condemned park actions as not conforming to official NPS policy.[9]

The ideological struggle over two management philosophies—light burning versus total suppression—opened a series of questions for debate: Should park officials manage protected areas in terms of scenery or ecology? Had Indigenous nations been responsible for stewarding the wildlands that were now part of the preserves? Did fire have a beneficial or detrimental effect on the sequoias? To find answers in its early years of formation, the National Park Service turned to the technocratic expertise of the US Forest Service. The Big Trees at Sequoia, Kings Canyon, and on other public lands of the Sierra Nevada paid the price of fire suppression.

Horace Albright's opposition to wildland fire stemmed from his understanding of what national parks were meant for. In the National Park Service's "dual mandate" of preservation and visitor use, nature protection was in service to tourist development, not the other way around. Born in 1890, Albright grew up in Owens Valley, California, in the eastern shadows of the Sierra Nevada and Sequoia National Park. He enrolled at the University of California, Berkeley, in 1908, studying economics and mining law while rubbing elbows with regional elites. Albright landed a job after graduation as a clerk for Secretary of the Interior Franklin Lane in Washington, DC, where he met Stephen Mather, the National Park Service's first director. Mather hired Albright as his private secretary, training the "assistant director" in the art of creating political support via business interests. An ardent capitalist and lifelong Republican, Albright took easily to the task building up national parks to serve the function of "national playgrounds." From 1919 to 1929, Albright held the superintendency of Yellowstone National Park, assisting Mather from afar until he took over the reins as the second NPS director. Albright believed light burning was "unsound" because, in his view, fire ruined the beauty that parks were trying to market and sell.[10]

Like Albright, White became a member of the dedicated cadre of park superintendents, collectively referred to as "Mather men." Coming from business, engineering, or military backgrounds, they used administrative talents to guide the new bureau through early management challenges. Born in England, White attended Oxford University before enrolling as a private in the US Army in 1899 at twenty years old. He served in the Philippine Constabulary, a paramilitary force established by US colonial officials to quell resistance on the archipelago, even participating in the 1906 Moro Crater Massacre against "Moro" (Muslim) peoples on the southwestern islands. White retired from the military due to chronic bouts with tuberculosis and malaria. He reenlisted during World War I, earning the rank of colonel, and then retired again in 1919. White was looking for work outdoors to rejuvenate his poor health; when his path crossed with Albright, he asked for a job in the National Park Service. Albright offered him a ranger position at Grand Canyon National Park in Arizona. Within a year, White took over as superintendent of Sequoia National Park.[11]

Unlike Albright, White developed a reputation as an obstructionist who favored ecological conservation over commercial developments. When automobiles enabled tourism at Sequoia to grow nearly tenfold, from 28,000 visitors in 1920 to 260,000 by 1938, White resisted the bureaucratic push for more roads, hotels, and other recreational amenities. While overseeing the completion of the Generals Highway, which linked Sequoia and General Grant, he opposed plans to build a southern route over the High Sierra to mirror the Tioga Road across Yosemite National Park. White also detested visitor congestion in the Giant Forest, especially after a plant pathologist showed that tourist traffic was harming the sequoias' shallow root structure, and he thus wanted the business operators known as concessionaires to move accommodations out of the grove to lessen human impact. White rejected Albright and Mather's capitalistic attitudes in a speech at the 1936 superintendents' conference. "We should boldly ask ourselves," he told the audience, "whether we want the national parks to duplicate the features and entertainments of other resorts, or whether we want them to stand for something distinct, and we hope better, in our national life."[12]

White's dissident positions toward park administration carried over to his views about fire. Two years into his superintendency, the NPS received its first congressional appropriation of $125,000 to be distributed to the parks for firefighting, but White doubted that total suppression would be feasible against major wildfires. In his 1926 superintendent's report, White commented that the "forest floor [was] thick laid with a mass of combustible pine needles, branches, logs, and snags, which [made] a conflagration almost impossible to control."[13] White referred to light burning as "protective burning" because, in a sense, the management practice protected forests from the prospect of catastrophic fires by reducing fuel loads.[14] White learned from fellow park employee Walter Fry that redwood bark was highly fire resistant and the tree's outer covering, a spongelike sheath one to three feet thick on mature sequoias, gave the woody plant an edge over other tree species. "California conifers, particularly the sugar and yellow pines, resist light or even moderately severe fires," Fry and White wrote. "But when the forest floor is littered with the dried accumulation of years the fire is so hot that it girdles the trees.... The pines, the firs, and the cedars die. But not the Big Trees."[15] They figured that light burning could be realized without harming old-growth sequoias; however, White held a minority stance that Washington officials denied or dismissed.

The light-burning controversy prompted park officials to consider the extent to which Indigenous nations had tended wild lands. A 1928 typescript report, left anonymous but likely prepared by the astute naturalist Fry, commented on the Native application of the firestick to giant sequoias. "Before the white man found these tree monarchs, they had been subjected to many fires," Fry reasoned. "The origins of the fires must have been either lightning or Indian."[16] Fry arrived at this conclusion two decades earlier when he had interviewed Hale Tharp, one of the first Euro-American newcomers to step foot in what became Sequoia National Park. Back in 1858, Tharp traveled up the Middle Fork of the Kaweah River to Hospital Rock, a granite storehouse located only a handful of miles from the Giant Forest, where he encountered a tribal elder called Chief Chappo and the Mono nation. Tharp told Fry, "The camp was never vacated during either winter or summer and the campfires were kept continually burning."[17] The Mono peoples who resided at this permanent village site were responsible for sustaining the open, park-like conditions within the sequoia groves—the very same ecological scenes valued by later park managers.

Combining ethnographic accounts and botanical science, it is feasible to present a view of how Indigenous nations used fire in sophisticated ways to enhance fiber and food production in the Sierra Nevada. In the southern foothills, for

Margaret Baty (Mono) presents a collection of California black oak acorns near Auberry, California, 1925. Acorn production depended on Native uses of cultural burning. Photo by George Holt. Courtesy of the Santa Rosa Junior College Museum's Flegal Collection.

example, Monos and Yokuts burned shrubs like redbud and sourberry to stimulate new vegetative growth and used their straight, flexible shoots for weaving and basketry. Higher up in the mountains, they burned around black oaks and sugar pines to improve acorn and pine nut collection. Both species need regular fire disturbances to maintain their leverage over white fir and incense cedar. Moreover, burning decreases oak tree susceptibility to many fungi- and insect-related diseases, which, in turn, increases acorn harvests. Melba Beecher recalled that her Mono forebears routinely kindled the landscape to achieve those management goals: "They'd burn and let the fire go where it was needed under the black oaks. They'd burn in October, November, or December every year but in different spots. Wherever it needed it."[18] Enhancing acorn production with the regular applications of fire was significant in providing for the bulk of Native diets in the Sierra: One family unit collected anywhere from four hundred to three thousand pounds (180 to 1,350 kilograms) of these protein-rich nuts each year. For the original inhabitants, burning sustained land and culture.[19]

When walking along the Kaweah River on a visit to Sequoia National Park, I came across a spread of bedrock mortars along the streambank. Native women had long used stone pestles to grind nuts into acorn and pine meal, wearing bowl-shaped holes into the bedrock. Archeologists date human detritus at Hospital Rock to 1300 CE and other artifacts of long-term occupation in the region to as early as two thousand to three thousand years ago.[20] Bedrock mortars can be found at many mid-altitude locations in the Sierra Nevada, such as on the edge of Giant Forest. In these peopled forests, Indigenous nations created their own names for sequoias, like *toos-pung-ish* in Wukchumni Yokut, *pusine* in Central Sierra Miwok, or *wah-ho'-nah* in Western Mono. Ron Goode of the North Fork Mono learned the language of living on the land from his mother, Owl Woman (Tahubnee), who in turn was mentored by her grandmother who knew a life before colonization.[21]

Indigenous burning practices left their mark on the Big Trees. While black oaks were only found on the margins of lower-elevation sequoia groves, sugar pines were typically part of the sequoia–mixed conifer forest. As a result of arboreal compositions, setting fires for acorn and pine nut production meant that flames often reached the Sierra redwood. Hector Franco, a citizen of the Wukchumni Yokuts, remembered, "The elders talked about burning in the giant sequoias. Burning was usually right after harvest of the acorns."[22] Franco reminisced that "old man Topna" was one of the last of his people to ignite fires within the Giant Forest during the early 1900s, at least a decade after Sequoia National Park was established. In doing so, Indigenous nations held enduring connections to areas now under US occupation.

Aside from setting flames where giant sequoias grew in association with black oaks and sugar pines, Native peoples burned under the Big Trees to enhance the forage for game species. Clara Charlie of the Chukchansi/Choynumni Yokuts recollected that family ancestors "burned in areas where the giant sequoias grew. The trees were sacred." Charlie emphasized, "They burned to keep things clear. They also burned for the animals—the deer, bear, rabbits, and squirrels. The new growth the following spring gave them better and higher quality foods."[23] Tree-ring studies reinforced that Native-ignited burning occurred regularly. Fire scars indicate that from 1480 to 1870, sequoias burned every eight to eighteen years. Given that lightning-caused wildfires happened every fifty years, the only plausible explanation for a higher fire frequency was the presence of Indigenous nations on what became park lands.[24]

Over the turn of the century, Indigenous fire management sparked the attention of white ranchers, woodcutters, and conservationists, some of whom adopted traditional burning to reduce fuel loads and regenerate alpine grasslands. One Madera County sheepherder hired Mono hands to tend flocks in the Sierra meadows and "learned some of the burning techniques from the Indians."[25] In 1910, California engineer G. L. Hoxie wrote in *Sunset* magazine that "the Indian for centuries, for his own convenience, no doubt, fired the forests."[26] Superintendent John Roberts White of Sequoia National Park learned how Indigenous nations conserved park-like features by talking with early settler communities. "Experienced mountaineers," White told the NPS director, "are unanimous in favor of 'light-burning' which, formerly conducted in spring or fall, cleansed the forest of accumulated debris [and] permitted new growth of vegetation."[27] While suppressing opposition to US empire in the Philippine Constabulary, White observed Native Filipinos who "burned timber to plant *palay* [rice] and *camotes* [sweet potatoes]." Encountering traditional knowledge in this transpacific context no doubt influenced his park management.[28]

Light burning held the potential to unsettle the dominant paradigms in US resource management, but White's calls for practicing the "Indian way of forestry" fell on deaf ears within the highest circles of NPS leadership. "It was suggested that money could be saved by just letting the fires burn themselves out as the Indians had done," NPS Director Horace Albright later scoffed. "But the Indians had been terrified of great fires and did everything in their power to keep them for starting."[29] From his elitist position in Washington, DC, it was easier for Albright to ignore the on-the-ground evidence of Indigenous management and to instead view parks as "pristine" wilderness untouched by humans.[30]

For policy, the National Park Service followed what was becoming the premier federal agency in wildland firefighting—the US Forest Service. After the 1910 Big Blowup, a series of massive wildfires that consumed three million acres in Idaho and Montana, professional foresters claimed the only solution to prevent the future loss of timber resources was total suppression. In California, forester Coert DuBois argued that federal involvement on public lands could serve as the handmaiden to capitalist extraction, writing in a manual, "The object of forest fire protection is to secure from each acre in the forest the maximum of all forest products which its soil is capable of producing."[31] NPS officials were not immune to this technocratic expertise when thinking about park management. "Throughout the federal government, the policy on fire was to fight it immediately and rigorously," Horace Albright wrote. "For the national forests, the reason was that valuable commercial timber could be burned. For the national parks, the idea was that the beauty of the landscape and the wildlife in them should be protected."[32] Just as the US Forest Service suppressed fire to sell lumber, the National Park Service did so to capitalize on scenery.

IN 1928, THE SAME YEAR that White ordered a backfire near Sequoia, the National Park Service required all park units to create fire control plans for standardizing responses to wildfire. John Coffman, formerly with the US Forest Service, became the NPS Fire Control Expert and doused the idea of light burning. Sequoia National Park's first fire plan began in adversarial terms: "Fire constitutes the greatest menace to the Park." It continued, "Therefore, the detection, suppression, and prevention of forest fires takes precedence for the Ranger Force over all other work in the Park except the preservation of human life."[33] Sequoia received $10,000 from US Congress for fire control measures, directing the money to park crews who built a lookout tower, fire trails, and several tool caches. From that year onward, regular federal funding provided for implementing exclusionary practices.[34]

Two transformative moments in mobilizing money, manpower, and machines for fire control came during economic depression and war. Authorized in 1933 under President Franklin Roosevelt's liberal New Deal programs, the Civilian Conservation Corps brought a labor force of millions of unemployed males to the national parks and other public lands across the country to conserve human and natural resources through landscape work. The corps hired young men, ages eighteen to twenty-five years old, for temporary, six-month enrollment periods. CCC enrollees worked six days a week, earning thirty dollars a month, all but five dollars of which was required to be sent back to their families. During its nine-year existence, the corps placed 198 camps within the national park system. By 1933, Sequoia maintained five CCC camps that were subjected to Jim Crow segregation—four for white enrollees, one for Black enrollees.[35]

The Civilian Conservation Corps more than doubled the firefighters and funds at Sequoia National Park, dedicating over 10,500 man-hours of labor and $16,000 in capital to suppression activities over the program's tenure. Besides trail building and road construction, fire-related tasks were the main responsibility for the corps. "CCC standby suppression crews will be retained at all times by each camp, available for dispatch and ready to leave within 5 minutes of any call," Chief Ranger Lawrence Cook reported. "Although they are doing constructive work, they will be continuously on call by the Fire Chief primarily for fire."[36] Park officials directed CCC workers to construct "a protection belt" of fuel breaks surrounding the Grant Grove and the Giant Forest; and these fire lines devoid of trees bounded the giant sequoias as fire exclusion zones.[37]

The World War II military buildup not only lifted the United States out of the Great Depression, but it also infused NPS firefighting with a martial ethic. Like the mindset for defeating a foreign enemy, Assistant Chief Ranger Irvin Kerr employed soldierly metaphors to describe work at Sequoia and Kings Canyon. "Fire suppression is a very difficult job and its direction is similar to the problems of war," Kerr commented. "The park policy demands that we reach and combat every fire that starts in the park, or that threatens the park, with such speed, strength, skill and equipment as to confine it to a satisfactory minimum of acreage."[38] Because wartime demands meant that nearly sixteen million men were drafted or voluntarily enlisted in the US armed forces, new park positions opened to women. Rose Vaughn became the first woman to work in fire prevention at Sequoia National Park when she was hired in 1943 as a lookout observer. An experienced mountaineer, Vaughn challenged gender stereotypes of what constituted women's labor being stationed at Cahoon Rock, a peak about fifteen miles from the nearest road and around 9,300 feet in elevation, where she could spot smoke at the Atwell Mill Grove of Big Trees. Fire demanded a wider gaze over park resources from lookout towers.[39]

Economic depression and global warfare had a profound impact on firefighting efficacy at Sequoia and Kings Canyon. Between 1928 and 1953, a reported 431 total fires began in the two parks, but the average size was no larger than ten acres burned. Strengthened by manpower and machines, park personnel virtually extinguished wildfires from the landscape, all in name of protecting their most valuable natural objects, the Big Trees.[40]

THE ENVIRONMENTAL COSTS of total suppression soon became apparent. On September 2, 1955, US Forest Service personnel noticed a column of smoke billowing from a portion of national forest lands close to Kings Canyon National Park. Blistering heat during the late summer months—a high temperature of 112 degrees Fahrenheit was recorded at the park's headquarters—kept a burn pile smoldering on the nearby McGee Ranch until it finally erupted into flame. The next day, an interagency dispatcher warned NPS administrators that the wildfire had already spread to four thousand acres and the blaze "would very shortly threaten National Park lands in the vicinity of the western boundary of [the] General Grant Grove Section."[41] To shield this iconic stand of Big Trees, park managers ordered all tourists to evacuate the area so they could prepare strategic defenses.

In cooperation with federal and state firefighters, Superintendent Eivind Scoyen mustered "every able-bodied NPS employee" to assist with the wildfire. Most blazes during the last decade required small teams consisting of three to five park rangers to put out flames; but for the McGee Ranch Fire, Scoyen called on forty men to join one thousand people from other agencies who were working on containing the inferno. Tactical measures included setting up thirty-five mobile pumping units with over sixteen thousand feet of hose from Sequoia Lake to form a "waterscreen" along state highway 180 and clearing strips of vegetation with bulldozers to act as fuel break lines around the Grant Grove. In paramilitary fashion, park officials equipped themselves for warfare against the wildfire.[42]

On September 5, powerful southerly gusts pushed the fire across the highway and closer to the national park. NPS crews battled the flames for three days until, fire control assistant Richard Boyer noted, the "fire situation had worsened considerably." The blaze blew up to over ten thousand acres upon reaching Converse Basin, where it torched slash and other combustible debris that had been left over from corporate logging more than a half century ago. Windborne embers ignited spot fires only a few hundred feet away from the sequoias, which fire teams swiftly extinguished. The inferno encircled Grant Grove; and park officials were about to lose these emblematic Big Trees. Fortunately for them, the wind died down on September 12, and so did the wildfire. The McGee Ranch Fire, a transboundary blaze that started outside the park, consumed 13,800 total acres before it let up just shy of the crimson giants.[43]

Nearly five years later, on June 26, 1960, visitor Henry Miwa was driving his station wagon on the Generals Highway with a carload of Boy Scouts to go on a camping trip in Sequoia National Park. After cruising through Tunnel Rock, a granite-topped underpass chiseled out by Civilian Conservation Corps laborers during the 1930s, Miwa noticed a burning smell emanating from his vehicle. He pulled onto the roadside to check things out. Miwa saw that his overheated brakes were smoking and his front tires were on fire. Miwa tossed sand on the wheels to try to smother the flames, but it was to no avail. One tire exploded—throwing burning rubber onto the grassy hillside and kindling a foliage pyre. Before park rangers arrived with their pumper trucks and water hoses, a gust of wind whipped the blaze up the oak-covered slopes toward the Giant Forest. Once again, wildfire threatened the park's most cherished natural feature.[44]

Two bulldozers plow the containment line along Ash Peaks Ridge within the oak-dominated woodlands to stop the Tunnel Rock Fire from reaching the Giant Forest of Big Trees, June 28, 1960. Courtesy of Museum Collections, Sequoia and Kings Canyon National Parks.

Sequoia's Fire Control Office called for emergency planning and cooperative assistance. Park officials set up an umbrella-shaped containment line north of Generals Highway, from Ash Peaks Ridge to Old Colony Mill Road, where they would attempt to hold the inferno. Federal and state personnel arrived at the park to supplement NPS fire teams. By July 1, over 1,400 firefighters were mobilized. Cold War–era military technologies flooded the scene: Ten North American B-25 Mitchell airplanes bombarded critical areas with a fire retardant; nine helicopters, including two Piasecki H-21 Shawnees, whisked men from the Potwisha and Marble Fork camps to the more inaccessible portions of the fire line; and four Caterpillar bulldozers cleared fuel breaks across a containment perimeter. "Maximum effort in men and supplies," Assistant Chief Ranger Wayne Howe observed, "was called for to keep the fire confined to the brushy lower areas and out of the Sequoia Tree groves."[45] Firefighters constructed new administrative lines in the forest.

220 | CHAPTER SIX

Since the previous fifty-five days registered no rainfall, dry conditions caused the Tunnel Rock Fire to flare up to 4,500 acres in less than a week. With the buildup of woody debris in the "highly inflammable forest" facing Deep Canyon, Superintendent John Davis stated, "a fire in this section has been dreaded."[46] Firefighters worked day and night to complete the fire line, especially back-up fuel breaks near the Marble Fork drainage where the Giant Forest began. On July 6, the wildfire was declared under control; in all, the blaze had scorched 4,960 acres and cost over $690,000. Considering that the inferno consumed 95 percent of Sequoia's annual fire-control budget, the Tunnel Rock Fire ended up as one of the costliest suppression efforts in the park's history.[47]

Resource managers at Sequoia and Kings Canyon National Parks found themselves caught between two fires. The 1955 McGee Ranch Fire almost risked the Grant Grove; the 1960 Tunnel Rock Fire jeopardized the Giant Forest—both served as wake-up calls to reevaluate land management practices in relation to the parks' biological treasures.[48] These wildfires revealed the severe consequences of excluding fire from the landscape.

But the real challenge to total suppression came from outside the National Park Service with university researchers, such as Starker Leopold and Harold Biswell. The Tunnel Rock Fire motivated Biswell to write a critical piece for *National Parks* magazine: "Would it not be worthwhile to select at least two groves of Big Trees—a dozen would be better—with their surrounding forests for some distance back," Biswell asked, "and manage them with light fires as a part of the environment, somewhat as they were managed by nature and the aborigines through thousands of years in the past? Soon it will become evident that a forest fire can be a 'friend,' it was in aboriginal times, and not our worst 'enemy' as it is today."[49] Only with the rise of ecological science—and its emphasis on forest ecosystems instead of hallowed objects—would burning the Big Trees be justified.

Restoring Prescribed Fire, Recovering Native People

The flames compliantly danced, never reaching higher off the ground than my kneecap. NPS crews applied a burning mix of diesel fuel and gasoline from drip torches to keep the blaze moving at an even pace. The embers consumed twigs, logs, and underbrush as the backing fire slowly crept downhill against the wind. I felt underwhelmed, as someone conditioned to seeing raging fires on sensationalized news, watching this controlled burn at Ash Mountain in Sequoia

National Park. Resource managers commonly refer to prescribed burning by a medical abbreviation, "Rx burning," since the practice is meant to improve forest health.

Among the Big Trees at higher elevations, fire brings salubrious changes down to their roots and up at their leaves. Combustion primes the soil with nutrients for a new generation of sequoias to grow just as the heat causes their cones to open and drop tiny seeds—ninety-one thousand of them to a pound. Crew leader Aidan McCormick told me that after setting a burn pile ablaze near the base of a sequoia tree, seeds rained down on her like an afternoon thundershower. I had a hard time wrapping my mind around how a seed similar to an oatmeal flake germinates into one of the largest living things on Earth.

The docile fire I watched at Ash Mountain was a long time in the making. Sequoia and Kings Canyon National Parks boast the second-longest active prescribed burning program in the United States. This management practice began on an experimental basis after Starker Leopold's *Wildlife Management in the National Parks*, more popularly known as the Leopold Report. Secretary of the Interior Stewart Udall had commissioned a scientific panel to speak on the issue of culling Yellowstone elk, but Leopold took the authorization as chairman to write one of the most wide-ranging, and influential, statements on park management. Published in 1963, the Leopold Report reoriented managerial policy toward ecological principles with the goal that national parks should represent a "vignette of primitive America," or the native plant and animal communities as first encountered by Euro-Americans.[50]

Restoring biological systems to preconquest benchmarks, Leopold believed, would be a Herculean task given the complex histories of environmental change. The adverse effects of fire suppression on Big Tree groves provided one such example. "When the forty-niners poured over the Sierra Nevada into California," Leopold wrote,

> those that kept diaries spoke almost to a man of the wide-spaced columns of mature trees that grew on the lower western slope in gigantic magnificence. The ground was a grass parkland, in springtime carpeted with wildflowers. Deer and bears were abundant. *Today much of the west slope is a dog-hair thicket of young pines, white fir, incense cedar, and mature brush—a direct function of overprotection from natural ground fires.* Within the four national parks—Lassen, Yosemite, Sequoia, and Kings Canyon—the thickets are even more impenetrable than elsewhere. Not only is this

accumulation of fuel dangerous to the giant sequoias and other mature trees but the animal life is meager, wildflowers are sparse, and to some at least the vegetative tangle is depressing, not uplifting.[51]

Leopold signaled that resource managers should take the stance of active manipulation based on science, instead of hands-off protection, to pursue the "maintenance of naturalness." For dealing with the "dog-hair thicket" in the Sierra forests, fire ecology and prescribed burning would offer the Promethean tools for the job.

The Leopold Report showed that groundbreaking knowledge was hardly the product of a sole genius but came from intellectual networks tied to specific places. The University of California, Berkeley, was one location smoldering with hot ideas. Starker Leopold, the oldest son of Aldo and Estella Leopold in their noted conservationist family, joined Berkeley's Museum of Zoology as a professor in 1947, after receiving a doctorate there. That same year, Harold Biswell arrived on campus to teach range management. Known by admirers as "Doc" Biswell and by detractors as "Harry the Torch," he offered field courses to his students in the Sierra foothills and showed his more open-minded colleagues how prescribed burning could improve grazing lands. Biswell had developed his pyromania during previous years in North Carolina, where private landholders kindled tracts of fire-dependent longleaf pine. With university offices just across the street from one another, Starker and Harry became lunchtime friends at Berkeley's Faculty Club, sharing thoughts about fire over coffee and cigarettes.[52]

Redwood Mountain was a unique location for grounding experimental knowledge in a physical environment—one giant sequoia grove with two sites of historic fire research. On the western side were Biswell's studies on fuels reduction. Doc Biswell turned his attention to higher up the Sierra Nevada at Berkeley's 320-acre experimental woodlands called Whitaker's Forest. Located just west of Kings Canyon National Park, Whitaker's Forest contained more than two hundred old-growth Big Trees. He recognized the Redwood Mountain Grove as a "powder keg" for holocaust fire because, according to his calculations, the forest floor was "littered with about 44,000 pounds of debris per acre."[53] In 1964, Biswell and his students began setting up four twenty-acre plots for low-intensity fires to reduce the hazard of catastrophic crown fires. After cutting down midsize trees, mostly shade-loving incense cedar and white fir that offered fuel ladders to large sequoias, they proceeded to set flame to forest. With a drip torch in hand, Biswell showed how broadcast burning effectively lowered fuel loads among the crimson giants.[54]

On the eastern side were Richard Hartesveldt and Thomas Harvey's investigations about sequoia regeneration. The two San Jose State University ecologists had received a research permit for Kings Canyon because park staff noticed that young sequoias failed to take root; that is, the reproduction of the species was at risk. Hartesveldt and Harvey demarcated two four-acre plots and two eight-acre plots in which logs and other large vegetative matter were mechanically removed to prepare for fire. In 1965, they burned half the area within all four plots. They returned the next year, and subsequent years after, to count the thousands of saplings that arose from the ashes. The survival ratio for seeds in fired versus unfired zones was two hundred to one. Hartesveldt and Harvey determined that burning readied the seedbed by exposing bare mineral soil and by increasing soil moisture, as well as offering sunnier conditions with a more open understory. Without regular fire disturbances clearing the ground, giant sequoias could eventually go extinct. At Redwood Mountain, smoke was rising for science.[55]

As ecologists challenged foresters in the National Park Service, the momentum against fire was shifting. In October 1967, Superintendent John McLaughlin traveled to Berkeley for a meeting hosted by Leopold about prescribed burning. Bruce Kilgore, a PhD student in fire ecology working with Leopold, recalled pushback among US Forest Service personnel who attended: "There was a lot of skepticism, a lot of questions."[56] Kilgore, however, was a fire acolyte who had helped Biswell with controlled burns at Whitaker's Forest. In March 1968, McLaughlin hired Kilgore as the parks' research scientist—the first position of its kind—to carry out the Leopold Report. That same year, park officials decided to place most areas above eight thousand feet in elevation within Sequoia and Kings Canyon under a "let burn" policy, where lightning-caused fires would be allowed to run their course. Let burn, later renamed "prescribed natural fire" to avoid misconceptions that no supervision occurred, would extend to three-quarters of the two national parks. They also conducted the first park-managed burning as a land management tool on an eight-hundred-acre plot in Rattlesnake Creek, a remote stand of lodgepole pine and red fir in the Kings River drainage.[57]

In 1969, resource managers gained enough confidence to bring fire to the Big Trees. Fire crews designated the 3,100-acre Redwood Mountain Grove as a prescribed burn unit, cautiously preparing smaller sections one thousand feet long by three hundred feet wide with fire lines and water hoses to keep blazes contained. From August through October, ten-person teams kindled tracts of land every week, totaling one hundred acres by the year's end. At study plots, Kilgore noted that burning drastically reduced the litter and duff fuels from fifty tons per acre to eight tons per acre. Yosemite National Park soon followed. With oversight from Biswell and Kilgore, resource managers ignited prescribed burns in the Mariposa Grove that autumn. While the Leopold Report inspired ecologists to substantiate fire's role among giant sequoias and in forest ecosystems, the guiding document left the incendiary politics of implementing the policy and educating the public open to practitioners.[58]

PARK STAFF held a careful, learn-as-you-go approach toward Sequoia–Kings Canyon's prescribed burning program during the early years. Between 1968 and 1974, NPS crews lit only fourteen fires that burned 1,421 total acres, or less than 1 percent of the two parks in total size.[59] "I think the public attitude may be one of 'wait and see,'" Superintendent McLaughlin observed. "In this respect, I suppose one could say we are playing with fire and at this stage there is no column on the score sheet for errors."[60] Only when park managers judged temperature, relative humidity, fuel moisture, and wind speed and direction to be absolutely suitable did prescribed burning occur. The bureaucratic inertia against fire took time and effort to redirect toward new fire practices.

Conducting prescribed burns demonstrates the cautious process of translating scientific research into management capabilities. Initially, fire crews performed strip burning on small plots of thirty to fifty acres in the sequoia–mixed conifer forest. Crew members created a wide fuel break, usually along a ridgetop, by scratching out a containment line with Pulaski tools and by burning alongside it with drip torches. They dropped down fifteen feet at a time and ignited ground fuels to burn upslope to the previously burned zone, where the fire died out. At Redwood Mountain, Fire Control Officer John Bowdler decided to drop down farther at the end of one day and burn out a wider area, but the fire got out of hand. "That strip started benignly enough, but as it moved upslope it picked up speed and intensity," Kilgore remembered. "Flame lengths nearly two to three times the height of previous flames leaped into the air beneath some giant sequoias, and John and his crew saw quickly this was not what they had intended to do!"[61] This management blunder, however, turned into a research success. Even though the "Bowdler burn" charred some lower sequoia branches, Kilgore documented how the larger prescribed fire had caused drying sequoia cones to release immense amounts of seeds.[62]

By the late 1970s, park managers recognized snags with prescribed burning procedures. The original objectives of reintroducing fire were to kill all the white fir less than eight inches in diameter and to remove at least 70 percent of dead and downed fuels. But the result was a homogeneous understory of even-aged

Baptized by Fire | 225

NPS fire crew member Ed Nelson lights a fire with a drip torch during a prescribed burn at Redwood Mountain, 1977. Courtesy of Museum Collections, Sequoia and Kings Canyon National Parks.

trees beneath the sequoia canopy, not the random, patchy forest as found in healthier ecosystems. In 1977, Tom Nichols arrived at Sequoia–Kings Canyon to take the position of environmental specialist, managing the prescribed burning program for the next twenty years. Nichols had recently graduated with an MS in fire ecology from San Diego State, and during a couple of forest management courses, he met Biswell and other scientists whom he called the "UC Berkeley fire mafia."[63]

Nichols made addressing woodland uniformity an administrative priority. He expanded the size of burn units from fifty to three hundred acres and changed their shape from rectangular blocks to the outlines of natural features so that fires could spread randomly. He also instructed fire crews to conduct spot burning, kindling fires at spots and letting them go where fuels and topography allowed within the burn unit. The outcome was an improved forest mosaic of varying age and vegetation types. Prescribed fire was beginning to break out of the administrative grid.[64]

When spot burning was implemented, the gender composition of fire crews started to change. Social movements of the 1960s and 1970s, including second-wave feminists, demanded federal legislation on workplace parity to open careers for women and minorities. In 1969, the Office Personnel Management removed many sex-specific qualifications on federal jobs; and the following decades saw an increase in female hiring within the National Park Service. Sandy Graban joined Sequoia–Kings Canyon in 1971 as a seasonal employee, helping record weather information for prescribed fires and watching over Redwood Mountain from the Park Ridge Lookout. Graban wanted to join her male counterparts in the physical labor, such as digging fire lines and lighting the burns, but she had to first convince them that she could do it. "I had to push on 'em a little bit," Graban recalled. By the mid-1970s, 20 percent of the thirty-person team conducting burns at Redwood Mountain were women.[65]

Communicating fire science and management to tourists and local communities was another difficult responsibility, especially as resource managers more generously applied fires. Between 1975 and 1983, NPS crews started ninety-three prescribed burns in Sequoia–Kings Canyon for over twenty-one thousand total acres, and they supervised another twenty-two thousand acres of prescribed natural fires.[66] Ranger-naturalist Bill Tweed, who informally oversaw public relations, recollected answering phone calls and getting screams from worried visitors who had seen smoke and asked why park personnel weren't doing anything about it. They tried to shift popular understandings about burns by saturating the parks' interpretation, or as Tweed put it, "We tried to build it into everything."[67] Local newspapers, visitor centers, wayside exhibits, campfire programs, and guided walks provided outlets for park employees to educate their patrons.

In general, public attitudes shifted toward a greater acceptance of fires in the parks. Don McGraw, a seasonal park ranger from 1971 to 1986, offered guided walks at Redwood Mountain where visitors could reassess their views of prescribed burning. After seeing the "ribbons of land that had been treated differently," McGraw remembered that even the more doubtful trail-goers "began to appreciate what [fires] did for giant sequoias."[68] Although public interpretation created trust in fire management, there were limits to what people would accept. The 1988 Yellowstone fires, managed as a prescribed natural fire before it got out of hand, showed park administrators that rational science could not always override public opinion.[69] That same year, Malinee Crapsey arrived at Sequoia–Kings Canyon to become the first designated public information officer. She handled media relations, including a barrage of questions asking if the

Yellowstone debacle signaled that the Sierra Nevada national parks were the next to go up in flames. Crapsey thought these concerns emerged from the Big Trees conveying "a sense of immortality and timelessness, so it was hard for visitors to grasp biological change."[70] It turned out that a human hand in biological change was hard for scientists, too.

BY THE EARLY 1980s, park managers at Sequoia–Kings Canyon began to have doubts about what they were doing. The guiding assumption, derived from the Leopold Report, was that decades of fire suppression had thrown giant sequoia ecosystems out of whack; and after park personnel reintroduced fire to reduce fuel loads, they would eventually be able to let lightning-caused fires take over because the balance of nature would be restored. But scientists and resource managers soon realized that wouldn't be the case.[71] As historian Stephen Pyne has commented, "Fire is not ecological pixie dust that can by itself magically transform the degraded into the wondrous."[72] The evolutionary relationship between humans and fire was an ongoing process, something that the US parks would find that Indigenous nations knew well.

Seeds of uncertainty were developing along two lines of evidence. The first seed sprouted from a greater awareness of the dynamism of nature. Back in April 1975, Starker Leopold gave a lecture on resource management to a gathering of NPS superintendents in San Francisco. With twelve years of perspective, Starker reflected on the Leopold Report's shortcomings and stated he would like to change one sentence, which read: "As a primary goal we would recommend that the biotic associations within each park be maintained or where necessary recreated as nearly as possible in the condition that prevailed when the area was first visited by the white man." The basic concept, Starker reasoned, "was too narrow and restrictive, and the phraseology was bad. In other words, it implied stopping the clock. And what we're talking about here are dynamic processes that don't stop. Biotic associations change, constantly, naturally."[73] His words confused fire managers at Sequoia, Kings Canyon, and Yosemite, who wondered if Leopold had abandoned the goal of restoring a product for recovering a process.[74]

The second seed of doubt germinated from tree-ring data that indicated that, before the parks were established, Indigenous peoples had set just as many fires as lightning, if not a good many more. Within Redwood Mountain and Bearskin Groves of Big Trees, Bruce Kilgore and a research technician recorded all the dates of fire scars found on the stumps of over two hundred conifers, mostly ponderosa pines. In a 1979 journal issue of *Ecology*, Kilgore claimed that tree rings from the last five hundred years had shown that 1870 was a benchmark year for fire frequency: Before that year, fires burned about once every decade, and after that year, fires only burned every half century. "The sharp decline in fire-scar occurrence," Kilgore wrote, "suggests that native Indians may have been a significant ignition source."[75]

Dave Parsons, a Stanford-trained plant ecologist who had replaced Kilgore as Sequoia–Kings Canyon's research scientist in 1973, remembered his mustached lip twitching and his eyes fluttering upon reading the study's conclusion. Parsons explained, "Native Americans have lived in that area for several thousand years and also did some burning, so are we trying to re-create the fire regime they had established? Do we need to mimic the intensities and intervals they had set?" Parsons felt uneasy because science couldn't answer the value-laden question, "How do we decide what we should be managing for?"[76]

Management objectives became murkier when fire crews ran into another problem. Tom Nichols and his team could not make prescribed burns big enough to take out larger white firs, the trees that had grown up during the years of fire suppression, unless they went in beforehand and cut them down, or they set a hefty fire that also killed mature sequoias. Nichols wanted some guidance from his colleagues Parsons and David Graber, another research scientist who directed bear management at the two parks. "I'd gotten to the point where I could do whatever we wanted with fire," Nichols boasted. "You want a small fire? I'll give you one. A big, intense fire? You got it. I could make fire jump through hoops. We had been thinking that just putting fire back in would take care of it. But what kind of fire did we want?"[77]

Nichols received scientific counsel, but not from Parsons and Graber. In August 1982, Tom Bonnicksen, a University of Wisconsin forestry professor, and Edward Stone, who had been his faculty mentor at Berkeley, argued that fire alone at Redwood Mountain wasn't getting the job done, so resource managers should opt for the chainsaw. Published in *Ecology*, their analysis revealed that the vegetative structure, in terms of tree sizes and species, did not match what the sequoia–mixed conifer forest had looked like before fire exclusion. Sequoia–Kings Canyon's prescribed burning program inhibited the growth of white fir saplings, but it typically did not burn down larger trees. Prescribed fires, they contended, were not restoring a "natural" condition but sustaining an entirely different, age-stratified forest. Bonnickson and Stone proposed that the only method to reestablish a presuppression sequoia grove was logging the understory.[78]

Amid this anxiety, Superintendent Boyd Evison asked Graber to write to his former PhD adviser, Starker Leopold, and ask him if he would be willing to update the now two-decades-old Leopold Report. Graber requested a meeting between Starker, himself, and a few park staff to discuss management issues, forwarding a copy of Bonnicksen's article with the letter. Graber even raised one of the problems bothering him: "If Indians played an important ecological role as burners," Graber wrote, "the Park Service will find itself with the task of simulating Indian burning in perpetuity, yet very possibly without the scientific data necessary for a realistic simulation."[79] Leopold replied that he would be more than willing to get together with Graber and his peers; and Starker arranged a lunch gathering at Berkeley's Faculty Club, the same campus building where he and Biswell had begun chatting about fire.

In the first week of June 1983, Graber, Nichols, Parsons, and Evison left Ash Mountain to meet the aging Berkeley sage. During the meal, Graber asked about a new Leopold Report, but, as he remembered, "Starker said there would be no second coming."[80] Discouraged by his response, the Sequoia–Kings Canyon scientists prodded Leopold to clarify what he meant by the phrase "vignette of primitive America," hoping an answer might provide direction for the parks' prescribed burning program. Parsons later recalled their appeal coming from a collective sense that "we're not really sure what we should be doing. Ecosystems are not static, so to what extent are we managing for change?" Starker laughed, looked them straight in the eyes, and then said something along the lines of "That's your job to figure out."[81] He essentially told them that scientific data would always be incomplete, mistakes would happen, and they should handle the management uncertainties with a good dose of flexibility and humility.[82]

Knowing the fire protégés left the meeting unsatisfied, Leopold wrote a follow-up letter to Superintendent Evison with his thoughts. "If the area is ready to burn, it makes little difference to me whether the fire is set by lightning, by an Indian, or by Dave Parsons, so long as the result approximates the goal of perpetuating a natural community," Leopold mused. "Our parks are too small in area to relegate to the forces of nature that shaped a continent."[83] Sadly, Evison was too late in following up; in August 1983, Leopold had a heart attack and died.

A couple messages stand out in Leopold's final letter. Most importantly, Starker realized that Indigenous peoples had powerfully shaped forest ecologies, including giant sequoia groves, through burning practices. Resource managers needed to concede that someone—whether Native inhabitants or NPS personnel—would always be part of that "natural" process; and they should recognize that management questions once beginning with "what" and "how" must transition to begin with "who" and "why." In other words, US agencies should consider collaborating with Indigenous nations.

And close behind in significance, Leopold judged that wildland fire was an elemental force operating far beyond the confines of any single protected area. National parks were too small in size, their boundaries too poorly drawn, for holistic treatment. Biological resources were endangered not only by what was going on *within* the parks or protected areas, but also by what was going on *around* them. The 1963 Leopold Report thus motivated thinking about natural resource management to shift from objects to ecosystems, whereas Leopold's closing remarks held the potential to move conservation from ecosystems to large landscapes.[84]

Managing for Global Change

Imagine the most ornate wooden tabletop you have ever seen. The huge slab, taken as a single piece of lumber, must sit atop sturdy legs because of its weight, say 150 pounds. Its surface has been polished with very fine sandpaper—400-grit for those woodworkers—to the point you can see your own reflection in the board and make out the contours of cell walls in the former tree. The orangish-red coloration, with black marbling along its edges, seems exotic. Its asymmetrical shape stands out as more modern looking than the conventional square, rectangle, circle, or oval. Its profile, however, was not directly shaped by human hands. Rather, intense burning whittled out these irregular cavities and darker markings in the wood. What you are visualizing is not an expensive piece of furniture for hosting dinner guests, but a one-of-a-kind sample for dating fire scars as far back as 10 CE. I saw hundreds of these beautiful specimens of Sierra redwood, prepared by Thomas Swetnam, in the basement archives of the University of Arizona's Laboratory of Tree-Ring Research.

Giant sequoias were fitting megaflora to study for Tom Swetnam, who looks part Paul Bunyan, part A. E. Douglass—a barrel-chested man sporting a thick beard, suspenders, and wire-rim glasses. Like the mythical lumberjack, Swetnam spent his life in the forest. Born in 1955, he recalled that "forest fires were part of growing up" because his father was a US Forest Service ranger stationed in northern New Mexico. After earning a bachelor's degree in biology from the University of New Mexico in 1978, Swetnam joined a helitack crew in the Gila Wilderness Area. Aside from aerial firefighting, Swetnam worked on horseback as a fire monitor when federal land managers reintroduced some of the first prescribed natural fires to the national wilderness area. Two years later, he left for Tucson to pursue graduate studies at the University of Arizona, receiving a PhD in forest ecology and management in 1987. Like the famous scientist, Swetnam was a dendrochronologist and became director of the Tree-Ring Lab from 2000 to 2015. In an interview, I asked Swetnam how he started down the path of using the Big Trees to investigate wildland fire and climate change. He remembered reading Douglass's book *Climatic Cycles and Tree-Growth* for his doctoral training and receiving a research request from the National Park Service.[85]

Studying fire and climate for the NPS began with a public complaint. In fall 1985, Eric Barnes returned to the Giant Forest for a day of sightseeing with his family. Growing up in Three Rivers, a gateway town at Sequoia National Park's entrance, Barnes spent much of his youth exploring the famous sequoia grove. He came to know individual Big Trees by name, like Pillars of Hercules, and fostered a deep emotional bond with them. During a homecoming trip, Barnes saw redwoods on the Soldiers Trail that outraged him: Park crews, he believed, had recently burned a cluster of crimson giants called the "Broken Arrow unit" to a crisp. Park officials admitted to no mismanagement, even if they acknowledged that fires burned a little too hot. In November, Barnes published a scathing broadside, "Sequoia in Flames," disputing why "the cinnamon-colored bark was scorched upwards of forty feet" and urging other concerned visitors to call their elected officials. Barnes also sent multiple letters of protest to Superintendent John Davis in which he asserted that prescribed burning amounted to "official vandalism" of park aesthetics.[86]

Wider criticism against the prescribed burning program began to mount when Barnes convinced leading environmentalists to get involved. David Brower, former director of the Sierra Club who in the 1950s led the successful campaign against building Echo Park Dam in Dinosaur National Monument, shared his contempt with Superintendent Davis. "I can think of no sound ecological justification for destroying the enjoyment of a forest in order to save the forest," Brower insisted. "It is a little like putting an opaque cover over the Mona Lisa in order to keep it from being damaged by light."[87] John Dewitt of the Save the Redwoods League added, "We must never lose sight of the fact that the Giant Sequoia forests are of world-wide fame and beauty. To charcoal broil them would be a tragic mistake."[88] A common assumption, running throughout objections to burning the sequoias, was that national parks should be valued for scenery over ecology. The so-called black bark controversy stirred up enough public indignation that NPS officials suspended prescribed burning at Sequoia–Kings Canyon in 1986 while conducting an external review.[89]

Norman Christensen, a Duke University ecologist, chaired the special advisory committee—consisting of three scientists and three landscape architects—to evaluate the scientific basis and management goals of the prescribed burning programs at Sequoia, Kings Canyon, and Yosemite. After studying the best available evidence, as well as receiving public comments, the committee members determined that park crews had appropriately reintroduced fire, not to return forest ecosystems to some historic *point* in time, but to maintain evolutionary *processes* over time. In context of the Leopold Report, they deemed that managing the national parks as a "vignette of primitive America" was less about capturing a "snapshot" and more about keeping the "motion picture" rolling—regular burns by Indigenous peoples, then government bureaucrats, kept biological systems rolling. "There is little doubt," the panelists noted, "that the fire regime of the sequoia-mixed conifer forests during the two millenia [sic] prior to the period of active suppression was dominated by Indian-set fires."[90] Although the final report suggested that resource managers in the future be more attentive to visual qualities when setting flame to forest within the "showcase" areas, or places of heavy visitation, it also affirmed that sequoia groves were not "living museums" and that prescribed burning should continue.[91]

The 1987 Christensen Report was a catalyst for large landscape conservation, recognizing that fire management within protected areas required a broader transboundary perspective: "The maintenance of natural disturbance cycles in wilderness ecosystems is a persistent reminder that our national parks are *tiny islands* in a sea of development and urbanization," the committee members wrote. "Fire regimes are landscape phenomena that operate on scales exceeding the size of most parks."[92] Resource managers knew giant sequoias needed to burn, that much was obvious from prior research, but they still had limited direction about when, how, and where to burn. For historical guidance, Dave Parsons and Jan van Wagtendonk, a research scientist at Yosemite, convinced the NPS administration to hire newly minted PhD Swetnam to research the frequency, intensity, and spatial extent of past burns through tree rings.[93]

A pilot study at Mountain Home State Park confirmed that obtaining fire-scar samples from dead stumps, logs, and snags of Big Trees was feasible. After these preliminary results, Swetnam turned to five additional giant sequoia groves for dendrochronological work along the Sierra Nevada range: Atwell Mill, Giant Forest, and Big Stump (in Sequoia–Kings Canyon); Mariposa (in Yosemite); and Calaveras. Beginning in 1988, Swetnam and his team—which consisted of research technician Chris Baisan and graduate students Peter Brown and Tony Caprio—devoted their summers to the onerous task of taking tabletop-like slices from sequoias in the field.

They commonly spent two to three hours on each sample, using chainsaws with six-foot guide bars (three times the length of standard equipment) to make two parallel cuts, four inches apart. One scientist sawed into the redwood while another spotted the blade path for precision: Sliced at too narrow of an angle, the section crumbled, but at too wide of an angle, it could not be removed. When spotting Baisan's chainsaw on one occasion, Swetnam almost lost his nose as the blade whizzed past his face due to a kickback. Because sampling five to six fire cavities per tree was mandatory to capture all the burn marks, they typically put in one to two days on a single sequoia, applying crowbars and hammers to pry the samples loose. "We looked like coal miners at the end of the day," Swetnam reminisced, "totally covered head to foot in sawdust."[94] After six summers of work, the researchers had filled long-haul rental trailers with redwood, taking samples from over one-hundred Big Trees.

Back at Tucson's Tree-Ring Lab, Swetnam and his team prepared sequoia samples for fire-scar dating by drying them out and then sanding them down. They wanted a surface that was polished enough to see, under a microscope, where fire damage had entered the plant's cell walls—and tree rings were less than a millimeter thick. Utilizing A. E. Douglass's timeline from the early twentieth century as the control against which all dates were cross-identified, they determined the years of fire events on the slices. Swetnam later emphasized that tree rings "have to be dated correctly and accurately, from that, all else follows. If you don't have accurate dating, you can't begin studying things like testing the association of fire regimes and drought events."[95] This process highlighted the fact that Douglass had been a meticulous scientist because, when they were finished, Swetnam found no errors in the original chronology.

Two scientific discoveries held significance for NPS resource managers. In their reconstructed fire history stretching back over three thousand years, Swetnam and the team noticed wide tree rings, indicating a period of rapid plant growth, in most samples after deep fire scars dated to 1297 CE. They interpreted this event as understory crown fire that killed most of the shade-tolerant white fir and incense cedar growing underneath the mature Big Trees. Their research suggested that high-intensity fire could now be understood as comparable to a shot of steroids into a bodybuilder's arm. Like a needle of performance-enhancing drugs, the inferno killed a few crimson giants but also contributed to what Swetnam has called a "growth release" effect because more water, sunlight, and nutrients were available to sequoias with less arboreal competition. The tree-ring studies reassured park officials that implementing hotter prescribed burns—even understory-replacing fires that might harm a few sequoias—were

Left: In the foreground, Chris Baisan (*left*) saws into a sequoia log while Tom Swetnam and Tony Caprio spot the blade, and in the background, Peter Brown records field data on a clipboard. Courtesy of Thomas W. Swetnam.

Below: A portion of a giant sequoia slice, displayed at Kings Canyon National Park, with its fire scars dated from 1236 to 1378 CE. Photo by Tony Caprio. Courtesy of Thomas W. Swetnam.

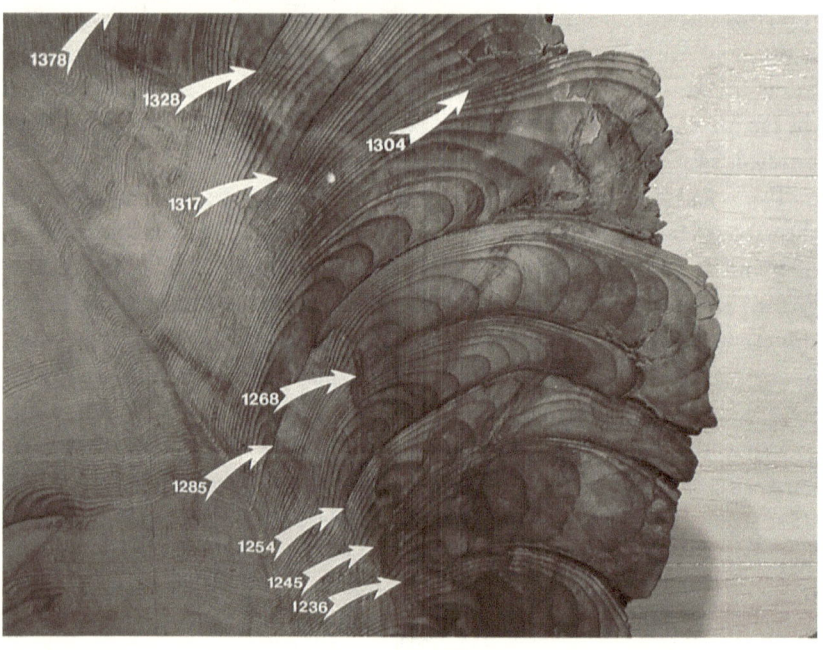

justified since intense blazes historically had opened the Sierra Nevada forest to resources for the Big Trees to prosper.[96]

During laboratory work, Swetnam observed a relationship between fire scars and ring widths. "We were dating these fires," Swetnam recalled, telling Baisan, "Hey, don't you notice that these fires are occurring on the narrow rings, the dry years?"[97] Since narrow and wide rings were the byproduct of climatic variability, the dendrochronologists sampled other conifers to capture a clearer picture of past meteorological conditions over the region. Using tools called increment borers, they took long, cylindrical cores from foxtail pines, located a few thousand feet above the Big Tree groves in Sequoia National Park, and from bristlecone pines, one of the oldest living things in the world situated in the White Mountains to the east of the Sierra Nevada. Because both tree species grew on edges of the timberline at higher elevations with very little rainfall, ring growth was more dependent on temperature than precipitation. In 1993, Swetnam published an article in the high-profile journal *Science* establishing a strong correlation between fire frequency and climatic changes within the sequoia groves. Fire scars over the last two millennia indicated that blazes were most prevalent—happening every two to three years—from 800 to 1300 CE, coinciding with a phase of hotter, drought-like conditions across parts of the Earth that scholars have called the Medieval Warm Period. In short, the Big Trees held a long record of more frequent fires during a warmer climate.[98]

This dendroclimatological finding informed management policies when global warming turned into a pressing issue. In 1988, the same year that Swetnam began investigating fire scars on sequoias, the Intergovernmental Panel on Climate Change (IPCC) formed to build a scientific consensus about the anthropogenic greenhouse effect. NASA scientist James Hansen also brought climate science to mainstream attention after giving testimony before the US Senate Energy and Natural Resources Committee that a buildup of carbon dioxide and other polluting gases was responsible for observed atmospheric warming over the twentieth century. By 1996, the same year that Swetnam was finishing up his redwood studies, the IPCC issued its second assessment report, concluding that multiple lines of evidence showed that CO_2 emissions from burning fossil fuels left a "discernable human influence" on the global climate system.[99]

For Swetnam, one strategy for climate change adaptation became apparent. "The last time the sequoias went through a warming period analogous to today," Swetnam stated, "was a period when there were lots and lots of fires." In 2009, Swetnam, Baisan, Caprio, and others contributed to a scientific paper for the journal *Fire Ecology*. Although recognizing that fire histories were not rigid templates, they proposed that "the prescribed burning program in the sequoia groves should continue and should probably be increased" because "the more fire, the more likely sequoias are to be resilient to recent and future warming."[100] To endure amid global change, the Big Trees must burn—and burn often.

SCATTERED INTO seventy-five groves, giant sequoias are restricted to a narrow belt, some 15 miles wide by 250 miles long, of the western Sierra Nevada. At present, Sequoia and Kings Canyon National Parks hold thirty-nine groves within their boundaries, which account for 40 percent of all sequoia forest by area in the world. Based on readings from five meteorological stations in or near the two parks, average yearly temperatures have risen by 1 degree Fahrenheit between 1975 and 2011. In the southern Sierra, projected warming of 4 degrees F is expected by 2050 and of 5–7 degrees F by the end of the twenty-first century.[101] The Big Trees and other charismatic megaflora will be on the front lines of climate change; their future boils down to three options: adapt, move, or die.

In 1991, NPS officials created the Global Change Program to understand how climate change would affect specific regions, not particular parks, because warmer conditions promised a major realignment of plant and animal geographies. For example, they realized that a majority of sequoias may one day be found outside of Sequoia National Park as new generations of trees find refuge in northern latitudes. Based on the biogeographical areas concept, in which "ecological boundaries transcend management boundaries," park managers increasingly understood Sequoia, Kings Canyon, and Yosemite as part of the Sierra Nevada bioregion. For a landscape-scale approach, park staff would have to coordinate their scientific research and resource management with the surrounding agencies.[102]

As early as 1993, park personnel at Sequoia–Kings Canyon began discussing the "global climate change implications for fire management" during their prescribed burning workshops.[103] The prospect of drought and megafires pressed resource managers to figure out how to get extra controlled fires onto the landscape. The challenge became more imperative when Tony Caprio, one of Tom Swetnam's former students, arrived in 1995 to take a job as the parks' fire research coordinator. Caprio analyzed the spatial coverage of prescribed burning over the years using GIS tools, finding that only half the giant sequoia groves and only one-sixth of all forested areas had been kindled since the program's inception in 1968. Caprio then calculated how many fire return intervals, the number of times each forest type burned on average, were missed during the suppression era to prioritize treatment areas on a regional scale.[104]

With an urgency to expand burning, Sequoia–Kings Canyon's program faced a mountain of obstacles for increasing fire's use. Foremost were budget priorities and budget cuts. Beginning in the early 1980s, prescribed burning was not funded directly by the National Park Service but by a general Department of the Interior appropriation—then divided among the various federal land agencies to use under their budget models. NPS officials allocated a significant portion to "fuels management," which ensured robust burning programs within national parks. This arrangement worked reasonably well until 2000, when Congress started to direct how and where funds were spent. More money went into firefighting efforts defending communities living on the forest's edge, or what fire experts have called the wildland-urban interface, and less money went into kindling prescriptive burns. Since 2000, the NPS fuels management budget has decreased 30 percent, from $33 million to $23 million (as of 2016). Austerity measures to reduce government spending found a home with neoliberal politicians, both Republicans and Democrats, who put greater trust in defending private property than public lands. Fighting flames was prioritized over lighting them.[105]

Despite the economic woes, prescribed burning was still more cost effective than fighting fires. As Tony Caprio pointed out, an entire 3,500-acre burn at Redwood Mountain cost less than $500,000, or about $150 per acre, to complete, and most other prescribed burns cost around $200 per acre. In contrast, the 2015 Rough Fire, a 151,000-acre inferno that burned on national forest lands just outside of Kings Canyon, was $900 per acre in firefighting expenses. This wildfire also demonstrated the potency of prescribed burning after it sent a bolt of flames up a valley toward Redwood Mountain. When the raging fire hit a patch of sequoia forest that park managers had burned with prescribed fires ten years earlier, the hundred-foot flames died down from thinner ground fuels and firefighters were able to stop the approaching fire.[106]

In 2012, when the global-change-type drought arrived, Sierra forests took a beating with more than 102 million trees dead or dying from a combination of insects, disease, heat stress, and decreasing snowpack. Like all trees, giant sequoias must tap into water at their roots, and the air surrounding their leaves causes water to draw up artery-like stem tissue until it evaporates from tiny leaf pores called stomata back into the atmosphere. The Big Trees can prevent water loss by constricting their stomata, but the tradeoff is less intake of carbon dioxide for photosynthesis—meaning if they are closed for too long, the mammoth plants will starve to death. Transpiration, as scientists formally call the whole process, creates tension within sequoias' water columns. The warmer the climate becomes and the less groundwater available, the higher the tension. "Under extreme drought conditions, when that tension grows too high, those columns of water snap like a rubber band," one nature writer has explained. "Gas bubbles form, creating an embolism that prevents the flow of water up the trunk. If this happens enough, a tree will shed its leaves and can, eventually, die."[107] In 2014, this self-pruning scenario was exactly what Dr. Sequoia (aka Nate Stephenson) was seeing among the groves.

Like emergency room physicians, Stephenson and other park scientists were forced to perform management triage: choosing which botanical patients needed to be treated first and where on the landscape limited resources would be applied. To answer these problems, Stephenson recruited federal, private, and university partners to form the Leaf to Landscape Project. For zooming in to the scale of leaves, Berkeley ecologists Anthony Ambrose and Wendy Baxter climbed up the crimson giants and cut small branches from high in their crowns. Back on the ground, Ambrose and Baxter immediately put the branches into a pressure chamber that squeezes out the water to approximate the level of drought stress in individual Big Trees. For zooming out to the scale of landscapes, Stanford scientist Greg Asner of the Carnegie Airborne Observatory flew over the Sierra Nevada in a Dornier 228 airplane carrying millions of dollars in remote sensing equipment. Two ex-military technologies, both of which peer out a hole cut in the belly of the plane, helped Asner to create maps of sequoia vulnerability. At a cruising altitude of 6,500 feet above thirty groves, a light detection and ranging system (LiDAR) sent radar-like laser pulses to capture the distance and shape of tree targets; and a high-fidelity imaging spectrometer (HiFIS) seized the full spectrum of light—from ultraviolent rays to short-wave infrared—to measure the forest canopy's water content. After flights in 2015 and 2016, Sequoia–Kings Canyon resource managers now had three-dimensional representations of the hotter drought.[108]

From the remote sensing, park personnel learned that grove areas with prescribed burn treatments had fared better during the global-change-type drought than those places that rarely experienced fire.[109] Because the two parks didn't have enough money to treat the entire sequoia forest, resource managers must pick areas to burn strategically. In 2017, NPS officials approved Sequoia–Kings Canyon's *Climate-Smart Resource Stewardship Strategy*, an administrative plan engaging in scenario planning to accommodate an uncertain future. Their management priorities included "continuing and expanding the use of fire and fuels treatment."[110] Prescribed burning was officially adopted as an adaptation strategy for human-driven climate change. "But that only buys us time," Nate Stephenson reminded me. If fossil fuel companies do not drastically mitigate their CO_2 emissions, then resource managers will begin experimenting with assisted migrations, planting sequoia trees at higher elevations or even outside the park boundaries.[111]

ANNIVERSARIES ARE opportunities to look back on the past and to project into the future. In 2016, the US National Park Service turned one hundred years old. During its tenure at Sequoia–Kings Canyon, the goal of fire management shifted three times among the giant sequoias. As treasured objects, park managers protected Big Trees from the aesthetic damage of wildfires because they desired to sell monumental scenery to the tourist public. As ecosystems, park personnel set flames to sequoia forests to restore the interactions between woody plants and natural processes. And as landscapes, park crews redirected burning across the Sierra Nevada to build resiliency in those sequoia groves most affected by global environmental changes. Human-driven climate change shook down the administrative barriers between parks and people.

It also had been a century since the last Indigenous person, a Yokut elder named Old Man Topna, had ignited fires within the Giant Forest. Although the Native firestick was largely responsible for upkeeping the Big Trees, NPS officials were slow to recognize Indigenous land management. In the debate over light burning and total suppression, park administrators smothered proposals for the "Indian way of forestry" as they pursued a policy of fire exclusion. Park ecologists finally realized that Native inhabitants had established the historic fire patterns that resource managers hoped to re-create through prescribed burning, but they were reluctant to abandon the Leopold Report mandate of maintaining "naturalness." When global warming required park managers to view sequoias as part of a "cross-jurisdictional landscape," they saw international cooperation with Indigenous nations under a new light.[112]

For the NPS Centennial, park archaeologist Jane Allen of Sequoia–Kings Canyon and tribal elder Ron Goode of the North Fork Mono wished to mark a new era in natural resource management based on shared authority. As tribal liaison, Allen reached out to the surrounding Indigenous nations to see if they wanted to collaborate on federal public lands in a tangible way. North Fork Mono members responded because, Goode explained, "for us, the forest lands and park lands are tribal homelands."[113] An idea evolved out of conversations to establish a Native demonstration garden in Sequoia National Park. "As part of the care and maintenance of that place," Allen noted, "the consensus was that we needed to use fire."[114]

Goode emphasized that "cultural burning," in contrast to prescribed burning, goes beyond the utilitarian goal of reducing fuel loads by setting fires to enhance targeted cultural resources. At Ash Mountain, cultural burning among the blue oaks would increase acorn yields not only to sustain Mono peoples with food and medicine but also to nourish animals as more-than-human relatives. Goode also stressed that cultural burning by his people never really ended, but the federal era of colonial oppression and fire suppression forcibly pushed traditional practices onto private and tribal lands. After more than one hundred years of park administration, the season was ripe for change. Ecological internationalism meant sharing management between US park managers and Indigenous fire practitioners.[115]

In 2017 and again in 2018, NPS crews burned twenty-six acres of blue oak woodlands near the park entrance with the advice and assistance of Sequoia's Indigenous neighbors. Within the plots, blue oak monitoring revealed a postburn rise in acorn production of 4 percent in Unit 6 and 14 percent in Unit 9. Goode and other practitioners, collectively called Keepers of the Flame, documented how improving forest health attracted a greater abundance of wildlife: from praying mantises and wild turkeys to western fence lizards and mountain lions. By proving that sophisticated manicuring can increase species diversity, Native peoples hope to extend Indigenous management to other parts of the park. Black oak, which historically grow at higher elevations on the periphery of giant sequoia groves, might one day benefit from Native-kindled fires, too.[116]

To be sure, cultural burning right now amounts to a very small fraction of land across the Sierra Nevada that needs to burn. Len Neilson, a prescribed fire expert who worked for the California Natural Resource Department, stated that five million acres need to be burned per year, but US agencies were only getting a million acres done. Transnational collaboration was significant for the possibilities it represents. "What are global warming, climate change, drought and dust bowls?" Goode asked. "These are terminologies, for humankind out of sync. They are not at-one with their own spirits let alone the spirits of the land."[117] If advocates of large landscape conservation take seriously the idea that no administrative boundaries separate humans from the rest of nature, then recovering the old ways and building resilience among the old trees should be one and the same.

JESSE VALDEZ handed me garden shears and we got to work clipping sourberry to the ground and stacking them on a burn pile. Valdez, a Mono fire practitioner who Goode had been mentoring as his protégé for the last fifteen years, was covered from head to toe in soot instructing us on how to conduct a cultural burn. The bushes were a dense, tangled mess, or "a dog-hair thicket" to use Leopold's phrase, that I could barely walk through—until they were piled up into the shape of an overturned rowboat for burning. Valdez tied his long, flowing mane into a ponytail, joking that nothing ruined the perfume of "good smoke" like burned hair, and set flame to the heap of sumac. The smoke rose through the overhanging branches of black oak trees and up into the sky. Fire consumed the sourberry quickly. We grabbed rakes to mix the ash with the topsoil until we couldn't tell that a fire had been there in the first place.

The goal, Ron Goode declared, was a "see through" forest. Goode watched over the operation and then showed a group of us from US agencies and universities the postfire effects of cultural burns. Sourberry sent up new, supple shoots for weaving materials like baskets and cradleboards; black oak produced better acorns because the smoke stopped parasitic dodder from robbing nutrients from the trees. Goode shared that we stood on Miwok territory, but anyone from the Keepers of the Flame—representing the Mono, Yokut, and Miwok Nations, among others—was welcome to gather cultural resources on the private estate. Ron welcomed youth, Native and non-Native alike, because he believed that imparting traditional knowledge with the next generation held the potential to transform US agencies when his fire acolytes entered them.

In the Sierra Nevada landscape, ecological internationalism was not only about a more equitable distribution of the biosphere between human and other than human, but it was also about repairing uneven geographies among people. Goode criticized the practice of industrial burning, as done by resource managers in the National Park Service, US Forest Service, and other federal entities, because it neglected how Indigenous nations tended the landscape. "What are you burning for? Who are you burning for?" the tribal elder asked. I knew that minimizing the risk of catastrophic wildfires, even though five of the largest infernos in California's history occurred in the year 2020 alone, was not the answer Goode was moving toward. Goode sought collaboration to revitalize lost treaty borders between the United States and Indigenous nations. "Without land, without fire, our traditions die."

Goode leaned into a cane with one hand, held a clapper stick in another, and sang a bear song to bless the ground and people of many nationalities who tended it. Flames from a burn pile jubilantly danced to his rhythmic clap and vocal cadence. Fire wanted to jump and prance across borders, less than thirty miles away was Yosemite National Park, where Goode's ancestors once stood beneath the shadows of the Mariposa Grove.

CONCLUSION

Barriers of Biodiversity

A Wilson's warbler on the move can easily be mistaken for a yellow warbler. Both are small, neotropical bird species with golden plumage that migrate from Central America to Colorado. Both can be found darting, diving, and snatching up insects around willow thickets at Rocky Mountain National Park. Only when the Wilson's warbler finally perched on a damaged shrub limb could Ben Bobowski and I spot the black-colored cap that differentiates it from a yellow warbler. North America's boreal forest, or taiga biome, a circumpolar belt of conifers stretching coast to coast from Newfoundland to Alaska, constitutes all the breeding range for Wilson's warblers, all except a thin strap of habitat that hangs down to Rocky. "This protected area serves as the southernmost limit of their summer territory." Bobowski hesitated. "For now."

But where habitat shifts, so do species. Imagine that band of boreal forest getting thinner and creeping north toward the polar region. At 2 degrees Celsius (or 3.6 degrees Fahrenheit) warmer than preindustrial times, climate modeling predicts that the Wilson's warbler will gain 17 percent of new breeding range toward the northern end, but it will also lose 40 percent at the southern end, including at Rocky. Willows have declined within the national park not only due to overgrazing elk but also from a complex interaction between parasitic fungi and climate change. The growth of riparian flora is contingent on temperature-dependent processes like the water table. Drought stress from below-average stream runoff during the 2000s exacerbated willows' inability to withstand beak gashes made by birds called red-naped sapsuckers, which introduced *Cytospora* infections to the plants.[1]

Beaver colonies can buffer against these climate impacts by raising groundwater through their dam construction, so long as elk have left vegetation alone for building materials. Riparian habitats make up a fraction of the landscape but hold the most avian species. Wilson's warblers build ground-level nests in dense willow cover to protect their eggs before hatching. At 3 degrees Celsius (or 5.4

degrees Fahrenheit) warmer than preindustrial times, their summer range will shrink by another 21 percent. With looming temperature increases in the future, the bird species in all likelihood will no longer settle down to brood offspring in the Lower 48 United States.[2]

Like the Wilson's warbler, Ben Bobowski pursued refuge at more northerly climes. In 2016, the park manager landed a leadership role as the superintendent of Wrangell-St. Elias National Park and Preserve in southeastern Alaska. In the new position, Bobowski remembered hearing reports about species that were found in Wrangell-St. Elias for the very first time in the protected area's history. Common yellowthroats were spotted during a Christmas Bird Count even though their historical summer range was restricted to Canada. Wood bison made a five-hundred-mile journey from the Ashihik herd in the Yukon Territory. Mule and whitetail deer have joined their smaller cousin, Sitka blacktail deer, in Alaska. "It's important to keep in mind that the park was only established in 1980," Bobowski stated, but after sorting out sampling bias to get "the signal versus the noise," these sightings do suggest a future of climate-induced shifts of species across the US-Canada border. "And it won't be the only border."[3]

The Earth is now warmer than at any time in the last 125,000 years. Climate breakdown has accelerated the movement of wildlife across the planet. Of four thousand species recently tracked, scientists documented that almost three-quarters of them shifted their ranges, mostly to cooler lands and waters. Terrestrial species, on average, were moving twenty kilometers (twelve miles) every decade toward the poles. As the world heats up, threatened biota needs more freedom of movement, greater flexibility with borders, to adapt and adjust. Survival on a hotter planet means the ability to move across international lines.[4]

Borders, with different qualities of habitat or different styles of management on one side or the other, can inhibit the biological imperative to move. They can act like a prison cell or fortress wall for living things. They are created and upheld by powerful nation-states to enforce social divides, but they easily become ecological divides too. The fall of the Berlin Wall in 1989 and the collapse of South African apartheid soon thereafter was supposed to signal the dissolution of borders for a more open world, but nation-states have pursued the exact opposite. At the end of the Cold War (when I was born), there were fifteen border barriers worldwide; now there are seventy-seven of them.[5]

Six out of every ten people today live in a country that has built one of these walls of its border, fragmenting habitats between United States and Mexico, Israel and Palestine, Bulgaria and Turkey, and Spain and Morocco at the Melilla and Cueta fences. What will the future of many borderland animals such as the ocelots of Tamaulipan Mezquital, mountain gazelles of the West Bank, or Eurasian lynx of Anatolia look like in a warming world? The answer will depend on how people of different nationalities resist and reformulate state-enforced borders. That's why tracing the roots of ecological internationalism is important.[6]

ANOTHER MIGRATION has been underway from Central America akin to the flight of neotropical birds. According to World Bank experts, as many as four million people could flee the Northern Triangle of Guatemala, Honduras, and El Salvador by 2050, dislocated because of climate-related disasters such as crop failures from intensifying droughts or storm surges from sea level rise. Climate disruption has exacerbated volatile weather patterns along the Dry Corridor, an agricultural region of the Central American Pacific Coast. Human beings, like all living things, are on the move.[7]

Beginning in 2014, rainfall dropped by 40 percent across the Dry Corridor, followed by an intense heat wave. Farmer Ronis Martínez remarked that his local community of El Rosario, Honduras, had a difficult time surviving: "We had a loss of almost 100 percent of the corn and beans. And that is the only income we have all year." Agriculture employs about two-thirds of the country's population, so poor harvests meant people lacked food and jobs. Nelson Mejia, also from El Rosario, prayed his rosary for guidance about the choices farmers faced: look for work in the cities like San Pedro Sula, where the prospect of gang violence from MS-13 was high, or move to the United States, where the risk of border-crossing apprehension was dangerous. In 2014, coinciding with the first year of a longer drought, at least ninety-one thousand Hondurans chose the latter. Migrant caravans headed north on a debilitating journey called Viacrucis del Refugiados (Refugees' Way of the Cross).[8]

One of the final border crossings was the Anzalduas International Bridge between Reynosa, Tamaulipas, and McAllen, Texas, an entry point where internationalist and nationalist responses were on full display. The National Butterfly Center, a one-hundred-acre private wildlife reserve founded by the North American Butterfly Association, sits on the US side of the Rio Grande–Río Bravo. The flagship center serves to educate the public on the role of butterflies and other pollinators in maintaining healthy ecosystems and sustainable foods. The borderlands institution does binational outreach by participating in the Lower Rio Grande Valley Conservation Corridor, a string of riverine habitats from Falcon Dam to the Gulf of Mexico, connecting mobile species on US- and Mexico-owned federal refuges, state parks, and private lands. The center's director,

Marianna Treviño-Wright, commented that her work came easy from growing up the adopted daughter of a binational family. Treviño-Wright remembered feeling the sting of Juan Crow laws in Texas during her 1970s childhood when she and her American mother entered a restaurant before her Mexican father did, who was a local physician from Monterrey, so they were not denied a table together. "Soy una fronteriza," Treviño-Wright stated. "I am a borderlander."[9]

On July 20, 2017, Treviño-Wright was walking the National Butterfly Center's grounds when she was surprised by five construction workers who were cutting down trees and mowing vegetation on the property. When she confronted them, they explained that the Trump administration had sent them here to "clear this land from the levee to the river ahead of the border wall." A day later, US Border Patrol agents showed up at the center, telling Treviño-Wright that what she had seen was not happening. In disbelief, Marianna showed them the heavy equipment that had been used to clear a 150-feet-wide "border enforcement zone," destroying five thousand milkweed and nectar plants as part of their Monarch Waystation in the process. "I did not sign up to be Plaintiff #1 against the Department of Homeland Security," Treviño-Wright stated, but they came here "with no notice, no authority, no due process for eminent domain." Marianna reached out to her elected officials for help, but neither Republicans nor Democrats were of any assistance. Senator Ted Cruz ignored her plea and Representative Henry Cuellar's staffer replied: "And? What do you want me to do about it?" With no redress, the National Butterfly Center sued the US government.[10]

The legal maneuver brought Treviño-Wright and the National Butterfly Center into the public spotlight. After a government shutdown over disputes about border funding in January 2018, US Congress passed the Consolidated Appropriations Act, which passed with bipartisan neoliberal support. Pundits and politicos claimed that President Donald Trump had been defeated on securing money for extending the border wall, but Representative Cuellar inserted a provision giving him $641 million for twenty-five new miles of "levee fencing." The conservative Texas Democrat had used a similar tactic to appease President George W. Bush a decade earlier. Treviño-Wright knew that levee fencing was a border wall in disguise. When she flew to Washington, DC, to meet with congressional offices, they all denied the new funding. The Secretary of Homeland Security then waived twenty-eight federal laws, including the National Environmental Policy Act, Clean Water Act, and Endangered Species Act, to expedite border wall construction on the Lower Rio Grande Valley and Santa Ana National Wildlife Refuges. On January 16, 2019, Treviño-Wright returned to Capitol Hill to testify before the House Committee on Natural Resources about how new border infrastructure will destroy transboundary ecologies.[11]

"The bulldozers will roll into the lower Rio Grande Valley wildlife conservation corridor, eliminating thousands of trees during spring nesting season for hundreds of species of migratory raptors and songbirds," Treviño-Wright declared under oath. "Wherever the land is cleared, vegetation that provides sensitive breeding and feeding areas for wildlife will disappear. Each butterfly species is intimately tied to a couple of plant species their caterpillars can consume to complete their metamorphosis; so if you eliminate the plant from the landscape, you eliminate the butterfly."[12]

Her activism caught the attention of We Build the Wall, an organization founded by US Air Force veteran Brian Kolfage with the goal of privately funding barrier construction through negotiations with landowners along the US-Mexico border area. Trump political strategist Steve Bannon jumped on as chairman of We Build the Wall to prevent what he xenophobically called the "invasion" of Latin American migrants. After raising $25 million through GoFundMe, Kolfage and Bannon began constructing a three-mile stretch of border fence on a private sugar cane farm near the National Butterfly Center. Treviño-Wright took boat rides with Father Roy Snipes, a local Catholic priest whose small historic chapel called La Lomita would be sealed off from parishioners by a border wall, along the river to raise public awareness about the atrocity. Kolfage retaliated by attacking Treviño-Wright and the center as "left wing thugs with a sham butterfly agenda" to his Twitter followers. White supremacist militias like the Three Percenters, Oath Keepers, and Boogaloo Bois started issuing death threats against Treviño-Wright almost daily.[13]

While Kolfage and Bannon stoked racist nationalism to fuel support for a border wall, Treviño-Wright believed the real reasons behind the entire project have been economic. A barrier would function, in Treviño-Wright's words, as "a hurdle and a harvester" by stalling migrants and allowing Border Patrol agents to shuttle them to private detention centers owned by for-profit prison companies like CoreCivic and GEO Group. A barrier would also deny public access to prospective oil and gas pipelines built by fossil fuel corporations like Kinder Morgan and Enbridge from the Permian Basin to the Port of Brownsville, Texas. President Trump made public remarks that he is "not a big believer in global warming," but he certainly is in private. His company cited global warming in a building permit as one of the rationales for erecting a coastal wall around his golf course in Ireland to protect it from more frequent storm surges. Trump, and the ultrawealthy around the world, have developed a strategy for dealing with climate chaos during the twenty-first century: Keep burning fossil fuels and hide from the consequences behind walls. Economic inequality between nation-states has been one of the strongest predictors of border hardening.[14]

Ecological internationalism forms the basis of combatting a closed world of borders, a "climate apartheid" between the rich and the rest. "We need to stand up for those who have no voice," Treviño-Wright remarked about her activism, which moved from nonhuman to human neighbors. The monarch butterfly has become a potent symbol for immigration justice, for abolishing the restrictions on mobility that come with state-imposed borders. After fighting the feds and far-right nationalists, Treviño-Wright commented, "I don't think I'll see a butterfly in the same way again."[15]

The border wall stood on shaky ground. In August 2020, Brian Kolfage and Steve Bannon of We Build the Wall were arrested on charges of fraud and money laundering for funneling hundreds of thousands of dollars in donations used for personal expenses. Kolfage spent $350,000 on a boat, luxury SUV, golf cart, jewelry, home renovations, and cosmetic surgery. Tommy Fisher, president of the North Dakota-based Fisher Industries, bragged that his company had built the "Lamborghini" of border walls along the Rio Grande for We Build the Wall. However, investigative reports from ProPublica uncovered that structural engineers and hydrology experts have already documented that water erosion was destabilizing the fence's groundwork because the eighteen-foot-tall structure was hastily erected on a sandy levee. The private border wall could fall over during a flood or hurricane. Until then, butterflies and birds will have to fly over it; land-dwelling animals will have to go around it. The National Butterfly Center will serve as a biological passageway.[16]

LARGE LANDSCAPE CONSERVATION holds the power to soften national divides. In October 2018, over 1,400 US scientists and six hundred Mexican scientists published a *BioScience* article condemning "the border wall's negative impacts on wildlife, habitat, and binational collaboration in conservation." Hundreds of animals, including sixty-two species on the edge of extinction like the Sonoran pronghorn, call the US-Mexico borderlands region home. For 34 percent of these creatures, a continuous border fence would cut their historical ranges in half. E. O. Wilson, Gerardo Ceballos, and other prominent conservation biologists expressed their collective opposition to constructing a wall because it "threatens some populations by degrading landscape connectivity."

Scientists from forty-three other countries endorsed the statement, speaking to the worldwide effects of border barriers on biodiversity.[17]

About 15 percent of the world's land, excluding Antarctica, now has some form of protected status. The percentage is equal to setting aside the combined territories of Australia, China, and India as habitat. Impressive as it may be, the figure was still 2 percent shy of the 2020 target established by signatories of the Convention on Biological Diversity a quarter century ago (and as of this writing, the United States government has yet to ratify the treaty). Like IPCC scientists setting goals to cut greenhouse gas emissions for achieving climate stabilization, conservation biologists also have created benchmarks to reach ecosystem stabilization: a doubling of protected lands to 30 percent of the total surface area by the year 2030 and getting to the 50 percent mark by the year 2050. If Half Earth—to use Wilson's phrase—has become a geopolitical objective, then more wildlife corridors will need to be created.[18]

Like Central America's Dry Corridor, the Eastern Mediterranean's Levant was also struck by drought. In Syria, where agriculture first originated in the Fertile Crescent, farmers experienced little rainfall from 2006 to 2011, the driest period for the region in the last nine hundred years. In the 1970s, the military regime led by Syrian President Hafez al-Assad pursued a policy of agricultural self-sufficiency, relying on groundwater wells to raise crops for the nation-state. In 2005, President Bashar al-Assad, Hafez's son and successor, made it illegal to dig new wells without a license issued by a governmental official. Syrian farmers, out of necessity due to water shortages, bribed civil servants for permits to drill more irrigation aquifers. The shrinking water table, exacerbated by drilling and drought, led to crop failures and food shortages. In 2011, before the Syrian civil war began, a million people had already left the country. Climate refugees fled the political violence by crossing into Türkiye and moving to Europe.[19]

The massive influx of Syrian migrants caused many European countries to close their borders. In November 2015, Slovenia constructed a razor wire security fence along its 670-kilometer-long border with Croatia. Newspapers featured images of lifeless red deer, after becoming entangled in the coils of wire, alerted conservationists to join other protesters in contesting the barrier on both humanitarian and environmental grounds. Before the Dinaric Alps were fenced, scientists had documented with GPS technologies the extent to which brown bears and Eurasian lynx regularly crossed the Slovenia-Croatia border. They demanded the border fence come down. "It is somewhat ironic," observed John Linnell of the Norwegian Institute for Nature Research, "that for the last

15 years, while conservation biologists have been largely promoting transboundary management and celebrating localised examples of fence removal, the global trend has been for an unprecedented increase in barriers preventing wildlife from moving across borders."[20] Internationalists must challenge hardening borderlines to conserve transboundary species during an age of climate dislocation.

"SWEET, SWEET, SWEET, I'm so sweet." That's the mnemonic device that English-speaking birders use to remember the song of a yellow warbler, Ben Bobowski told me. He had originally learned the phrase in Spanish from Costa Rican naturalist Johnny Pérez Artavia, who alternated between saying "Dulce, dulce, dulce, soy tan dulce" and "Dulce, dulce, más dulce que dulce." I closed my eyes for a moment to focus on listening to the series of six whistled notes. The simple, high-pitched chirps of this international birdsong seemed to invite a reply.

I have come to believe that it's incredible how such a small bird inspired such grand connections among people of different nationalities. Like Treviño-Wright battling over borders because of the monarch butterfly, I hope to never see the yellow warbler in the same way again. Our future comes down to solidarity or extinction.

I opened my eyes to watch two yellow warblers zoom overhead, chasing each other with twists and turns like a performance of aerial acrobatics along the Big Thompson River. The pair of birds settled momentarily on the olive-green branches of a willow bush, before rising off again in flight. *Zweet, zweet*, the yellow warblers called out to each other, moving downstream to pursue a cloud of insects until they slipped out of my view.

ACKNOWLEDGMENTS

Like any ecological relationship, producing this book connected me to numerous people and institutions. Mark Fiege has been a personal and scholarly model to me for more than a decade. I must thank him for getting me interested in environmental history in the first place: the irrigated fields of Idaho's Snake River Plain were both the subject of his first book and the substance of my childhood home. Stimulating conversation, bottomless cups of coffee, and yet another book recommendation were always ready. Mark nudged me to think about what story I wanted to tell rather than what topic to write—and that shift in perspective has made a big difference. I also appreciate how Mark showed me that the best writing was done not by the line of argument reasoned, but by the emotions those words evoked.

This project has its origins in doctoral study at Montana State University. I thank my adviser, Mark Fiege, for his passionate guidance, plus my committee members Catherine Dunlop, Tim LeCain, Michael Reidy, Molly Todd, Emily Wakild, and Brett Walker for their feedback. While graduate students come to MSU for its strengths of environmental history and public history, my training was enriched by a network of scholars: Crystal Alegria, Jacey Anderson, Micah Chang, Shane Doyle, Jill Falcon Ramaker, Kerri Keller Clement, Jen Dunn, Kirke Elsass, Will Griffiths, Amanda Martin, James Martin, and Janet Ore.

In 2021, the History Department at Augustana University took a chance on a newly minted PhD and offered me a job. I am thankful for my history colleagues Cory Conover, Peg Preston, and Mike Mullin who make Augustana a welcoming and vibrant place to work. I owe a special debt to Peg for reading and critiquing the entire manuscript when I was headed toward a dead end on revision. Getting to teach public history in Sioux Falls is a constant reminder that creating knowledge should be in service to a community. I thank Zach Van Harris Jr. of the South Dakota African American History Museum, Rick Tupper of the USS South Dakota Battleship Memorial, Adam Nyhaug of the Siouxland Heritage Museums, and Diane deKoeyer of the Sioux Falls Board of Historic Preservation for their collaborations. I enjoy the privilege of being at a small, liberal arts institution where professors from various departments get to work together. I thank faculty who contribute to the Environmental Studies

Program, particularly Dave O'Hara, Julie Loveland-Swanstrom, Carrie Olson-Manning, and Emily Roehl.

Before Augustana, I was employed as a postdoctoral fellow at the University of British Columbia, where I worked on a project that combined the animal history and public attitudes toward gene editing. During this joint endeavor between UBC's History Department and Animal Welfare Program, I am grateful to Heidi Tworek, Katie Koralesky, Nina von Keyserlingk, and Dan Weary for exhibiting what interdisciplinary research should be like. I am convinced that the most impactful historical investigations turn hindsight into foresight.

I cannot underestimate the intellectual encouragement that came from the WEST Network gathering, a retreat held every summer at a scientific field station rotating among host institutions of Colorado State University, Montana State University, University of Arizona, and University of Colorado. Because the historian's craft traditionally prioritizes reading archival documents to understand the human experience, I valued how the retreat emphasized the need to "read" a landscape to understand the relationships between people and place. To those who came to the Southwestern Research Station in Arizona; Flathead Lake Biological Station, Judson Mead Geologic Field Station, and Lubrecht Experimental Forest in Montana; and Mountain Research Station and CSU Mountain Campus in Colorado, your company was invaluable. I shared conversations with many people on the trails, including those MSU folks mentioned above, but I additionally want to thank Ruth Alexander, Mark Boxell, Nichelle Frank, Amado Guzman, Marcus Hernandez, Adrian Howkins, Lora Key, Karen Lloyd-D'Onofrio, Jared Orsi, Sarah Payne, Sara Porterfield, Ariel Schnee, Sheu, Sierra Standish, Paul Sutter, Dane Vanhoozer, and Jeremy Vetter.

For the book, I relied on countless interviews to bring stories to life. I tried emulating the journalistic techniques of science writers Elizabeth Kolbert and David Quammen who tangibly brought audiences along with them to far-flung places by serving as their eyes, ears, and other bodily senses. I am grateful for the conservation practitioners, land managers, and scientists who acted as my guide, allowing me to tag along in the field. I want to thank Malou Anderson-Ramirez, Homero Aridjis, Chris Baisan, Ben Bobowksi, Diane Boyd, Tony Caprio, Don Davis, Ron Goode, Carlos Gottfried, Jodi Hilty, Joel Moreno, Karen Oberhauser, Bob Pyle, Ellen Sharp, Doug Smith, Nate Stephenson, Tom Swetnam, Chip Taylor, Bill Tweed, and Jesse Valdez. These people generously shared their knowledge of the places they care for.

I could not have completed my work without generous financial assistance, and I want to acknowledge the institutions that made research trips all over

Canada, Mexico, and the United States possible: the Forest History Society's Walter S. Rosenberry Fellowship, Brigham Young University's Charles Redd Center, Montana State University's Ivan Doig Center, MSU History and Philosophy Department, Augustana University's History Department, and Augustana's Research and Artist Fund.

I would like to thank the members of my family for their love and support: my parents, Andrea and Russ Wright; my parents-in-law, Robie and Brad Culver; my brothers, Ryan and Adam Wright; my brothers-in-law, Cleve and Clint Culver; sisters-in-law, Maddy and Tay Wright, Tohlina and Rachel Culver; and grandmother, Patricia Knoblauch. All my family, immediate and extended, have been sources of both kindness and generosity.

My sons, Bradley, Anders, and Halden Wright, deserve special mention. They have forced me to view things from a more youthful perspective, seeing the world anew in awe and wonder. Meeting their needs are constant reminders of leaving behind my self-centeredness. Their laughter brings me back to living each moment.

I owe the deepest gratitude to my wife, Carly Culver Wright, who sustained and loved me through it all. Her good judgment, constant encouragement, and unshakable faith have been the pillars on which our family rests. Her passion for helping and healing other people inspires me. Carly brings forth the best in me, and to her I dedicate this book.

Will Wright
Sioux Falls, South Dakota

NOTES

Abbreviations

AGN	Archivo General de la Nación
CSU-Schutt	Harold G. Schutt Collection, Madden Library, California State University, Fresno
Douglass Papers	Papers of Andrew Ellicott Douglass, Special Collections, University of Arizona Libraries
FHS-OH	Oral History Collection, Forest History Society Library and Archives
LAC	Library and Archives Canada, Ottawa
MWL	Monarch Watch Library, University of Kansas
NARA-RG22	Record Group 22: Records of the US Fish and Wildlife Service, National Archives and Records Administration
Pyle Papers	Robert Michael Pyle Papers, Special Collections Library, Texas Tech University.
Ream Papers	Robert Ream Papers, Special Collections, Mansfield Library, University of Montana
SA-FWS	Record Unit 7176: Field Reports of the U.S. Fish and Wildlife Service, Smithsonian Institution Archives
SEKI Fire Effects	Fire Monitoring and Fire Effects Records, Sequoia and Kings Canyon National Parks Museum Collection
SEKI Fire Records	Fire Records, Sequoia and Kings Canyon National Parks Museum Collection
SEKI Park Files	Park Files, Sequoia and Kings Canyon National Parks Museum Collection
Stewart Papers	George W. Stewart Papers, Tulare County Public Library
UC-Glenbow	Glenbow Western Research Centre, University of Calgary
WU-Monsanto	Monsanto Company Records, Washington University Archives
Whyte-PC	Parks Canada Fonds, Archives and Library, Whyte Museum of the Canadian Rockies
Whyte-Y2Y	Yellowstone to Yukon Collection, Archives and Library, Whyte Museum of the Canadian Rockies

Introduction

1. Brown, "Place in Biography for Oneself"; Fiege, *Republic of Nature*, 3–13; US National Park Service, "Elk and Vegetation Management Plan—Rocky Mountain National Park—Colorado," Final Environmental Impact Statement, December 2007, www.nps.gov/romo/learn/management/elk_vegetation_planning.htm.
2. "Birds Without Borders," *Walk in the Park with Nick Mollé*; Wilson, *Seeking Refuge*.
3. Ben R. Bobowski, interview by author, 20 November 2020.
4. Pérez as quoted in "Birds Without Borders," *Walk in the Park with Nick Mollé*; Evans, *Green Republic*; Todd, "'We Were Part of the Revolutionary Movement There.'"
5. Villalobos, Núñez, and Padrón, "Cuidades Hermanas Estes Park"; "Sister Park Arrangement Between Rocky Mountain National Park, National Park Service (NPS), United States and Arenal-Tempisque Conservation Area ... Sistema Nacional de Áreas de Conservación (SINAC), Ministry of Environment and Energy, Republic of Costa Rica," 16 March 2015, in author's possession.
6. Mazerolle et al., "Effects of Large-Scale Climatic Fluctuations"; Albright et al., "Effects of Drought"; Strode, "Implications of Climate Change."
7. "Birds Without Borders," *Walk in the Park with Nick Mollé*; Ben Bobowski, Jeff Connor, Ryan Monello, Bill Monahan, Jim Thompson, Summer Olsen, and Chelsea Frost, "Species Relationships of Rocky Mountain National Park, Colorado, with Monteverde, Costa Rica," 2011, in author's possession.
8. I explain "ecological internationalism" as a political consciousness that is shaped by mobile organisms or biophysical processes that traverse national borders. This awareness of shared species, in turn, creates a sense of solidarity between people of different nationalities. See Taylor, "Boundary Terminology." Taylor defines "international" as anything dealing with state-to-state compacts or the collaboration of state officials, "transboundary" as political barriers that arise when natural processes cross state borders, and "transnational" as material or cultural exchanges that show interconnections among separated national spaces.
9. Wilderness Act, Pub. L. No. 88-577, 78 Stat. 890 (1964): 887–90; Turner, *Promise of Wilderness*.
10. Richards, *Unending Frontier*, 463–546; Stein, *Plumes*; Andrews, *Coyote Valley*, 46–51; Price, *Flight Maps*, 1–110; Jones, *Empire of Extinction*, 1–20, 138–69; Bolster, *Mortal Sea*, 72–73.
11. Gissibl, Höhler, and Kupper, *Civilizing Nature*; Kupper, *Creating Wilderness*; Tyrrell, *Crisis of the Wasteful Nation*; Howkins, Orsi, and Fiege, *National Parks Beyond the Nation*; Tyrrell, "Ian Tyrrell Replies"; Sutter, "Trouble with 'America's National Parks'"; Dunlap, "Beyond the Parks"; Swenson, "Response to Ian Tyrrell"; Pawson and Brooking, *Making a New Land*, 141–73; Frank, *Making Rocky Mountain National Park*, 7–21; Sellars, *Preserving Nature*.
12. Tweed, *Uncertain Path*, 45; Carruthers, *Kruger National Park*; Simonian, *Defending the Land*; Warren, *Hunter's Game*; Spence, *Dispossessing the Wilderness*; Jacoby, *Crimes Against Nature*; Brockington, *Fortress Conservation*; Jacobs, *Environment, Power, and Injustice*; Shetler, *Imagining Serengeti*; Wakild, *Revolutionary Parks*; Beattie, *Empire and Environmental Anxiety*; Boyer, *Political Landscapes*; Von Hardenberg et al., *Nature State*.
13. Grove, *Green Imperialism*; Quammen, *Song of the Dodo*; Hennessey, *On the Backs of Tortoises*.

14. Farnham, *Saving Nature's Legacy*, 22–23, 129–31; MacArthur and Wilson, *Theory of Island Biogeography*, 181.
15. Frank, *Making Rocky Mountain National Park*, 114–42; Boxell, "Rapid Ascent," 85–118.
16. Andrews, *Coyote Valley*, 216–48; US National Park Service, "Elk and Vegetation Management Plan."
17. Tonya Bina, "Wolf Sighting Doesn't Mean Animals Have Returned to Rocky Mountain National Park," *Sky-Hi Daily News* (Grand County, CO), 7 January 2008; Jiminez et al., "Wolf Dispersal."
18. Dorsey, *Dawn of Conservation Diplomacy*; Dorsey, *Whales and Nations*; Boardman, *International Organization and the Conservation of Nature*; Brüggemeier, "Ecological Movement and Internationalism," 75–92; De Bont, *Nature's Diplomats*; Franklin, "Biosphere Reserve Program."
19. Bobowski interview; Ben Bobowski to Will Wright, email, 10 December 2020, in author's possession; Maier, "Consigning the Twentieth Century to History"; Graybill, *Policing the Great Plains*; Benton-Cohen, *Borderline Americans*; Hernández, *Migra!*; St. John, *Line in the Sand*; Dunlop, *Cartophilia*; Geiger, *Subverting Exclusion*; Hogue, *Metis and the Medicine Line*; Rensink, *Native But Foreign*; Walia, *Border and Rule*; Baud and Van Schendel, "Toward a Comparative History of Borderlands"; Adelman and Aron, "From Borderlands to Borders"; Hämäläinen and Truett, "On Borderlands"; Orsi, "Construction and Contestation."
20. Quammen, *Song of the Dodo*, 409–91; Diamond, "Island Dilemma"; Newark, "Land-Bridge Island Perspective."
21. Bobowski interview; Barnosky et al., "Has the Earth's Sixth Mass Extinction"; Ceballos et al., "Accelerated Modern Human-Induced Species Losses"; Dirzo et al., "Defaunation in the Anthropocene"; Kolbert, *Sixth Extinction*, 173–92.
22. Beever et al., "Successes and Challenges"; Kati et al., "Challenge of Implementing"; Aycrigg et al., "Completing the System."
23. Wilson, *Half-Earth*, 3.
24. Bingham et al., "Sixty Years of Tracking"; Locke, "Nature Needs Half"; Lovejoy, "Extinction Tsunami Can Be Avoided"; Locke, "International Movement"; Baillie and Zhang, "Space for Nature"; Helen Briggs, "COP15: Nations Reach 'Historic' Deal to Protect Nature," BBC News, December 19, 2022, www.bbc.com/news/science-environment-64019324.
25. Evans, *Green Republic*, 1–93; Villalobos, Arias Núñez, and Medina Padrón, *Proyecto de Sistematización*; Bobowski interview.
26. Ellis et al., "Anthropogenic Transformation"; Lindenmayer, Thorn, and Noss, "Countering Resistance"; Hilty et al., *Corridor Ecology*.
27. Sánchez-Bayo and Wyckhuys, "Worldwide Decline of the Entofauna"; International Union for the Conservation of Nature, "Table 1a: Number of Species Evaluated in Relation to the Overall Number of Described Species, and Numbers of Threatened Species by Major Groups of Organisms," *Red List*, December 10, 2020, www.iucnredlist.org/resources/summary-statistics#Summary%20Tables.
28. Dunbar-Ortiz, *Indigenous Peoples' History*, 1–14; Harvey, *Social Justice and the City*; Jo Guldi, "What Is the Spatial Turn?," Spatial Humanities: A Project of the Institute for Enabling Geospatial Scholarship, accessed 21 January 2021, https://spatial.scholarslab.org/spatial-turn/; Hunt, *Writing History in the Global Era*; Baud and Van Schendel, "Toward a Comparative History of Borderlands."

29. Worster, "World Without Borders"; White, "Nationalization of Nature."
30. Wadewitz, "Scales of Salmon"; R. M. Wilson, "Mobile Bodies"; Ritvo, "On the Animal Turn"; Walker, "Animals and the Intimacy of History"; Isenberg, *Destruction of the Bison*; Anderson, *Creatures of Empire*; Weisiger, *Dreaming of Sheep*; Smalley, *Wild by Nature*.
31. Jones, *Wolf Mountains*; Coleman, *Vicious*; Robinson, *Predatory Bureaucracy*; Walker, *Lost Wolves of Japan*; Van Nuys, *Varmints and Victims*; Wise, *Producing Predators*; Rutherford, *Villain, Vermin, Icon, Kin*.
32. Gottfried, *Mariposa Monarca*; Urquhart, *Monarch Butterfly: International Traveler*; Brower, "Understanding and Misunderstanding"; Merino Pérez and Hernández Apolinar, "Destrucción de instituciones comunitarias"; Brenner, "Aceptación de políticas de conservación ambiental"; Van der Meer, "Butterfly Effect"; D. A. Rose, "Politics of Mexican Wildlife," 280–85; Agrawal, *Monarchs and Milkweed*; Gonzalez-Duarte, "Butterflies, Organized Crime, and 'Sad Trees'"; Sharp and Wright, "'We Were in Love with the Forest'"; Wright, "Monarch Butterfly Conservation (Mexico)."
33. Johnston, *They Felled the Redwoods*; Lori Vermaas, *Sequoia*; Anderson, *Tending the Wild*; Farmer, *Trees in Paradise*; Tweed and Dilsaver, *Challenge of the Big Trees*; Tweed, *King Sequoia*; Miller, *This Radical Land*.
34. Scott, *Seeing Like a State*, 11–52; Brown, "Gridded Lives," 17–48.
35. Haraway, "Situated Knowledges"; Finnegan, "Spatial Turn," 369–88; Kohler, *Landscapes and Labscapes*; Lewis, *Inventing Global Ecology*; Reidy, *Tides of History*; Benson, *Wired Wilderness*; Edwards, *A Vast Machine*; Vetter, *Knowing Global Environments*; Sayre, *Politics of Scale*; De Bont and Lachmund, *Spatializing the History of Ecology*; LeCain, *Matter of History*, 112–14; Wohlleben, *Hidden Life of Trees*.
36. Demuth, *Floating Coast*; Black, *Global Interior*; Raby, *American Tropics*; Cushman, *Guano and the Opening of the Pacific World*; Ekbladh, *Great American Mission*; Cullather, *Hungry World*.
37. Chakrabarty, "Climate of History"; Malm and Hornborg, "Geology of Mankind?"; LeCain, "Against the Anthropocene"; Haraway, "Anthropocene, Capitalocene, Plantationocene, Chthulucene"; Wright, "Geophysical Agency in the Anthropocene"; Davis and Todd, "On the Importance of a Date."
38. Fiege, "Weedy West"; Wakild, "Border Chasm"; Chester, *Conservation Across Borders*.
39. Dogliani, "Fate of Socialist Internationalism"; Clavin, "Men and Markets"; Herren, "Fascist Internationalism"; Vik, "Indigenous Internationalism"; Estes, *Our History Is the Future*, 201–45; Tompkins, "'Ecological Internationale?'"
40. Anderson, *Imagined Communities*; Hahn, *Nation without Borders*; Grandin, *End of the Myth*; Miller, *Empire of Borders*.

Chapter 1

1. Elva Wineman, "Famous White Killer Wolf Lived an Almost Unbelievable Career," *Democrat News* (Lewistown, MT), 11 May 1930; Walter, *Montana Campfire Tales*, 84–86.
2. Limerick, *Legacy of Conquest*, 17–32; Coleman, *Vicious*, 1–15.
3. Young, *Last of the Loners*; T. P. Strode to R. E. Bateman, 17 April 1926, in R. E. Bateman, "Annual Report, Fiscal Year 1925, U.S. Biological Survey Cooperating with the

State Game and Fish Department of Montana in the Control and Eradication of Predatory Animals," Container 21, Entry # P 179, NARA-RG22.
4. 1900 U.S. Census, Choteau County, Montana, John E. Oswood; R. E. Bateman, "Narrative Report for Month of March 1926," Container 21, Entry # P 179, NARA-RG22.
5. T. P. Strode to R. E. Bateman, 17 April 1926, Container 21, Entry # P 179, NARA-RG22.
6. Benton-Cohen, *Borderline Americans*, 1–17.
7. Gomer D. Thomas to R. E. Bateman in R. E. Bateman, "Narrative Report for Month of December 1926," Container 25, Entry # P 179, NARA-RG22.
8. Jones, *Wolf Mountains*; Robinson, *Predatory Bureaucracy*.
9. Rutherford, *Villain, Vermin, Icon, Kin*, 51, 66.
10. McDougall, *Pathfinding on Plain and Prairie*, 114.
11. The Scribbler [James Willard Schulz], "The Eagle Creek Wolfers," *Forest and Stream*, 5 January 1901, 2.
12. Kirk Munroe, "A War of Extermination," *Harper's Weekly*, 2 August 1884, 499–500; Isenberg, *Destruction of the Bison*, 26; Hampton, *Great American Wolf*, 22; Wayne et al., "Mitochrondrial DNA Variability of the Gray Wolf," 565; Wise, *Producing Predators*, 18.
13. Wise, "Killing Montana's Wolves," 53–54.
14. Taylor, *Twenty Years on the Trap Line*, 69.
15. Taylor, *Twenty Years on the Trap Line*, 70–71; Hampton, *Great American Wolf*, 108; Robinson, *Predatory Bureaucracy*, 18.
16. William McD. Tait, ed., "'I Remember': Recollections of 'Kootenai' Brown Related Verbatim to W. D. T.," Glenbow Foundation typescript copy, 5, UC-Glenbow.
17. Tait, "'I Remember,'" 4, 9; Rodney, *Kootenai Brown*; MacKinnon, *The Identities of Marie Rose Delorme Smith*, 48–52.
18. Tait, "'I Remember,'" 15–38; Hogue, *Metis and the Medicine Line*, 66–68.
19. Tait, "'I Remember,'" 46, 53.
20. McManus, *Line Which Separates*, 7–16.
21. Tait, "'I Remember,'" 54.
22. Curnow, "History of the Eradication of the Wolf," 28–30; US Department of Labor, *History of Wages*, 328–29; "The Bounty Question—Views of a Prominent Montana Stockgrower," *River Press* (Fort Benton, MT), 2 January 1889.
23. "A Bloody Deed," *Benton Record* (Fort Benton, MT), 4 May 1877.
24. Rodney, *Kootenai Brown*, 109–17.
25. "Local News," [Fort] *Benton Record* (Fort Benton, MT), 16 November 1877; Tait, "'I Remember,'" 59.
26. Dempsey, *Firewater*, 2.
27. Sharp, *Whoop-Up Country*, 55–77; Hogue, *Metis and the Medicine Line*, 72–73, 83–84.
28. Graybill, *Policing the Great Plains*, 12–16, 29–33, 37–43; Batty, *How to Hunt and Trap*, 214–17; Royal North-West Mounted Police, "Dominion Wolf Bounty Return," January-December 1913, File No. 47, Vol. 455, Record Group 18: Records of the Royal Canadian Mounted Police, LAC; Maxwell Grant [Graham], "Revelstoke Park," 15 October 1915, File No. 401-1, Vol. 1201, Record Group 85: Records of the North Affairs Program, LAC.
29. Marie Rose Smith, "Kootenai Brown," 5, M-1154-5, Marie Rose Smith fonds, UC-Glenbow.

30. Tait, "'I Remember,'" 84–85; Rodney, *Kootenai Brown*, 166–75; Lothian, *Brief History of Canada's National Parks*, 29.
31. Tait, "'I Remember,'" 59.
32. Brown, "Report of the Superintendent of Waterton Lakes Park," 45.
33. Rodney, *Kootenai Brown*, 201–2; Jones, *Wolf Mountains*; MacDonald, *Where the Mountains Meet the Prairies*, 41–69.
34. Tim Golden, "Big-Game Hunter's Gift Roils the Smithsonian," *New York Times*, 17 March 1999; Post, *Who Own's America's Past?*, 226–61.
35. Philip, "Institutionisation of Poison," 129–39; Paddle, *Last Tasmanian Tiger*; Walker, *Lost Wolves of Japan*, 165–67; Beinart, "Night of the Jackal"; Weaver, *Great Land Rush*; Richards, *Unending Frontier*; Virginia DeJohn Anderson, *Creatures of Empire*.
36. McManus, *Line Which Separates*, 66.
37. McManus, *Line Which Separates*, 83–105; Robbins, *Colony and Empire*; Spence, "Crown of the Continent, Backbone of the World"; Brado, *Cattle Kingdom*, 37.
38. Carl, "Montana Wolves and Panthers," *Forest and Stream*, 22 July 1886, 508.
39. *Report on the Statistics of Agriculture*, 75; Stelfox, "Wolves in Alberta," 19; Hampton, *Great American Wolf*, 113.
40. Elofson, *Cowboys, Gentlemen, and Cattle Thieves*, 8–16; Brado, *Cattle Kingdom*, 15.
41. White, "Animals and Enterprise"; Malone, Roeder, and Lang, *Montana*, 153–63; Gressley, *Bankers and Cattlemen*, 107; Craig, *Ranching with Lords and Commons*; Brisbin, *Beef Bonanza*.
42. Robinson, *Predatory Bureaucracy*, 27–28; Cronon, *Nature's Metropolis*, 218–24.
43. Elofson, *Cowboys, Gentlemen, and Cattle Thieves*, 6, 161n17.
44. "By-Laws" and "Report, Committee on Bulls" (April 1885), 13, 31, Montana Stock Growers Association Minute Book, Vol. 8, Montana Stockgrowers Association Records, Montana Historical Society Research Center, Helena.
45. "Meeting of the Executive Committee," 19 August 1897, 30–31, Folder 3: Minutes, 1896–1972, Box 1, Western Stockgrowers Association fonds, UC-Glenbow.
46. Elofson, *Frontier Cattle Ranching in the Land and Times of Charlie Russell*, 44–45; W. F. Cochrane to father, 28 January 1885, 47–48, Series 1: Letterbook, Cochrane Ranche Company Ltd. fonds, UC-Glenbow; "Minutes of the Semi-Annual Meeting of the Montana Stock Growers Association," 17 April 1888, 178–79, Montana Stockgrowers Association Records, Montana Historical Society Research Center, Helena.
47. "Fifth Annual Meeting of the Western Stock Growers' Association," 11 April 1901, 105, Folder 3: Minutes, 1896–1972, Box 1, Western Stockgrowers Association fonds, UC-Glenbow; Elofson, *Cowboys, Gentlemen, and Cattle Thieves*, 94; W. H. McKay, "Recollections as a Wolf Hunter," *Calgary Herald*, 15 January 1944.
48. White, "Animals and Enterprise," 252–69; Netz, *Barbed Wire*, 1–49; Nimmo, *Report in Regard to the Range*, 42–44.
49. Warren, *Conrad Kohrs*, 80–81; Curnow, "History of the Eradication of the Wolf."
50. Wise, *Producing Predators*, 33–45; Kelly, *Range Men*, 177.
51. "Bounty on Wolves," *River Press* (Fort Benton, MT), 12 December 1888.
52. Robinson, *Predatory Bureaucracy*, 37; Wise, "Killing Montana's Wolves," 56–65; *Second Annual Report of the Bureau of Agriculture, Labor, and Industry*, 124–27; *Fifth Annual Report of the Bureau of Agriculture, Labor, and Industry*, 128–29.
53. *Annual Report of the Department of Agriculture of the North-West Territories 1898*, 86.
54. *Annual Report of the Department of Agriculture of the North-West Territories 1899*,

77–78; "Wolf Bounties," *Macleod (Alberta) Gazette*, 26 May 1899; Wetherell, *Wildlife, Land, and People*, 260–64.
55. *Annual Report of the Department of Agriculture of the Province of Alberta . . . 1905, to . . . 1906*, 37; Curnow, "History of the Eradication of the Wolf," 88.
56. Morton, *Short History of Canada*, 166–67; *Annual Report of the Department of Agriculture of the Province of Alberta 1907*, 24.
57. *Annual Report of the Department of Agriculture of the Province of Alberta 1907*, 24; *Annual Report of the Department of Agriculture of the Province of Alberta 1908*, 26.
58. Van Nuys, *Varmints and Victims*, 40–44.
59. Robinson, *Predatory Bureaucracy*, 45–55; Pinchot, *Breaking New Ground*, 86, 177.
60. Smith et al., *Yellowstone Wolf Project: Annual Report 2011*, 3; Smith et al., *Yellowstone National Park Wolf Project: Annual Report 2017*, 3–5; Cassidy et al., "Territoriality and Inter-Pack Aggression in Gray Wolves," 37–42; Mech and Boitani, *Wolves*, 19–34; Bryan et al., "Heavily Hunted Wolves"; Stahler et al., "Wolf Turf."
61. Radkau, *Nature and Power*.
62. Smalley, *Wild by Nature*, 81–118; Bailey, *Wolves in Relation to Stock*, 5–7; Brett, *Report of the Acting*, 16; Bailey, *Animal Life of Yellowstone National Park*, 135–37; *Annual Reports for Yellowstone National Park, 1916–1919*, 38, Montana History Portal, Montana State Library, www.mtmemory.org/nodes/view/12997; Smith, *Yellowstone and the Smithsonian*.
63. S. W. McClure, "The Wolf at the Stockman's Door: Sheep and Cattle Killers Breed in the National Reserves," *Country Gentleman*, 14 November 1914, 1845–46.
64. U.S. Senate, "Agricultural Appropriation," *Congressional Record* 52 (25 February 1915): 4568–83.
65. Van Nuys, *Varmints and Victims*, 54–55; Robinson, *Predatory Bureaucracy*, 78–80.
66. Wallis Huidekoper, "The Wolf Question and What the Government Is Doing to Help," address at the Annual Convention of the Montana Stock Growers' Association, 18 April 1916, Montana Stockgrowers Association Records, Montana Historical Society Research Center, Helena.
67. J. F. Gillings, *Narrative Report of Predatory Animal Work*, October 1919, Container 5, Entry # P 179, NARA-RG22.
68. Schullery, *Yellowstone Wolf*, 37–47.
69. *Annual Report of the Department of Agriculture of Alberta 1916*, 131–33.
70. "Minutes of the Western Stock Growers Association," 11 May 1916, 17–18, Folder 3: Minutes, 1896–1972, Box 1, Western Stockgrowers Association fonds, UC-Glenbow.
71. Fish and Wildlife Historical Society, *Fish, Fur, and Feathers*, 158, 160.
72. Bailey and Merriam Bailey, *Wild Animals of Glacier National Park*, 82–83; Vernon Bailey, "Mammals: Glacier Park, Mont.," 5 July–30 August 1917, Folder: Montana -Bailey, V.o. Special Reports, 1917, Box 65, SA-FWS.
73. Stephen T. Mather to W. W. Payne, 19 July 1918; Luther J. Goldman to W. W. Payne, 9 September 1918; W. W. Payne to W. J. Dorrington, 16 December 1918; all in Microfilm Reel #1, Glacier National Park Collection, Archives and Special Collections, Mansfield Library, University of Montana, Missoula; Jones, *Wolf Mountains*, 72–77.
74. Vernon Bailey, "Glacier Park Region, Mont." 2 November 1930, Folder: Montana -Bailey, V.o. Predatory Animal Control in Montana, Washington, and Oregon, Box 65, SA-FWS.
75. Warren Sibbald, "Re. destruction of Wolves, Waterton Lakes," 18 February 1919; and

J. B. Harkin to W. W. Cory, memorandum, 25 August 1922, both in Microfilm Reel T-12918, LAC.
76. Wise, *Producing Predators*, 38–39; R. E. Bateman, "Narrative Report for Month of December 1926," Container 25, Entry # P 179, NARA-RG22.
77. Tsing, *Mushroom at the End of the World*.
78. Snow, *These Mountains Are Our Sacred Places*, 9; Philip H. Godsell, ed., *The R. N. Wilson Papers* (Calgary, AB: Glenbow Foundation, 1958), 93–95, 322–24, UC-Glenbow; Hampton, *Great American Wolf*, 32–35; Secretary Tom Lean to Dept. of Agriculture, 20 July 1908, File No. 195485, Vol. 1073, Record Group 17: Records of the Department of Agriculture, LAC; Treaty Inspector to Deputy Superintendent General, memorandum, 10 February 1908, File No. 278, Vol. 4019, Record Group 10: Records of Indian Affairs, LAC.
79. Maxwell Graham to Commissioner of the Dominion Parks Branch, memorandum, 15 October 1915, File No. 1915–988, Vol. 1163, RG 25: Department of External Affairs, LAC.
80. Maxwell Graham, "Re. Wolf Situation in Waterton Lakes Park," 15 February 1921; and Henry Riviere, "Report on Wolves in the South End of Waterton Lakes Park," 1 June 1921, both in Microfilm Reel T-12918, LAC.
81. Maxwell Graham to Commissioner of the Dominion Parks Branch, memorandum, 15 October 1915, File No. 1915–988, Vol. 1163, RG 25: Department of External Affairs, LAC.
82. Curnow, "History of the Eradication of the Wolf," 75–81; Green, *Wolves of Banff National Park*, 41–44; Jones, *Wolf Mountains*, 70–71.
83. J. S. Jackson to William Jennings Bryan, 23 April 1915, File No. 401-1, Vol. 1201, Record Group 85: Records of the North Affairs Program, LAC.
84. William Jennings Bryan to Cecil Arthur Spring-Rice, 10 May 1915; Lt. Gov. George H. V. Bulyea of Alberta to the Sec. of State in Ottawa, 10 August 1915; D. F. Houston to Secretary of State, 13 December 1915; Cecil Spring Rice to Governor General of the Dominion Canada, 9 October 1916; Cecil Spring Rice to Lord Duke of Devonshire, 22 November 1916, all in File No. 1915-988, Vol. 1163, RG 25: Department of External Affairs, LAC; Fiege, "Weedy West."
85. H. A. Craig to J. H. Grisdale, 6 June 1919, File No. 268610, Vol. 1350, Record Group 17: Records of the Department of Agriculture, LAC.
86. C. Gordon Hewitt, "The Control of Wolves and Coyotes in Western Canada," 26 April 1919, File No. 268610, Vol. 1350, Record Group 17: Records of the Department of Agriculture, LAC.
87. Norman Criddle, *The Habits and Economic Importance of Wolves*, Bulletin No. 13 (Ottawa: Department of Agriculture, 1925), 17–18, UC-Glenbow.
88. J. B. Harkin to W. W. Cory, memorandum, 25 August 1922, Microfilm Reel T-12918, LAC.
89. E. W. Nelson, "Report of Chief of Bureau of Biological Survey," Fiscal Year 1925, Entry # E 232, Box 2, NARA-RG22.
90. E. W. Nelson, "Report of Chief of Bureau of Biological Survey," Fiscal Year 1924, Entry # E 232, Box 2, NARA-RG22.
91. R. E. Bateman, "Narrative Report of Predatory Animal Work," September 1919, Container 5, Entry # P 179, NARA-RG22.
92. Young, *Last of the Loners*.

93. E. Bateman to Dr. Bell, 11 October 1919, Container 5, Entry # P 179, NARA-RG22.
94. R. E. Bateman to Dr. Bell, 11 October 1919, Container 5; R. E. Bateman to E. W. Nelson, 18 July 1921, Container 8; R. E. Bateman, "Annual Report for Fiscal Year. July 1.1921. To June 30.1922," Container 10; R. E. Bateman, "Annual Report, Fiscal Year 1923, U.S. Biological Survey Cooperating with the State Game and Fish Department of Montana in the Control and Eradication of Predatory Animals," Container 13, all in Entry # P 179, NARA- RG22.
95. Vernon Bailey, "Predatory Animal Problems in Montana," 1930; Vernon Bailey, "Field Inspections in Montana, Washington, and Oregon from October 17 to December 16, 1930"; and Vernon Bailey, "Big Belt Mountains (Lingshire), Mont." 2 November 1930, all in Folder: Montana -Bailey, V.o. Special Reports, 1924–30, Box 65, SA-FWS; R. E. Bateman, "Annual Report for Fiscal Year. July 1.1921. To June 30.1922," Container 10, Entry # P 179, NARA-RG22; Netz, *Barbed Wire*, xi–xiv.
96. Vernon Bailey, "Big Belt Mountains (Lingshire), Mont.," 2 November 1930, Folder: Montana-Bailey, V.o. Special Reports, 1924–30, Box 65, SA-FWS.
97. R. E. Bateman, "Narrative Report for Month of September 1926," Container 21, Entry # P 179, NARA-RG22.
98. Vernon Bailey, "Big Belt Mountains (Lingshire), Mont." 2 November 1930, Folder: Montana-Bailey, V.o. Special Reports, 1924–30, Box 65, SA-FWS; Dunlap, *Saving America's Wildlife*.
99. R. E. Bateman, "Progress Report Calendar Year 1925," Container 21, Entry # P 179, NARA-RG22; R. E. Bateman, "Narrative Report for Month of March 1927," Container 25, Entry # P 179, NARA-RG22; King, *Reasons for the Decline of Game*, 103–5; R. E. Bateman, "Report of Leader Predatory Animal Control of the Montana District," 10 July 1930, Container 30, Entry # P 179, NARA-RG22.
100. Vernon Bailey, "Predatory Animal Problems in Montana," 1930, Folder: Montana-Bailey, V.o. Special Reports, 1924–30, Box 65, SA-FWS.
101. Stelfox, "Wolves in Alberta."
102. Wonderley, "Most Utopian Industry"; Harding, *Steel Traps*, 50–64; Gerstell, *Steel Trap in North America*, 196–98, 203, 209; Hampton, *Great American Wolf*, 137.
103. R. E. Bateman, "Annual Report for Fiscal Year 1925," Container 18, Entry # P 179, NARA- RG22; Netz, *Barbed Wire*.
104. Van Nuys, *Varmints and Victims*; Coleman, *Vicious*; US Fish and Wildlife Service, *Northern Rocky Mountain Wolf Recovery Plan* (1987), 9.
105. Douglas W. Smith, interview by author, 18 July 2018.
106. Smith interview, July 18, 2018.

Chapter 2

1. Hilty et al., *Corridor Ecology*; Benson, *Wired Wilderness*, 47–48.
2. Cornelia Dean, "Wandering Wolf Inspires Project," *New York Times*, 23 May 2006.
3. Boyd et al., "Transboundary Movements."
4. Jodi A. Hilty, interview by author, 11 September 2018.
5. Lewis, *Inventing Global Ecology*.
6. Banff Superintendent, "Re: Wolves, Banff National Park," 5 January 1945, Microfilm Reel T-10422, LAC.
7. Jones, *Wolf Mountains*, 116–23; Banff National Park, "Re: Wolves, Banff National

Park;—Winter 1944–5," 16 April 1945, Folder 27, Whyte-PC; "Notes on the Occurrence of Wolves in the Banff Nat. Park," 29 January 1948, Folder 27, Whyte-PC.

8. Chief of Dominion Wildlife Service to Marshall McLean, 24 January 1948, Microfilm Reel T-9724, LAC.
9. W. C. Fisher, "Sabotage in the National Parks," *Hunting and Fishing in Canada*, October 1942; McTaggert Cowan, "Timber Wolf in the Rocky Mountain National Parks of Canada"; Alan MacEachern, "Ian McTaggart-Cowan in Banff & Jasper: Bringing Wildlife Science to the National Parks," NiCHE, 29 April 2016, http://niche-canada.org/2016/04/29/ian-mctaggart-cowan-in-banff-jasper-introducing-wildlife-science-to-the-national-parks/.
10. MacEachern, "Sentimentalist"; Tony Lascelles, "Predator vs. Game in National Parks," *Hunting and Fishing in Canada*, February 1943, 13; "The Case for the Predator," "The Predator—Friend or Foe," and "Wildlife in National Parks" in Hubert Green fonds, Whyte Museum of the Canadian Rockies.
11. Hampton, *Great American Wolf*, 150–51; Leopold, *Sand County Almanac*, 137–40; Leopold, *Game Management*, 34–36, 54–57, 230–52.
12. MacEachern, "Sentimentalist," 30; Banff National Park, "Wild Life Notes," April 1956, Whyte-PC; Banff National Park, "Wild Life Notes," January 1957, Whyte-PC.
13. Colpitts, "Howl," 219–53.
14. Colpitts, "Howl," 219–53.
15. Jones, *Wolf Mountains*, 133; Banff National Park, "Wild Life Notes," December 1954, Folder 24, Whyte-PC; Banff National Park, "Wild Life Notes," May 1955, Folder 24, Whyte-PC.
16. "FGA Officials Protest 'Protection' of Wolves," *Calgary Herald*, 7 December 1959.
17. Carson, *Silent Spring*, 6; Lear, *Rachel Carson*, 306–46, 373–74, 428–38.
18. Guha, *Environmentalism*, 93–132; Rome, *Genius of Earth Day*.
19. MacDowell, *Environmental History of Canada*, 243–85; Monte Hummel and Erin James-Abra, "Environmental and Conservation Movements," *Canadian Encyclopedia*, accessed October 10, 2019, www.thecanadianencyclopedia.ca/en/article/environmental-and-conservation-movements.
20. Mowat, *Never Cry Wolf*, 76; Rutherford, *Villain, Vermin, Icon, Kin*, 97–101.
21. Karen Jones, "Never Cry Wolf," 81.
22. Robert Guest, *In Defence of Wolves* (Edmonton, AB: Canadian Wolf Defenders, 1970), UC-Glenbow; Canadian Wolf Defenders newsletters, Fall 1969–Fall 1970, LAC.
23. Kerr in *Calgary Herald*, 8 November 1971, quoted in Canadian Wolf Defenders newsletter, Spring 1972, LAC.
24. Province of British Columbia, *Preliminary Wolf Management Plan*, 13–21; Loo, *States of Nature*, 149–82.
25. Van Nuys, *Varmints and Victims*, 182–97; Hampton, *Great American Wolf*, 177–80.
26. Gordon Aalborg, "Pipeline from the Peace: He's a Fanatic Fan of Wolves," *Edmonton Journal*, 2 August 1971.
27. Douglas H. Pimlott, "Wolves in the World," *National Parks and Conservation Magazine*, August 1972, 18; Pimlott, *Wolves*, 89–102, 113–19.
28. Flight Sheet—Wolf Ecology Project, 28 June 1980, and 1 July 1980, Folder: Untitled, Box 2, Ream Papers.
29. Brian Kennedy, "Wolf Stars on Radio," *Hungry Horse News*, 14 June 1979; Robert Ream, "Postscript to Quarterly Report!!!," 11 April 1979, Folder: WEP Correspondence

(letters from Bob), Box 1, Ream Papers; "Wolf Sign Data," 10 October 1979, Folder: 1979 B. C., Box 2, Ream Papers.
30. Jack Nisbet, "The Track of the Wolf," *Seattle Weekly*, 9–15 July 1986, Folder: Wolf Articles, Misc. 86–87, Box 1, Ream Papers; Ream, "Bob Ream—Before and After the Millennium," unpublished autobiography, 2016, in author's possession.
31. Ream, "Bob Ream."
32. Ursula Mattson, interview by author, 21 August 2019; Brian Kennedy, "Elusive Is Common Term in Wolf Study," *Hungry Horse News*, 19 April 1979; Robert Ream, Richard Harris, and Ursula Mattson, "Wolf Ecology Project Progress Report," January–September 1977; Ursula Mattson, "Wolf Ecology Project," 1 October 1977–15 April 1978 progress report; and Ursula Mattson and Robert R. Ream, "Wolf Ecology Project," 1978 Annual Report, all in Folder: Progress Reports '77 & '78, Box 1, Ream Papers; Carly Holmquist, "Scientists Like Cry of 'Wolf!'" *Miles City (MT) Star*, 4 December 1981.
33. Diane K. Boyd, interview by author, 4 September 2019; Myers Reece, "The Jane Goodall of Wolves," *Flathead Beacon* (Kalispell, MT), 27 March 2017; Perri Knize, "The Woman Who Runs with the Wolves," *Sports Illustrated*, 18 October 1993; John Aloysius Farrell, "The Woman and the Wolf," *Empire: A Magazine of the West*, 15 December 1985, Folder: Wolf Articles, Misc. 86–87, Box 1, Ream Papers; Diane Boyd to Robert R. Ream, 5 November 1978, Folder: Personnel File, Box 2, Ream Papers.
34. Boyd interview.
35. Boyd interview; Boyd, "Food Habits and Spatial Relations"; "Statewide Endangered Species Research," Folder: Wolf Project, Box 2, Ream Papers.
36. Mech and Boitani, "Wolf Social Ecology."
37. Robert Ream and Joseph Smith, "Wolf Ecology Project," 1 January–31 March 1980 progress report, Folder: Winter '80 Quarterly Report, Box 1, Ream Papers; Ursula Mattson, "A Review of the Wolf Population in the Rocky Mountain Region of the Canadian/U.S. Border," unpublished report, Folder: Ursula's Orig. Wolf Pop, Box 1, Ream Papers; Robert Ream to John Spinks, 29 July 1980, Folder: WEP Correspondence (letters from Bob), Box 1, Ream Papers.
38. Robert Ream to Max Baucus, 6 September 1978, Folder: WEP Correspondence (letters from Bob), Box 1, Ream Papers; Turner, *Promise of Wilderness*.
39. Province of British Columbia, *Preliminary Wolf Management Plan*, 13–21; Robert Ream to Tom Coston, 24 March 1980, Folder: WEP Correspondence (letters from Bob), Box 1, Ream Papers; Robert Ream, "Enhancement of an International Wolf Population in the Northern Rockies," date unknown, Folder: Border Wolf Technical Committee, Box 2, Ream Papers; Mattson interview.
40. Boyd interview; Farrell, "Woman and the Wolf"; Bob Ream, "The Wolf Is at Our Door: Population Recovery in the North Rockies," *Western Wildlands*, Summer 1984, 2–7.
41. Ursula Mattson, "Montana Wolf Magic," unpublished essay, 1987, in author's possession.
42. Mattson, "Montana Wolf Magic"; Jones, *Wolf Mountains*, 86–91; Robbins, "Wolves Across the Border."
43. Mattson, "Montana Wolf Magic"; Jones, *Wolf Mountains*, 86–91; Jim Robbins, "Wolves Across the Border," *Natural History*, May 1986, 6–16.
44. Boyd et al., "Transboundary Movements"; Dave Cooper, "Rancher's Bullet Ends Wolf's Trek North," *Edmonton Journal*, 24 July 1987.
45. Robert R. Ream to Superintendent Gilbert Lusk, 24 July 1987, Folder: New Yorker article with attached correspondence, Box 4, Ream Papers.

46. US Fish and Wildlife Service, *Northern Rocky Mountain Wolf Recovery Plan* (1980); U.S. Fish and Wildlife Service, *Northern Rocky Mountain Wolf Recovery Plan* (1987), 10–33.
47. Ursula Mattson, "Canadian Wolves 'Packing' to Go South?" *Equinox Magazine* [draft], 1982, Folder: Correspondence to Bob, Box 2, Ream Papers.
48. Hampton, *Great American Wolf*, 205; Jones, *Wolf Mountains*, 88.
49. Robbins, "Wolves Across the Border," 14.
50. John Howe, *Return of the Wolves*, aired on Montana PBS, recorded 27 January 1989, Box 4, Ream Papers.
51. Howe, *Return of the Wolves*.
52. George Wuerthner, "Ranchers Fear the Wolf at Their Door," *High Country News*, 15 September 1986.
53. Hampton, *Great American Wolf*, 180; Van Nuys, *Varmints and Victims*, 189; "Gray Wolf History," Montana Fish, Wildlife, and Parks, accessed November 21, 2019, http://fwp.mt.gov/fishAndWildlife/management/wolf/history.html.
54. Hampton, *Great American Wolf*, 180; Van Nuys, *Varmints and Victims*, 189; McIntyre, "1992."
55. Fischer, *Wolf Wars*, 101–2.
56. Fischer, "1995"; Defenders of Wildlife, "Wolf Compensation Trust," 2009, in author's possession.
57. Bangs, "Return of a Predator"; Bangs, "Operation Wolfstock"; Hampton, *Great American Wolf*, 210–20; Van Nuys, *Varmints and Victims*, 195–219; Jones, *Wolf Mountains*, 50–55, 195–98.
58. Jones, *Wolf Mountains*, 222.
59. Soulé, "What Is Conservation Biology?"; Robin, Sörlin, and Warde, *Future of Nature*, 405–6; Fraser, *Rewilding the World*, 1–34; Dave Foreman, John Davis, David Johns, Reed Noss, and Michael Soulé, "The Wildlands Project: Mission Statement," *Wild Earth*, Special Issue, June 1992, 1–6, https://wildlandsnetwork.org/resource-topics/wild-earth-journal/page/4/.
60. Soulé and Terborgh, *Continental Conservation*; Hannibal, *Spine of the Continent*, 195–209; Fancy, *Satellite Telemetry*, 1–3. ARGOS stands for Advanced Research and Global Observation Satellite.
61. Paul C. Paquet, interview by author, 30 August 2019; Trish Pryce, "Paul Paquet—A Catalyst in Action," *Wilderness Alberta*, June 1992, 8–9, 16, Whyte Archives and Library.
62. Paquet interview; Diane K. Boyd et al., "Transboundary Movements."
63. Paquet interview.
64. Boyd interview.
65. Boyd-Heger, "Dispersal Genetic Relationships," 1–40; Don Thomas, "Alberta Wolf Control Hurting U.S. Project," *Edmonton Journal*, 28 April 1990; Chris Dawson, "Accidents, Shootings Cripple Research Work," *Calgary Herald*, 4 November 1990.
66. Paquet interview; Boyd, "1995," 365, emphasis added.
67. Hannibal, *Spine of the Continent*, 79–81; Chester, *Conservation Across Borders*, 143–48.
68. Chester, *Conservation Across Borders*, 143.
69. Harvey Locke, "Yellowstone to Yukon," 1993, Document 300, Box 2, Whyte-Y2Y. Later published as Harvey Locke, "Preserving the Wild Heart of North America: The Wildlands Project and the Yellowstone to Yukon Biodiversity Strategy," *Borealis*, 1994, 18–24.

70. Locke, "Yellowstone to Yukon."
71. Hilty interview; Hilty, "Use of Riparian Corridors"; Fraser, *Rewilding the World*, 38–42.
72. Hilty interview; Hilty, "Use of Riparian Corridors."
73. Chester, *Conservation Across Borders*, 189.
74. Gary M. Tabor, "Yellowstone-to-Yukon: Canadian Conservation Efforts and Continental Landscape/Biodiversity Strategy," July 1996, Document 270, Box 2, Whyte-Y2Y.
75. Meeting summary of the Yellowstone to Yukon Conservation Initiative, Kananaskis, Alberta, 13–14 April 1996; Meeting minutes of the Yellowstone to Yukon Network, Kananaskis Field Station, Alberta, 1–2 February 1997; Meeting minutes of the Yellowstone to Yukon Conservation Initiative, Kalispell, Montana, 26–27 April 1997; all in Document 190, Box 2, Whyte-Y2Y.
76. "Connections; The First Conference of the Yellowstone to Yukon Conservation Initiative," 2–5 October 1997, Document 188, Box 2, Whyte-Y2Y; Howard Schneider, "Conservationists Take Stock of the Land," *Washington Post*, 27 October 1997.
77. Stephen Legault, "Parks May Contribute to Genetic Isolation," *Wild Lands Advocate*, September 1997, 4, Document 181, Box 1, Whyte-Y2Y.
78. Chester, *Conservation Across Borders*, 140; Debarbieux and Rudaz, *Mountain*, 265–83.
79. Louisa Willcox, Bart Robinson, and Ann Harvey, eds., *A Sense of Place: An Atlas and Resources in the Yellowstone to Yukon Ecoregion* (Canmore, AB: Yellowstone to Yukon Conservation Initiative, 1998), 31–50, 65–69, Document 169, Box 1, Whyte-Y2Y; Tom Lee and Robert G. Stanton, "Memorandum of Understanding between the NPS and Parks Canada on Cooperation in Management, Research, Protection, Conservation, and Presentation of National Parks and National Historic Sites," May 1998, Document 170, Box 1, Whyte-Y2Y.
80. "The Yellowstone to Yukon Biodiversity Strategy," 1993, VHS Tape #1, produced by the Canadian Parks and Wilderness Society and the Wildlands Project, Whyte-Y2Y.
81. Harvey Locke, "Yellowstone to Yukon: Eight Years Later," 8 February 2003, Document 399, Box 3, Whyte-Y2Y.
82. "A Y2Y Network Handbook of Principles, Policies, and Guidelines," 27 April 1997, Document 251, Box 3, Whyte-Y2Y; David Johns, "Protecting the Wild Heart of North America: The Politics of Y2Y"; "Y2Y Workplan: Overview"; and "Yellowstone to Yukon Conservation Initiative—Meeting Summary"; all in Document 190, Box 2, Whyte-Y2Y.
83. Lothian, *Brief History of Canada's National Parks*, 33–35; Gary M. Tabor, "Yellowstone-to-Yukon: Canadian Conservation Efforts and Continental Landscape/Biodiversity Strategy," July 1996, Document 270, Box 2, Whyte-Y2Y; Cathy Ellis, "Wolf Expert May Sue Parks Canada," *Banff (Alberta) Crag and Canyon*, 26 May 1999; and Scott Crowson, "Banff's Wildlife Corridors Attract U.S. Researchers," *Calgary Herald*, 11 September 2002; both in Newspaper Clippings: Wolves, UC-Glenbow.
84. Paul C. Paquet et al., "Wildlife Corridors in the Bow River Valley, Alberta: A Strategy for Maintaining Well-Distributed, Viable Populations of Wildlife," report to the Bow River Valley Corridor Task Force, 1994, 1–38; Tony Clevenger, "Highways Through Habitats: The Banff Wildlife Crossings Project," *TR News*, Transportation Research Board, March–April 2007, 14–17, https://onlinepubs.trb.org/onlinepubs/trnews/trnews249hwyhabitats.pdf; Cathy Ellis, "Wildlife Crossing Data Deciphered After 17 Years," *Rocky Mountain Outlook*, 19 February 2015, all in Newspaper Clippings: Banff National Park, 2010–2019, UC-Glenbow; Fraser, *Rewilding the World*, 35.

85. Chester, "Responding to the Idea of Transboundary Conservation," 111–16.
86. Chester, "Responding to the Idea of Transboundary Conservation," 118.
87. Karsten Heuer and Erica Heuer, *Yellowstone to Yukon Hike: Final Report*, December 1999, Document 177, Box 2, Whyte-Y2Y; "Yellowstone to Yukon Conservation Initiative, Meeting of the Council," Kalispell, Montana, 26–27 April 1997, Document 190, Box 2, Whyte-Y2Y.
88. *The West Wing*, season 1, episode 5, "The Crackpots and These Women," directed by Anthony Drazen, aired 20 October 1999 on NBC.
89. Chester, *Conservation Across Borders*, 173.
90. National Parks and National Historic Sites of Canada, "Nahanni National Park Reserve of Canada: Park Expansion Memo of Understanding," June 2003, https://dehcho.org/wp-content/uploads/2022/03/DFN.24.06.03.NahanniParkMOU-5.pdf; Norwegian in Locke, "International Movement to Protect Half the World," 3; Weaver, *Big Animals and Small Parks*.
91. National Parks and National Historic Sites of Canada, "Nahanni National Park Reserve of Canada: Interim Park Management Arrangement," June 2003, https://dehcho.org/wp-content/uploads/2022/03/DFN_NEG_NS_NahanniInterimParkManagementArrangement_2016_06_03-1.pdf, Nahanni National Park Reserve of Canada: Management Plan," June 2010, https://dehcho.org/docs/DFN_CR_Nahanni_MP_2010_01_30.pdf.; International Union for the Conservation of Nature, "Expansion of the Nahanni National Park Reserve: Parks Canada," June 24, 2009, www.iucn.org/content/expansion-nahanni-national-park-reserve-parks-canada.
92. Yellowstone to Yukon Conservation Initiative, *Yellowstone to Yukon Vision*, 18–19; Wilson, *Half-Earth*.
93. "The Yellowstone to Yukon Biodiversity Strategy," 1993, VHS Tape #1, produced by the Canadian Parks and Wilderness Society and the Wildlands Project, Whyte-Y2Y.
94. Smith and Ferguson, *Decade of the Wolf*, 140, 190, 221.
95. Van Nuys, *Varmints and Victims*, 237–44; Frank Clifford, "Wolves and the Balance of Nature in the Rockies," *Smithsonian Magazine*, February 2009; Defenders of Wildlife et al. v. H. Dale Hall et al. 565 F. Supp. 2d 1160 (D. Mont. 2008).
96. Rob Chaney, "Judge: How Do We Know Whether Wolf Population Is Recovered?" *Helena (MT) Independent Record*, 16 June 2010; Ray Ring, "How Green Is Judge Molloy?," *High Country News*, 25 August 2010; Christopher Ketchum, "Wolves to the Slaughter," *American Prospect*, 13 March 2012; Nuys, *Varmints and Victims*, 241–42.
97. Montana Code Annotated 2023, 87-5-132 (MCA 2023).
98. Jim Robbins, "As It Goes High-Tech, Wildlife Biology Loses Its Soul," *High Country News*, 17 December 2012; "USDA-Collared Judas Wolves Used over and over to Lead Killers to Their Families," Timber Wolf Information Network, accessed February 16, 2020, www.timberwolfinformation.org/usda-collared-judas-wolves-used-over-and-over-to-lead-killers-to-their-families/ (page removed).
99. Fraser, *Rewilding the World*, 51; DeCesare et al., "Wolf-Livestock Conflict," 712; Ramler et al., "Crying Wolf?"; Richard Conniff, "Getting Ranchers to Tolerate Wolves—Before It's Too Late," TakePart, 31 January 2014, https://strangebehaviors.wordpress.com/2014/02/01/getting-ranchers-to-tolerate-wolves-before-its-too-late/; Grant, "Ranches with Wolves."
100. Lisa Upson, interview by author, 4 February 2020.

101. Wilson, Bradley, and Neudecker, "Learning to Live with Wolves"; Douglas H. Chadwick, "Wolf Wars," *National Geographic*, March 2010.
102. Wilson, Bradley, and Neudecker, "Learning to Live with Wolves"; Chadwick, "Wolf Wars."
103. Suzanne Goldenberg, "How America Is Learning to Live with Wolves Again," *Guardian*, 8 December 2010.
104. Malou Anderson-Ramirez, interview by author, 25 February 2020.
105. Anderson-Ramirez interview.
106. Anderson-Ramirez interview; Nathan Rott, "Montana Ranchers Learn Ways to Live with Wolves," *Weekend Edition Saturday*, NPR, 8 February 2014, www.npr.org/2014/02/08/273577607/montana-ranchers-learn-ways-to-live-with-wolves; Kristina Johnson, "These Montana Ranchers Are Helping Grizzlies and Cattle Coexist," Food and Environment Reporting Network, 28 November 2017, https://thefern.org/2017/11/montana-ranchers-helping-grizzlies-cattle-coexist/; Joseph Bullington, "Montana Ranchers Share Landscape with Grizzlies," *Great Falls (MT) Tribune*, 11 May 2019.
107. Anderson-Ramirez interview; "Livestock and Predators Can Coexist," Natural Resources Defense Council, 28 June 2019, www.nrdc.org/stories/livestock-and-predators-can-coexist.

Chapter 3

1. Patricio Moreno Rojas, interview by Ellen Sharp, 7 April 2020, FHS-OH.
2. "Mystery of Migration," *Washington Post*, 17 September 1911; 1910 US Census, Saint Paul, Minnesota, Lillie T. George.
3. Williams, *Americans and their Forests*, 233; Cronon, *Nature's Metropolis*, 148–206.
4. Vane-Wright, "Columbus Hypothesis," 179–88.
5. Comstock, *Comstocks of Cornell*, 59–115; Armitage, *Nature Study Movement*, 66–70, 195–209.
6. Comstock, *Handbook of Nature Study*, 1.
7. Kohlstedt, *Teaching Children Science*, 1–10, 78–84.
8. Comstock, *Ways of the Six-Footed*, 50.
9. Anna Botsford Comstock, "Butterflies and Moths," *Country Life in America*, June 1902, 50–51.
10. Comstock and Comstock, *How to Know Butterflies*, 171, 206–7; Comstock, *Ways of the Six-Footed*, 39–54; "Insects: He Fools Them," *Washington Post*, 2 September 1923.
11. Comstock, *How to Know Butterflies*, 205.
12. *Annual Report of the Entomological Society of Ontario, 1917*, 22.
13. Jennie Brooks, "The Migration of the Monarch Butterfly," *Country Life in America*, August 1911, 48, 62.
14. C. B. Hutchings, "A Note on the Monarch or Milkweed Butterfly with Special Refence to Its Migratory Habits," *Canadian Field-Naturalist*, November 1923, 150; Brower, "Understanding and Misunderstanding," 304–18.
15. Williams, *Migration of Butterflies*, 141–54, 378, 411.
16. Seitz, *Macrolepidoptera of the World*, 113.
17. José Leonel Moreno Espinoza, interview by Patricio Moreno Rojas and Ellen Sharp, 22 May 2020, FHS-OH.

18. "Resolución en el expediente de ampliación de ejidos al poblado El Capulín, Estado de México," *Diario Oficial*, 19 June 1937, 11–13; Wakild, *Revolutionary Parks*, 1–14.
19. Boyer, *Political Landscapes*, 36; Pérez Talavera, *La expotación*, 62; Moreno Espinoza interview.
20. Emiliano Zapata and Otilio Montaño, "Plan of Ayala," 28 November 1911, https://laii.unm.edu/info/k-12-educators/assets/documents/mexican-revolution/plan-of-ayala.pdf.
21. Boyer, *Political Landscapes*, 74–77.
22. Moreno Espinoza interview.
23. Juan Correa et al. to Lazaro Cardenas, 2 September 1938, 4 September 1938; and David Riqelme to Pedro Pizá Martínez, 26 September 1938, all in File Q/021/4585, Box 1442, Secretaría Particular de la Presidencia, Lázaro Cárdenas, AGN; Boyer, *Political Landscapes*, 104; Boyer, "Contested Terrain," 27–48.
24. Elidió Moreno de Jesús, interview by Patricio Moreno Rojas and Ellen Sharp, 27 April 2020, FHS-OH; Merino Pérez and Hernández Apolinar, "Destrucción de instituciones comunitarias."
25. Wakild, "Border Chasm"; Gallegos, *Áreas naturales*, 29–31; Wakild, *Revolutionary Parks*, 70–91.
26. Aridjis, *El poeta niño*, 55.
27. Beutelspacher, *Las mariposas*; Peter Menzel, "Butterfly Arms Now Under Guard in Annual Bivouac," *Smithsonian Magazine*, November 1983, 174–81; Rodriquez, "Mariposa Monarca"; Ortíz Monasterio et al., "Magnetism as a Complementary Factor."
28. Halpern, *Four Wings and a Prayer*, 67–68; Aridjis, *La montaña de las mariposas*.
29. Rzedowski, "Nota sobre un vuelo migratorio."
30. De la Maza E. and Calvert, "Investigations of Possible Monarch Butterfly Overwintering Areas," 295–98; Melquiades Moreno de Jesús, interview by Patricio Moreno Rojas, 1 July 2020, FHS-OH.
31. Lincoln P. Brower, interview by Christopher Koehler, 14 March 1994, University of Florida Libraries; Lincoln P. Brower, interview by Mark Madison, 21 May 2015, https://digitalmedia.fws.gov/digital/collection/videos/id/142/rec/1; Rome, *Bulldozer in the Countryside*, 119–52.
32. Brower interview by Koehler.
33. Brower interview by Koehler; Van Zandt Brower, "Experimental Studies of Mimicry."
34. Brower interview by Koehler.
35. Brower interview by Koehler; Parsons, "Digitalis-Like Toxin."
36. Brower, Van Zandt Brower, and Corvino, "Plant Poisons in a Terrestrial Food Chain"; Reichstein et al., "Heart Poisons in the Monarch Butterfly"; Agrawal, *Monarchs and Milkweed*, 34–39.
37. Brower interview by Koehler.
38. Lincoln P. Brower, "Ecological Chemistry," *Scientific American*, February 1969, 22–29.
39. Brower interview by Koehler.
40. Donald A. Davis, interview by author, 20 November 2018; Gustafsson et al., "Monarch Butterfly Through Time and Space."
41. Frederick A. Urquhart, interview by Paul A. Bator, 8–12 January 1979, Special Collections, University of Toronto; Frederick A. Urquhart, "Monarch Migration Studies: An Autobiographical Account," *News of the Lepidopterists' Society*, July–August 1978, 3–4; Vetter, *Field Life*, 76–137.

42. Urquhart, interview; Urquhart, "Ecological Study."
43. Urquhart interview; Urquhart, "Proposed Method for Marking."
44. "Norah Patterson," *Torontonensis* (Victoria College yearbook), 1941, 62; Urquhart interview; *Insect Migration Studies: A Newsletter to Research Associates*, 1964, MWL.
45. Urquhart, "Marked Monarchs"; Agrawal, *Monarchs and Milkweed*, 73–75.
46. "Attempt to Solve Insect Mystery by Tagging Butterflies," *Brandon (Manitoba) Daily Sun*, 18 June 1955; "Canadians Ask Aid with Butterflies," *Rocky Mount (NC) Evening Telegram*, 21 April 1955, "Butterfly Flys South for Winter," *New York Times*, 13 November 1955; "Migratory Butterflies," *Time*, 31 December 1956, 46; "Banding Butterflies," *Winona (MN) Daily News*, 20 September 1957; Urquhart, *Report on the Studies of the Movements*, iii–ix.
47. Halpern, *Four Wings and a Prayer*, 83.
48. *Insect Migration Studies*, 1964–1965, MWL.
49. *Insect Migration Studies*, 1964–1965, MWL; Gary Webster, "This Butterfly Gets Around," *Corpus Christi (TX) Caller Times*, 21 September 1958; Urquhart, *Monarch Butterfly*, 88–89, 252–321; Urquhart, "Monarch Butterfly (*Danaus Plexippus*) Migration."
50. *Insect Migration Studies*, 1968, MWL.
51. *Insect Migration Studies*, 1970, MWL.
52. *Insect Migration Studies*, 1970, MWL; Urquhart interview; Urquhart, *Monarch Butterfly: International Traveler*, 154.
53. *Insect Migration Studies*, 1972, MWL; Fred A. Urquhart, "The Migrating Monarch," *The News* (Mexico City), 25 February 1973, Hemeroteca Nacional de México, Mexico City.
54. Frederick A. Urquhart, "Found at Last: The Monarch's Winter Home," *National Geographic*, August 1976.
55. *Insect Migration Studies*, 1974, MWL; Herberman, *Great Butterfly Hunt*; Robert McG. Thomas Jr., "Kenneth C. Bugger, 80, Dies," *New York Times*, 12 December 1998.
56. Monika Maeckle, "Founder of the Monarch Butterfly Roosting Sites in Mexico Lives a Quiet Life in Austin, Texas," Texas Butterfly Ranch, 10 July 2012, https://texasbutterfly ranch.com/2012/07/10/founder-of-the-monarch-butterfly-roosting-sites-in-mexico -lives-a-quiet-life-in-austin-texas/.
57. Sheryl Smith-Rodgers, "Maiden of the Monarchs: Discoverer of Butterfly Wintering Site Breaks Decades of Silence to Tell Her Story," *Texas Parks and Wildlife*, March 2016, https://tpwmagazine.com/archive/2016/mar/LLL_catalina/.
58. Smith-Rodgers, "Maiden of the Monarchs"; Brower, "Understanding and Misunderstanding," 331–32; Carlos F. Gottfried Joy, interview by author, 25 November 2020.
59. Maeckle, "Founder of the Monarch Butterfly Roosting Sites"; Smith-Rodgers, "Maiden of the Monarchs"; Lindsey Cherner, "Austin Woman Remembers Finding Monarch Sanctuary," *Austin (TX) American-Statesman*, 2 February 2013.
60. Maeckle, "Founder of the Monarch Butterfly Roosting Sites"; Albert Moldvay, "In Focus," *Westways*, May 1982, 21–23, 70.
61. *Insect Migration Studies*, 1975, MWL; Fred A. Urquhart, "Found at Last."
62. *Insect Migration Studies*, 1976, MWL; Urquhart and Urquhart, "Overwintering Site"; Nicholas C. Chriss, "Butterflies Wing It—as Far as Mexico," *Los Angeles Times*, 11 November 1976.

63. Urquhart, "Found at Last"; Ellen Sharp, "Who Gets to Be a Citizen-Scientist?" *Medium*, 14 October 2016, https://medium.com/invironment/who-gets-to-be-a-citizen-scientist-e7d52a6239d2.
64. Jørgensen, Jørgensen, and Pritchard, *New Natures*, 13–17, 179–94.
65. Brower, "Understanding and Misunderstanding," 333–34; "National Science Foundation Refuses a Grant," *Insect Migration Studies*, 1975, MWL; Donald Davis to Will Wright, email, 27 April 2020.
66. Fred Urquhart and Norah Urquhart, "A Special Report to the Research Associates Who Have Been Involved in the Studies of the Monarch Butterfly Migrations," October 1977, Donald Davis Personal Collection, Toronto, Ontario.
67. Robert Thomas Allen, "Why Dr. Fred Urquhart Sometimes Wishes He'd Never Eaten a Monarch Butterfly in the First Place," *Maclean's*, 1 July 1961.
68. *Insect Migration Studies*, 1976, MWL.
69. Halpern, *Four Wings and a Prayer*, 93.
70. Brower interview by Koehler.
71. Brower, "Understanding and Misunderstanding," 335; Brower interview by Koehler.
72. Bayard Webster, "2D Group Uncovers Butterflies' Secret," *New York Times*, 29 May 1977; "Lepidopterist Leak," *New York Times*, 14 June 1977.
73. Brower, "Monarch Migration," 40–53; Brower et al., "Biological Observations of an Overwintering Colony."
74. Urquhart and Urquhart, "Special Report to the Research Associates"; *Insect Migration Studies*, 1978, MWL; Lynne Baranski, "Butterfly Biologists Are Aflutter (and a-Feudin') Over a Royal Hangout," *People*, 3 April 1978; Fink and Brower, "Birds Can Overcome,"; Ramachandra Guha, "The Authoritarian Biologist and the Arrogance of Anti-Humanism," *The Ecologist*, January–February 1997, 14–20.
75. Urquhart and Urquhart, "Special Report to the Research Associates."
76. Brower, "Monarch Migration," 51.
77. Gottfried, *Monarcas*, 12; Ortíz Monasterio and Ortíz Monasterio Garza, *Mariposa monarca*, 51; Sharp and Wright, "'We Were in Love with the Forest,'" 4–16.
78. Moreno Espinoza interview; Emilio Velázquez Moreno, interview by Ellen Sharp, 9 May 2020, FHS-OH; Urquhart, "Conservation Areas," 109.
79. Moreno Espinoza interview.
80. Elidió Moreno de Jesús interview.
81. Moreno Espinoza interview; Elidió Moreno de Jesús interview; Melquiades Moreno de Jesús interview.
82. Rice, *Making Machu Picchu*, 1–44.
83. Ellen Sharp, "Butterfly Trailblazer: An Interview with Cerro Pelon's First Female Guide," *Medium*, 23 March 2018, https://medium.com/@ellensharp/butterfly-trailblazer-an-interview-with-cerro-pelons-first-female-guide-a1c1abef4d2a.
84. Robert McG. Thomas, Jr., "Kenneth C. Brugger, 80, Dies; Unlocked a Butterfly Mystery," *New York Times*, 12 December 1998; Chip Taylor, "In Pursuit of a Little History," *1998 Season Summary*, Monarch Watch, May 1999, p. 24, https://monarchwatch.org/download/pdf/season-summary-1998.pdf; Catalina Aguado Trail to Will Wright, email, 10 November 2018.
85. Robert Michael Pyle, interview by author, 29–30 June 2020, 4 September 2020; Robert Michael Pyle, "Las monarcas," *Orion*, Spring 2001.

86. Pyle interview; Pyle, *Watching Washington Butterflies*, Pyle, "History of Lepidoptera Conservation."
87. Pyle interview; "The 1972 Annual Meeting," *News of the Lepidopterists' Society*, 15 September 1972, 1–2; Pyle, "Eco-Geographic Basis."
88. Fraser, *Rewilding the World*, 80–81; Reinalda, *Routledge History of International Organizations*, 512–13; Christopher Andreae, "Sir Peter Scott: Save the World by Enjoying It," *Christian Science Monitor*, 20 April 1989; Peter Scott to Dr. Pyle, 28 March 1976, Folder 12, Box 34, Pyle Papers; Peter Scott to Dr. Pyle, 19 July 1976, Folder 12, Box 34, Pyle Papers; Pyle interview.
89. Robert M. Pyle to Thomas Lovejoy III, 19 March 1976, Folder 1, Box 33, Pyle Papers; Robert M. Pyle to Thomas E. Lovejoy, 24 May 1976, Folder 1, Box 33, Pyle Papers; Pyle, "Lepidoptera Specialist Group," 29.
90. Pyle interview; Robert M. Pyle and Sarah A. Hughes, *Conservation and Utilization of the Insect Resources of Papua New Guinea* (Port Moresby, Papua New Guinea: Department of Natural Resources, 1978), Folder 29, Box 75, Pyle Papers.
91. Pyle, "International Problems in Insect Conservation"; Susan Sleeth Mosedale, "Imperiled Monarchy," *Northwest Magazine*, 1984, Folder 2, Box 38, Pyle Papers.
92. "Palabras de Rodolfo Ogarrio," *Acciones para la protección de la mariposa Monarca, Sierra Chincua, Michoacán,* 22 August 1986, Box 4, Secretaría de Desarrollo Urbano y Ecologia, Unidad de la Crónica Presidencial, Administración Presidencial de Miquel de la Madrid Hurtado, AGN; Ogarrio, "Development of the Civic Group."
93. Peter Scott to Bob Pyle, 10 January 1979, Folder 12, Box 34, Pyle Papers; Pyle, "Recent IUCN Activity in Insect Conservation"; Pyle, "International Efforts for Monarch Conservation."
94. "Decreto por el que por la causa de utilidad pública se establece zona de reserva y refugio silvestre los lugares donde la mariposa conocida con el nombre de 'Monarca' hiberna y se reproduce," *Diario Oficial*, 9 April 1980; R. M. Pyle, "IUCN Report of the Lepidoptera Specialist Group," 28 March 1980, Folder 20, Box 82, Pyle Papers; Robert M. Pyle, phone call notes with Lincoln Brower, 18 February 1980, Folder 20, Box 82, Pyle Papers.
95. R. M. Pyle, "IUCN Report of the Lepidoptera Specialist Group," 28 March 1980, Folder 20, Box 82, Pyle Papers; Robert M. Pyle, phone call notes with Lincoln Brower, 18 February 1980, Folder 20, Box 82, Pyle Papers.
96. Lincoln P. Brower to Leonila Vazquez Garcia, 22 October 1980, Folder 1, Box 38, Pyle Papers; Lincoln P. Brower to Bob Pyle, 22 October 1980, Folder 1, Box 38, Pyle Papers; "Leonila Vázquez García: In Memoriam"; Jo Brewer, "A Visit with 200 Million Monarchs," *Defenders*, 1982, 13–16; Pyle interview.
97. Calvert et al., "Monarch Butterfly Conservation."
98. Robert Michael Pyle to Cuauhtémoc Cárdenas, 8 May 1981, Folder 10, Box 31, Pyle Papers; Ricardo Barthelemy to Dillon Ripley, 9 February 1983, Folder 2, Box 38, Pyle Papers; "Palabras de Cuauhtémoc Cárdenas Solórzano," *Acciones para la protección de la mariposa Monarca, Sierra Chincua, Michoacán*, 22 August 1986, Box 4, Secretaría de Desarrollo Urbano y Ecologia, Administración Presidencial de Miquel de la Madrid Hurtado, AGN.
99. "Congreso Mexico Americano de Lepidopterología," *News of the Lepidopterists' Society*, September–October 1981, 57–63; Pyle, "Symposium on the Biology and Conservation."

100. Vázquez G. and Perez R., "Monarch Butterfly as a Resource"; Rodriquez, "Mariposa Monarca"; Lincoln P. Brower to F. A. Urquhart, 7 August 1981, Folder 1, Box 38, Pyle Papers; Lincoln P. Brower et al. to Secretario de Agricultura y Recursos Hidráulicos, 6 November 1981, Folder 2, Box 38, Pyle Papers; Gallegos and López García, "Contribucion Geográfica al Programa," 9–26.
101. Simonian, *Defending the Land of the Jaguar*, 182–89; Robert Pyle to Peter Scott, 4 March 1983, Folder 12, Box 34, Pyle Papers; Lincoln P. Brower to Thomas Lovejoy, 25 August 1981, Folder 1, Box 38, Pyle Papers.
102. Robert Pyle, "Urgent Message to Prince Philip via Peter Scott," Folder 12, Box 34, Pyle Papers; Pyle, "Conclusion"; Prince Philip, foreword to Gottfried, *Monarcas*.
103. Pyle, "International Red Data Book of Invertebrates," 60; Wells, Pyle, and Collins, *IUCN Invertebrate Red Data Book*, 463–66.
104. Robert Michael Pyle, interview by Mark Madison, US Fish and Wildlife Service, 2015, https://digitalmedia.fws.gov/digital/collection/videos/id/202/rec/2.
105. "Monarch Project Steering Committee History," Folder 2, Box 38, Pyle Papers; Robert Michael Pyle, "A Proposal for the Conservation of the Monarch Butterfly Migratory Phenomenon," Folder 1, Box 38, Pyle Papers; Robert M. Pyle to Lincoln Brower, Melody Mackey Allen, and Richard Lindley, 13 February 1983, Folder 1, Box 38, Pyle Papers; Melody Mackey Allen, "Monarch Project Memorandum," June 1984, Folder 2, Box 38, Pyle Papers; Pyle interview.
106. Peter Sauer, "The Monarch Versus the Global Empire," *Orion*, April 2001.

Chapter 4

1. Nathan Bomey, "Monsanto Shedding Name: Bayer Acquisition Leads to Change for Environmental Lightning Rod," *USA Today*, 4 June 2018.
2. Orley R. Taylor, interview by author, 23 July 2020. See "Monarch Waystation Registry" for an updated count of pollinator gardens at https://monarchwatch.org/waystations/.
3. Taylor interview.
4. Van der Meer, "Butterfly Effect," 38; Snook, "Conservation of the Monarch Butterfly Reserves," 368; Merino Pérez and Hernández Apolinar, "Destrucción de instituciones comunitarias."
5. Russell, *History of Mexico*, 480–85, 514–16.
6. Sharon Sullivan, "Guarding the Monarch's Kingdom," *International Wildlife*, November–December 1987, 4–10; Aridjis, "Conspiración contra la monarca"; Peter Menzel, "Butterfly Arms Now Under Guard in Annual Bivouac," *Smithsonian Magazine*, November 1983, 174–81.
7. Gottfried, *Mariposa monarca*; Rose, "Politics of Mexican Wildlife," 280–85; Rodolfo Ogarrio to Robert Michael Pyle, 29 January 1985, Folder 2, Box 38, Pyle Papers.
8. Aridjis, "Los pájaros"; Ardjis, "Declaración de 100 intelectuales y artistas"; Aridjis, "La mariposa monarca."
9. Ogarrio, "Conservation Actions Taken by Monarca"; Joy, "Monarch Conservation in Mexico"; Camas de Castro, "Operative Programs in the Monarca A.C. Project"; González de Castilla, "Importance of Alternative Sources of Income to 'Ejidatarios.'"
10. "Decreto por el que por razones de orden público e interés social, se declaran áreas naturales protegidas para los fines de la migración, invernación y reproducción de la mariposa Monarca . . ." *Diario Oficial*, 9 October 1986.

11. "Palabras de Curtis Freese," "Palabras de Rodolfo Ogarrio," "Palabras de Manuel Camacho Solís," and "Palabras de Valentina Ortíz Monasterio," *Acciones para la protección de la mariposa Monarca, Sierra Chincua, Michoacán,* 22 August 1986, Box 4, Secretaría de Desarrollo Urbano y Ecologia, Administración Presidencial de Miquel de la Madrid Hurtado, AGN.
12. Gottfried, "One of Nature's Most Incredible Phenomenon"; Rose, "Politics of Mexican Wildlife," 282; Lincoln P. Brower to Robert Pyle, 26 August 1987, Folder 1, Box 38, Pyle Papers; Carlos Federico Gottfried Joy, interview by author, 25 November 2020.
13. Gottfried Joy, interview.
14. Rose, "Politics of Mexican Wildlife," 283; Sullivan, "Guarding the Monarch's Kingdom," 4; Brenner, "Áreas naturales protegidas y ecoturismo"; Pro-Monarca, *Mariposa monarca: Instrucciones de comportamiento* (Mexico City: Monarca, A.C., 1988), Folder 12, Box 39, Pyle Papers.
15. Hoth, "Mariposa monarca," 25; Tapia Torres, "Retos para la conservación," 335.
16. Halpern, *Four Wings and a Prayer*, 41; Aridjis, "Conspiración contra la monarca"; Ramírez et al., "Threats to the Availability of Overwintering Habitat."
17. Brower et al., "Quantitative Changes in Forest Quality"; Alex Shoumatoff, "Flight of the Monarchs," *Vanity Fair*, November 1999.
18. Homero Aridjis and Lincoln P. Brower, "Twilight of the Monarchs," *New York Times*, 26 January 1996; Aridjis, "La nueva reserva de la mariposa monarca."
19. Melquiades Moreno de Jesús, interview by Patricio Moreno Rojas, 1 July 2020, FHS-OH.
20. Aridjis, "Grandeza y miseria de la mariposa monarca."
21. Russell, *History of Mexico*, 486–90.
22. Van der Meer, "Butterfly Effect," 13; Elidió Moreno de Jesús, interview by Patricio Moreno Rojas and Ellen Sharp, 27 April 2020, FHS-OH.
23. Gottfried Joy, interview.
24. Grandin, *End of the Myth*, 233–36; Robin, *World According to Monsanto*, 139–88.
25. Elmore, "Commercial Ecology of Scavenger Capitalism."
26. Charles, *Lords of the Harvest*, 25; Elmore, "Roundup from the Ground Up."
27. Karen Keeler Rogers, "Fields of Promise: Monsanto and the Development of Agricultural Biotechnology, Part I," *Monsanto Magazine*, Winter 1996, Folder 18, Box 13, Series 8, WU-Monsanto.
28. Karen Keeler Rogers, "Fields of Promise: Monsanto and the Development of Agricultural Biotechnology, Part II," *Monsanto Magazine*, Spring 1997, Folder 19, Box 13, Series 8, WU-Monsanto; Charles, *Lords of the Harvest*, 2–21. The 1980 case allowing living organisms to be patented was *Diamond v. Chakrabarty*.
29. Robin, *World According to Monsanto*, 138–39; Charles, *Lords of the Harvest*, 60–68.
30. Robin, *World According to Monsanto*, 230; Charles, *Lords of the Harvest*, 41–54; Vaeck et al., "Transgenic Plants Protected from Insect Attack."
31. "Chemical Bonds" section of Jon R. Luoma, "Pandora's Pantry," *Mother Jones*, January/February 2000; Executive Office of the President, Office of Science and Technology Policy, "Coordinated Framework for Regulation of Technology," 26 June 1986, www.aphis.usda.gov/brs/fedregister/coordinated_framework.pdf; Food and Drug Administration, "Statement of Policy: Foods Derived From New Plant Varieties," 29 May 1992, www.fda.gov/regulatory-information/search-fda-guidance-documents/statement-policy-foods-derived-new-plant-varieties; Robin, *World According to Monsanto*, 147–56.

32. Dan Glickman, "New Crops, New Century, New Challenges: How Will Scientists, Farmers, and Consumers Learn to Love Biotechnology and What Happens If They Don't?" C-Span, 13 July 1999, www.c-span.org/video/?150302-1/crops-century-challenges.
33. "Availability of Determination of Nonregulated Status for Corn Line Genetically Engineered for Insect Resistance," *Federal Register* 61 (29 January 1996): 2789–90, in US Department of Agriculture, "Petition 95-195-01 for Determination of Nonregulated Status for Bt11 Corn," January 1996, International Service for the Acquisition of Agri-Biotech Applications, www.isaaa.org/gmapprovaldatabase/event/default.asp?EventID=128&Event=Bt11%20(X4334CBR,%20X4734CBR.
34. Losey, Rayor, and Carter, "Transgenic Pollen Harms Monarch Larvae," 214; Charles, *Lords of the Harvest*, 243–48.
35. Wassenaar and Hobson, "Natal Origins of Migratory Monarch Butterflies"; "Corn Area Planted and Harvested, Yield, Production, Utilization, Price, and Value—United States: 1866–2017," US Department of Agriculture, *Crop Production Historical Track Records*, April 2018, 31; Rick Weiss, "Biotech vs. 'Bambi' of Insects? Gene-Altered Corn May Kills Monarchs," *Washington Post*, 20 May 1999; Carol Kaesuk Yoon, "Altered Crop May Imperil Butterfly, Study Says," *New York Times*, 20 May 1999.
36. Nigel Hawkes and Nick Nuttall, "Modified Maize 'Killing Butterflies,'" *Times* (London), 20 May 1999.
37. Hodgson, "Monarch Bt-Corn Paper Questioned," 627; Michael Fumento, "The World Is Still Safe for Butterflies," *Wall Street Journal*, 25 June 1999.
38. Memorandum for the President, "Decision on Agricultural Biotechnology Issues," 10 November 1999, Subject Files: GMOs (Genetically Modified), Clinton Digital Library, https://clinton.presidentiallibraries.us/items/show/31336.
39. Memorandum for the President, "From Secretary Dan Glickman: Report for Week of May 9, 1999," Subject Files: History of USDA—Archival Documents—Chapter 11:00 Weekly Reports, 1999, Clinton Digital Library, https://clinton.presidentiallibraries.us/items/show/5648.
40. Lincoln Brower, "The Monarch and the Bt Corn Controversy," *Orion*, 2 April 2001.
41. Brower, "Monarch and the Bt Corn Controversy."
42. "Scientific Symposium to Show No Harm to Monarch Butterfly," Biotechnology Innovation Organization, 2 November 1999, https://archive.bio.org/media/press-release/scientific-symposium-show-no-harm-monarch-butterfly; Carol Kaesuk Yoon, "No Consensus on Effect of Genetically Altered Corn on Butterflies," *New York Times*, 4 November 1999; Robert Steyer, "Scientists Discount Threat to Butterfly from Altered Corn," *St. Louis Post-Dispatch*, 2 November 1999; Robin, *World According to Monsanto*, 228–31.
43. Hansen Jesse and Obrycki, "Field Deposition of Bt Transgenic Pollen"; Sears et al., "Impact of Bt Corn Pollen on Monarch Butterfly Populations."
44. Chip Taylor, "Transgenics and Monarchs," *2001 Season Summary*, Monarch Watch, 2002, pp. 58–59, accessed 19 February 2021, www.monarchwatch.org/read/.
45. Roger Segelken, "Biologists Invent Gun for Shooting Cells with DNA," *Cornell Chronicle* (Cornell University, Ithaca, NY), 14 May 1987; Charles, *Lords of the Harvest*, 75–87; Klein et al., "High-Velocity Microprojectiles."
46. In 1996, Monsanto sold 20 percent more Roundup than the previous year, which the

Notes to Chapter 4 | 277

company credited to the introduction of Roundup Ready soybeans. Monsanto Company, *1996 Annual Report*, Folder 10, Box 6, Series 8, WU-Monsanto; Charles, *Lords of the Harvest*, 109–25.
47. Charles, *Lords of the Harvest*, 177.
48. Monsanto Company, *1998 Annual Report*, Folder 12, Box 6, Series 8, WU-Monsanto.
49. Monsanto Company, *2001 Annual Report*, Folder 14, Box 6, Series 8, WU-Monsanto.
50. Monsanto Company, "Roundup: Into the Twenty-First Century," Folder 5, Box 3, Series 4, WU-Monsanto; Monsanto Company, "We're in the Roundup Ready Zone," Folder 7, Box 3, Series 4, WU-Monsanto; Franz, Mao, and Sikorski, *Glyphosate*, 2–9; Pleasants, "Monarch Butterflies and Agriculture," 173.
51. Malcom, Cockerell, and Brower, "Spring Recolonization of Eastern North America."
52. Oberhauser, "Material Investment in Mating"; Oberhauser and Frey, "Coercive Mating by Overwintering Male Monarchs"; Karen S. Oberhauser, interview by author, 16 June 2020.
53. Karen Oberhauser, "Larval Monitoring," *1996 Season Summary*, Monarch Watch, May 1997, p. 16, www.monarchwatch.org/read/; Prysby and Oberhauser, "Temporal and Geographic Variation in Monarch Densities."
54. Oberhauser interview; Oberhauser et al., "Temporal and Spatial Overlap Between Monarch Larvae and Corn Pollen," 11918.
55. Chip Taylor, "Transgenics and Monarchs," *1999 Season Summary*, Monarch Watch, Summer, 2000, p. 8, www.monarchwatch.org/read/.
56. Pleasants, "Monarch Butterflies and Agriculture"; U.S. Department of Agriculture, "Recent Trends in GE Adoption," 17 July 2020, www.ers.usda.gov/data-products/adoption-of-genetically-engineered-crops-in-the-us/recent-trends-in-ge-adoption.aspx.
57. David A. Wurdeman to Orley Taylor, as quoted in Chip Taylor, "Effects of Transgenic Crops on Milkweeds," Monarch Watch Update, 22 June 2004, www.monarchwatch.org/update/2004/0622.html.
58. Chip Taylor, "Conservation: Monarch Waystations," Monarch Watch Update, 30 March 2005, https://monarchwatch.org/update/2005/0330.html#2; "Monarch Waystation Program Launched," Monarch Watch Update, 25 April 2005, https://monarchwatch.org/update/2005/0425.html#2; "Monarch Waystation Registry," Monarch Watch, accessed September 19, 2025, https://monarchwatch.org/waystations/index.html#registry.
59. Taylor interview; Thogmartin et al., "Restoring Monarch Butterfly Habitat in the Midwestern U.S.," 1–10.
60. Hurt, *American Agriculture*, 397–98; "CRP Ending Enrollment by Fiscal Year, 1986–2018," Conservation Reserve Program Statistics, US Department of Agriculture, accessed 2 March 2021, www.fsa.usda.gov/programs-and-services/conservation-programs/reports-and-statistics/conservation-reserve-program-statistics/index; Thogmartin et al., "Monarch Butterfly Population Decline in North America," 1–16.
61. Cameron and Tomlin, *Making of NAFTA*, 188; Knox and Markell, "Innovative North American Commission for Environmental Cooperation."
62. "North American Agreement on Environmental Cooperation between the Government of Canada, the Government of the United Mexican States, and the Government of the United States of America," signed 14 September 1993, https://ustr.gov/sites/default/files/naaec.pdf.
63. Homero Aridjis, "Last Call for Monarchs," *Huffington Post*, 7 February 2014.

64. Howard LaFranchi, "Monarchs' Reign in Mexico Is Threatened," *Christian Science Monitor*, 8 November 1995; Jeffrey P. Crolla and J. Donald Lafontaine, *Status Report on the Monarch Butterfly (Danaus plexippus) in Canada*, 29 March 1996, www.monarch watch.org/read/article_canmon.html.
65. Homero Aridjis and Lincoln P. Brower, "Twilight of the Monarchs," *New York Times*, 26 January 1996; Anderson and Brower, "Freeze-Protection of Overwintering Monarch Butterflies in Mexico," 107–16.
66. Aridjis, "El tamaño de la tala."
67. Karen Oberhauser, "Trip to Mexico," 8 March 1996, Monarch Watch, www.monarch watch.org/read/articles/trip.htm; O. R. Taylor and William Calvert, "Conservation of Monarch Butterflies: An Informal Meeting to Discuss Issues and Options," Monarch Watch, 16 June 1996, https://monarchwatch.org/read/articles/houston.htm; "Trilateral Agreement. 2 August 1996," *1996 Season Summary*, Monarch Watch, May 1997, p. 29, www.monarchwatch.org/read/.
68. Hoth, "Monarch"; Brower, "Biological Necessities"; Halpern, *Four Wings and a Prayer*, 41.
69. Sigala Páez, "La conservación de la monarca"; Caro and Sigala quoted in US Fish and Wildlife Service, *1997 North American Conference on the Monarch Butterfly: Roundtable Discussions*, 8, 15. Caro's agency was Secretaría del Medio Ambiente, Recursos Naturales y Pesca, or SEMARNAP.
70. US Fish and Wildlife Service, *1997 North American Conference on the Monarch Butterfly: Roundtable Discussions*, 16; Yeager, "Roundtable Address," 225; Halpern, *Four Wings and a Prayer*, 45.
71. Merino, "Reserva especial," 239.
72. "Decreto por el que por área natural protegida, con el carácter de la reserva de la biosfera, la región denominada Mariposa Monarca . . .," *Diario Oficial*, 10 November 2000; "Conservación de la Naturaleza, Un Nuevo Decreto de la Mariposa Monarca", 9 November 2000, Box 80, Dirección de Operación de Eventos, Secretaría Particular II, Administración Presidencial de Ernesto Zedillo Ponce de León, AGN.
73. Jordi Honey-Rosés et al., "Monitoreo Forestal del Fondo Monarca 2003," World Wildlife Fund Mexico, www.wwf.org.mx/nuestro_trabajo/ecosistemas_terrestres /conservacion_de_la_mariposa_monarca/; Honey-Rosés et al., "To Pay or Not to Pay?" 120–28; Honey-Rosés, Baylis, and Ramírez, "Spatially Explicit Estimate of Avoided Forest Loss," 1032–43.
74. Van der Meer, "Butterfly Effect," 73; Merino, "Reserva especial," 240–43.
75. Toledo Manzur, "Estrategia integral para el desarrollo sustentable," 42; Vicente Guzmán Reyes to Presidente Vicente Fox Quesada, 16 June 2003, File 52429, Box 176, Red Federal de Servicio a la Cuidadania 2000–2005, Coordinación General de Administración, Administración Presidencial de Vicente Fox Quesada I, AGN.
76. World Wildlife Fund Mexico, *La tala ilegal*, 3.
77. Abel Castro Posadas to Presidente Vicente Fox Quesada, 16 June 2003, File 52240, Box 175, Red Federal de Servicio a la Cuidadania 2000–2005, Coordinación General de Administración; Cuauhtémoc González Pacheco to Doroles Huante Mares, 4 June 2004, File 136721, Box 458, Red Federal de Servicio a la Cuidadania 2002–2005, Coordinación General de Administración, both in Administración Presidencial de Vicente Fox Quesada I, AGN.

78. Emilio Velázquez Moreno, interview by Ellen Sharp, 9 May 2020, FHS-OH; Merino, "Reserva especial," 243.
79. Velázquez Moreno interview.
80. Gonzalez-Duarte, "Butterflies, Organized Crime, and 'Sad Trees'"; Columba Gonzalez-Duarte and Manuel Ureste, "Indigenous Communities in Mexico Take Up Arms to Defend the Monarch Forest," North American Congress on Latin America, 24 March 2021, https://nacla.org/mexico-monarchs-organized-crime.
81. Patricio Moreno Rojas, interview by Ellen Sharp, 7 April 2020, FHS-OH.
82. Velázquez Moreno interview.
83. Fletcher and Büscher, "Conservation Basic Income," 1–7; Wright, "Monarch Butterfly Conservation (Mexico)."
84. Haddad, *Last Butterflies*, 187–201.
85. Pleasants and Oberhauser, "Milkweed Loss in Agricultural Fields," 134–44; Homero Aridjis, "Last Call for Monarchs," *Huffington Post*, 7 February 2014.
86. Aridjis et al., "Letter to President Barack Obama."
87. Howard LaFranchi, "A Win for Monarch Butterflies at the Mexico Summit?" *Christian Science Monitor*, 20 February 2014; Monica Maeckle, "Monsanto: 'We Are Absolutely Committed' to Monarch Butterfly Conservation," Texas Butterfly Ranch, 27 January 2015, https://texasbutterflyranch.com/2015/01/27/monsanto-we-are-absolutely-committed-to-monarch-butterfly-conservation/.
88. White House Office of the Press Secretary, "Presidential Memorandum—Creating a Federal Strategy to Promote the Health of Honey Bees and Other Pollinators," 20 June 2014, https://obamawhitehouse.archives.gov/the-press-office/2014/06/20/presidential-memorandum-creating-federal-strategy-promote-health-honey-b.
89. Janet Marinelli, "Can the Monarch Highway Help Save a Butterfly Under Siege?" *Yale Environment 360*, 11 July 2017, https://e360.yale.edu/features/can-the-monarch-highway-help-save-a-butterfly-under-siege; Chip Taylor, "Creating a Monarch Highway," *Monarch Watch Blog*, 1 December 2015, https://monarchwatch.org/blog/2015/12/01/creating-a-monarch-highway/; Kasten et al., "Can Roadside Habitat," 1047–57; Memorandum of Understanding, "Agreement for the Support of a Monarch Highway," May 2016, American Association of State Highway and Transportation Officials, https://downloads.transportation.org/Monarch_Highway_MOU_052616.pdf.
90. Center for Biological Diversity et al., "Petition to Protect the Monarch Butterfly"; Maeckle, "Monsanto: 'We Are Absolutely Committed.'"
91. "The Monarch Butterfly," *Monsanto Blog*, 24 February 2014, https://web.archive.org/web/20140605060415/http://monsantoblog.com/2014/02/24/the-monarch-butterfly; "Myth #6: Monsanto Is Killing the Monarch Butterflies," *Mosanto Blog*, 20 May 2014, https://web.archive.org/web/20140523005517/http://monsantoblog.eu/myth-6-monsanto-is-killing-the-monarch-butterflies; David F. comment on Monika Maeckle, "Endangered Species Act Petition: Wrong Tool for Monarch Butterfly Conservation?," Texas Butterfly Ranch, 10 November 2014, https://texasbutterflyranch.com/2014/11/10/endangered-species-act-petition-wrong-tool-for-monarch-butterfly-conservation/.
92. Nicole Greenfield, "For a Family in Mexico, a Mission to Protect Monarchs," Natural Resources Defense Council, 9 February 2021, www.nrdc.org/stories/family-mexico-mission-protect-monarchs; "About Us," J. M. Butterfly B&B Welcome Packet, in possession of author.

93. Greenfield, "For a Family in Mexico"; "About Us," J. M. Butterfly B&B Welcome Packet.
94. Greenfield, "For a Family in Mexico"; "About Us," J. M. Butterfly B&B Welcome Packet.
95. Darlene A. Burgess, interview by author, 1 March 2021.
96. Patricio Moreno Rojas interview.
97. Monika Maeckle, "Success! Petition Spurs Rangers' Reinstatement at Monarch Butterfly Roosting Forest," Texas Butterfly Ranch, 10 July 2017, https://texas butterflyranch.com/2017/07/10/success-petition-spurs-reinstatement-at-monarch-butterfly-roosting-forest/.
98. Osvaldo Esquivel Maya, interview by Ellen Sharp, 9 May 2020, FHS-OH.
99. Francisco Moreno Hernández, interview by Ellen Sharp, 9 May 2020, FHS-OH.
100. Ellen Sharp, "Letter from Ellen Sharp: Early Departure," Journey North, 10 March 2021, https://journeynorth.org/monarchs/resources/article/03102021-letter-ellen-sharp-early-departure.
101. Agrawal, *Monarchs and Milkweed*.

Chapter 5

1. Effie (Comstock) Simmons, interview transcript, 10 August 1950, Folder 1, Box 258, SEKI Park Files; Hartesveldt et al., *Giant Sequoias*, 10.
2. Hartesveldt et al., *Giant Sequoias*, 10; White, *"It's Your Misfortune and None of My Own,"* 282–85. The grove's name comes from the felling of the Mark Twain Tree in 1891 for the American Museum of Natural History. With a trunk diameter of thirty feet across, all that remained was a "big stump." See Willard, *Giant Sequoia Groves of the Sierra Nevada*, 93.
3. Theodore Wagner to J. D. Hyde, 7 January 1880, Folder 5, Box 176, SEKI Park Files.
4. Tweed and Dilsaver, *Challenge of the Big Trees*, 73–79.
5. Johnston, *They Felled the Redwoods*; McGraw, *A. E. Douglass and the Role of the Giant Sequoia*.
6. Belich, *Replenishing the Earth*, 307; Isenberg, *Mining California*, 75–98; Orsi, *Sunset Limited*, 17–21.
7. Madley, *American Genocide*, 1–15, 67–102.
8. "Treaty Made and Concluded at Camp Fremont, State of California, March 19, 1851 . . .," in "1851–1852 Eighteen Unratified Treaties Between California Indians and the United States," Hornbeck Collection, California State University, Monterey Bay, Digital Commons, accessed January 27, 2021, https://digitalcommons.csumb.edu/hornbeck_usa_2_b/5/.
9. "Treaty Made and Concluded at Camp Keyes, on the Cah-Wia River, in the State of California, May 30, 1851 . . ."; and "Treaty Made and Concluded at Camp Burton, on Paint Creek, State of California, June 3, 1851 . . .," both in "1851–1852 Eighteen Unratified Treaties Between California Indians and the United States," Hornbeck Collection, California State University, Monterey Bay, Digital Commons, accessed January 27, 2021, https://digitalcommons.csumb.edu/hornbeck_usa_2_b/5/.
10. Larisa K. Miller, "The Secret Treaties with California's Indians," *Prologue: Magazine of NARA*, Fall/Winter 2013, 38–45; Aldern, "Native Sustainment," 37–44.

11. Tweed, *King Sequoia*, 78–79; Otter, *Men of Mammoth Forest*, 14–19; Madley, *American Genocide*, 244–46.
12. Whitney, *Geology*, 368–69; Alfred Bannister, *Map of Tulare County—State of California* [map] (San Francisco: Britton and Rey, 1884), Library of Congress; Brewster, *Life and Letters of Josiah Dwight Whitney*, 231–37; Farquhar, *History of the Sierra Nevada*, 129–43; Moore, *Capitalism in the Web of Life*.
13. Timber and Stone Act, Pub. L. No. 151, 20 Stat. 89 (1878); Tweed and Dilsaver, *Challenge of the Big Trees*, 64–65; "Timber Land Frauds," *San Francisco Chronicle*, 2 April 1892.
14. 1870 U.S. Census, San Francisco County, California, Austin D. Moore; Biographical Information: Moore, Austin D., Folder 6, Box 7, CSU-Schutt; Millard, *History of the San Francisco Bay Region*, 311–12; Johnston, *They Felled the Redwoods*, 24–25; "Kings River Lands," *San Francisco Chronicle*, 3 April 1892; L. A. Winchell, "History of Fresno County," 85, Logging-Lumber Folder, Box H, June English Forestry Collection, Madden Library, California State University, Fresno.
15. Kaweah Colony, *By-Laws of the Joint-Stock Company*; O'Connell, *Co-Operative Dreams*; Miller, *This Radical Land*, 161–212.
16. Haskell, *Pen Picture of the Kaweah Co-Operative Colony Co.*, 4.
17. Tweed, *Kaweah Remembered*; O'Connell, *Co-Operative Dreams*; Miller, *This Radical Land*; Otter, *Men of Mammoth Forest*, 70.
18. "The Timber Belt," *Fresno Daily Republican*, 27 June 1890; "Great Lumber Mills," *Fresno Daily Republican*, 1 January 1892; Johnston, *They Felled the Redwoods*, 25–35; Tweed, *King Sequoia*, 82–83.
19. Lizzie McGee, "Mills of the Sequoias," unpublished manuscript, 1952, Folder 17, Box 267, SEKI Park Files; "Millwood Musings," *Fresno Weekly Republican*, 18 May 1899; 1910 US Census, Mill Creek, California, Cilicy M. Kanawyer.
20. "Creditors Called," *San Francisco Chronicle*, 15 September 1893; "Heavy Failure," *San Francisco Morning Call*, 15 September 1893; Johnston, *They Felled the Redwoods*, 49; "Hoisting Machinery," *Fresno Weekly Republican*, 6 April 1894.
21. "H. C. Smith Insolvent," *Fresno Morning Republican*, 22 August 1897; "Report of Timber Sales," 1896–1905, Folder 29, Box 15, Charles H. Hackley and Thomas Hume Papers, Archives and Historical Collections, Michigan State University.
22. Hume-Bennett Lumber Company, Articles of Association, 22 November 1905, Folder 30, Box 15, Charles H. Hackley and Thomas Hume Papers, Archives and Historical Collections, Michigan State University; Cox, *Lumberman's Frontier*, 125–48; 263–90; "Hume, New Lumber Town," *Los Angeles Times*, 17 October 1909; Willard, *Giant Sequoia Groves of the Sierra Nevada*, 117–19.
23. "California Lumber to Take the World," *Los Angeles Times*, 15 June 1903.
24. "Makers of Los Angeles—Wheatley, Wilkes," *Out West Magazine*, January–June 1910, 332; Williams, *Deforesting the Earth*, 263–317; Tyrrell, *True Gardens of the Gods*, 56–102.
25. Johnston, *They Felled the Redwoods*, 60; Otter, *Men of Mammoth Forest*, 73–76.
26. "Logging in the Sierra Nevada Mountains, California," *Scientific American*, 19 December 1896, 444.
27. Johnston, *They Felled the Redwoods*; Otter, *Men of Mammoth Forest*, 77.
28. Photographs in Folders 1–5, Box 12, and Folders 6–7: Indexed Photographs: Logging,

1851–1912, 1947–1955, & Undated, b Box 13, CSU-Schutt; "Porterville a Center for the Lumber Business," *San Francisco Chronicle*, 7 February 1903.
29. "The Lumber Flume," 1–4, Flumes-Logging Folder, Box H, June English Forestry Collection, Madden Library, California State University, Fresno; "Hectic, Rugged Was Ride Down Famed Sanger Flume," *Fresno Bee*, date unknown, Flumes-Logging Folder, Box H, June English Forestry Collection, Madden Library, California State University, Fresno; Marilyn Scott, "Riding the Flumes," Folder 19, Box 267, SEKI Park Files.
30. "Great Lumber Mills," *Fresno Daily Republican*, 1 January 1893.
31. "Sanger Yard Inventory," 13 December 1905, Folder 35, Box 15, Charles H. Hackley and Thomas Hume Papers, Archives and Historical Collections, Michigan State University; "Sanger Doing Much Business," *San Francisco Chronicle*, 20 February 1904; Haugland et al., *Images of America: Sanger*, 45–57.
32. Jepson, *Trees of California*, 48; Sudworth, *Forest Trees of the Pacific Coast*, 140.
33. Piirto, "Wood of Giant Sequoia," 19–23; Farquhar, *History of the Sierra Nevada*, 87.
34. Neubrech, *California Redwood and Its Uses*, 17–21; Luxford and Markwardt, *Strength and Related Properties of Redwood*; Farmer, *Trees in Paradise*, 44–49.
35. "Foreign Markets," *American Lumberman*, 30 December 1899, 20.
36. W. M. Barr, "Sanger, California," *Out West Magazine*, July–December 1908, 83.
37. Shepard to Grosvenor, 16 June 1929, Folder 4, Box 65, Douglass Papers; Williams, *Deforesting the Earth*, 317.
38. "Coast Industrial Notes," *Mining and Scientific Press*, 8 December 1894, 366; "For Russian Use—A Shipment of California Redwood," *San Francisco Chronicle*, 9 May 1895; "California Redwoods," *Los Angeles Times*, 1 May 1896; "A New Enterprise—Fresno County Shipping Sequoia to Germany," *Fresno Bee*, 20 November 1894; Walter J. Ballard, "California Redwood Lumber Exports," *Los Angeles Times*, 12 September 1907.
39. Sudworth, *Forest Trees of the Pacific Slope*, vii–xv.
40. Sudworth, *Forest Trees of the Pacific Slope*, 138.
41. Sudworth, *Forest Trees of the Pacific Slope*, 138; Sudworth, "Present Condition of the California Bigtrees," 227–36.
42. Johnston, *They Felled the Redwoods*, 138.
43. Willard, *Giant Sequoia Groves of the Sierra Nevada*, 88–89.
44. Worster, *Passion for Nature*, 118–80.
45. Muir, *Our National Parks*, 276.
46. John Muir, "The New Sequoia Forests of California," *Harper's Monthly*, November 1878, 821.
47. Tweed, *King Sequoia*; Tweed and Dilsaver, *Challenge of the Big Trees*; Farmer, *Trees in Paradise*; Vermaas, *Sequoia*; Berland, "Giant Forest's Reservation"; Strong, "History of Sequoia National Park."
48. LeCain, *Matter of History*; Dunlop, *Cartophilia*.
49. Muir, "On the Post-glacial History of *Sequoia Gigantea*," 12. Later scientists have shown that glaciation was not a primary factor in sequoia grove distribution.
50. Muir, "On the Post-glacial History of *Sequoia Gigantea*," 12.
51. Chaney, "Revision of Fossil Sequoia and Taxodium"; Anderson, "Paleohistory of a Giant Sequoia Grove"; Hartesveldt et al. *Giant Sequoias*, 14–16, 42–43.
52. Piirto, "Factors Associated with Tree Failure of Giant Sequoia," 32–40.
53. Koehler, *Causes of Brashness in Wood*, 1–4, 8.

54. Muir, *Our National Parks*, 299.
55. Piirto, "Wood of Giant Sequoia," 19–23.
56. US Division of Forestry, *Report on the Big Trees of California*, 29.
57. "Return of Annual Net Income," and "Corporation Income Tax Return," 1910–1925, in Folders 1–16, Box 15, Charles H. Hackley and Thomas Hume Papers, Archives and Historical Collections, Michigan State University.
58. D. G. Wood, photo and description of "feather bed" near General Grant Park, undated, Folder 7: Indexed Photographs: Logging, 1851–1912, 1947–1955, & Undated, Box 13, CSU-Schutt.
59. Malinee Crapsey, "Walter Fry: Ambassador of Nature," *Sequoia Bark*, Summer 1994; "'Grand Old Man of the Sequoias' Laid to Rest," *Terra Bella News*, 1941, Folder 5, Box 1, SEKI Park Files; "Judge Walter Fry Taken by Death at Woodlake; Funeral Rites Tomorrow," newspaper clipping (newspaper unknown), Folder 16, Box 1, SEKI Park Files.
60. "Walter Fry—United States Commissioner, Sequoia and General Grant National Parks," 1925, Folder 14, Box 179, SEKI Park Files; Hays, *Conservation and the Gospel of Efficiency*; White, "'Are You an Environmentalist'"; Tweed, *Uncertain Path*, 45.
61. George Stewart to Anna Hays, 16 May 1924; and George Stewart to Enos Mills, 29 August 1916, both in File 1: Correspondence Related to the History of the Park, 1890–1927, Stewart Papers.
62. John Touhy, "Tulare County Parks: An Address to the Tulare Grange," *Tulare City Register*, 17 October 1890, Folder: Tulare County, Place Names, Sequoia National Park, Founding, 1890, Box 10, CSU-Schutt.
63. John Touhy to Honorable Senator Cockrell, 23 August 1890, Folder: Tulare County, Place Names, Sequoia National Park, Founding, 1890, Box 10, CSU-Schutt.
64. George Stewart to John R. White, 8 June 1929, File 1: Correspondence Related to the History of the Park, 1890–1927, Stewart Papers.
65. Strong, "History of Sequoia National Park," 93–99.
66. "An Act to Set Apart a Certain Tract of Land in the State of California as a Public Park," Pub. L. No. 51-926, 26 Stat. 478 (1890); "An Act to Set Apart Certain Tracts of Land in the State of California as Forest Reservations," Pub. L. No. 51-1263, 26 Stat. 650 (1890); Tweed and Dilsaver, *Challenge of the Big Trees*, 74–75; Tweed, *King Sequoia*, 96–99.
67. Berland, "Giant Forest's Reservation," 68–82; Orsi, *Sunset Limited*, 370–71; Strong, *Trees—Or Timber?*; *Map showing wagon roads and trails in territory embraced by Sequoia and General Grant National Parks* [map] (Washington, D.C.: Department of Interior, 1893), SEKI Park Files; Isaac N. Chapman, *Plat of the exterior boundaries of Sequoia National Park, California* [map] (San Francisco: U.S. Surveyor General's Office, 1901?), Library of Congress.
68. J. J. Martin, "A Co-Operative Commonwealth: The Kaweah Colony," *The Nationalist*, October 1889, 208.
69. Miller, *This Radical Land*, 201–6; Strong, "History of Sequoia National Park," 116; Tweed, *King Sequoia*, 100; *Report of Select Committee on Forest Reservations in California*.
70. Dorst, *Report of the Acting Superintendent*, 9; John Muir, "A Rival of Yosemite: The Cañon of the South Fork of King's River, California," *Century Magazine*, November 1891, 97.

71. Albright, Albright Schenck, and Tweed, *Mather Mountain Party of 1915*, 4–10.
72. Fry, *Report to the Secretary of the Interior . . . 1915*, 10–11; Albright, Albright Schenck, and Tweed, *Mather Mountain Party of 1915*, 13–24, quote 25.
73. Runte, *National Parks*, 43–56.
74. Mills, *National Parks for All the People*, 11–14; "Speech of Hon. Henry E. Barbour . . .," *Congressional Record* 62 (21 December 1922): 842–43, Folder 7, Box 159, SEKI Park Files; Strong, "History of Sequoia National Park," 235, 240–42; Robert Sterling Yard, "The Tehipite Valley and the Kings River Cañon, Greater Sequoia," *Sierra Club Bulletin*, 1918, 352–56; Henry H. Slayor, "The Proposed Roosevelt National Park," *Sierra Club Bulletin*, 1920, 29–33; William E. Colby, "The Proposed Enlargement of Sequoia National Park," *Sierra Club Bulletin*, 1924, 76–77.
75. "Quick Facts: Susan Thew," National Park Service, accessed 30 June 2018, www.nps.gov/people/susan-thew.htm); Susan P. Thew to George W. Stewart, 4 March 1926, File 6: Correspondence Related to Park Enlargement, 1913–1928, Stewart Papers; Susan P. Thew to George Stewart, 8 June 1926, File 6: Correspondence Related to Park Enlargement, 1913–1928, Stewart Papers; George Stewart to Susan P. Thew, 20 June 1926, File 6: Correspondence Related to Park Enlargement, 1913–1928, Stewart Papers.
76. Thew, *Proposed Roosevelt-Sequoia National Park*, 1.
77. Stephen T. Mather to Colonel Stewart, 20 March 1926, File 6: Correspondence Related to Park Enlargement, 1913–1928, Stewart Papers; H. E. Barbour to Geo. W. Stewart, 19 March 1926, File 6: Correspondence Related to Park Enlargement, 1913–1928, Stewart Papers.
78. Strong, "History of Sequoia National Park," 289.
79. See multiple letters dated March 1926 in File 6 of Stewart Papers for congressional praise of Thew's publication. "Roosevelt-Sequoia National Park," *Congressional Record* 67 (26 May 1926): 10043–48, File 0: Acts of Congress, Circulars, Committee Reports, Maps, Stewart Papers; "An Act to Revise the Boundary of Sequoia National Park, California," 16 U.S.C. § 45a; Tweed and Dilsaver, *Challenge of the Big Trees*, 147.
80. Freemuth, *Islands Under Siege*.
81. White and Pusateri, *Sequoia and Kings Canyon National Parks*, ix, 104; emphasis added.
82. Bernice Cosulich, "Huge Slice of Giant Sequoia Will Be Sent to Dr. Douglass," *Arizona Daily Star*, 22 November 1931, Folder 1, Box 95, Douglass Papers.
83. Webb, *Tree Rings and Telescopes*, 13–21, 102–3.
84. Webb, *Tree Rings and Telescopes*, 52, 84, 105; Douglass, "Weather Cycles in the Growth of Big Trees," 225–37.
85. Douglass, *Climatic Cycles and Tree Growth*, 15–29.
86. Douglass, *Climatic Cycles and Tree Growth*, 15–29.
87. A. J. Potter to European contacts, multiple letters, 7 October 1912, Folder 6, Box 75, Douglass Papers; Douglass, *Climatic Cycles and Tree Growth*, 29–41.
88. Tyrrell, *Reforming the World*, 106.
89. Fleming, *Historical Perspectives on Climate Change*, 95–106; Martin, *Ellsworth Huntington*, 5–26.
90. Huntington, "Geological and Physiographic Reconnaissance," 208; Martin, *Ellsworth Huntington*, 33–43; Davis, "Coming Desert," 23–43.
91. Martin, "Robert Lemoyne Barrett, 1871–1969," 29–31; Huntington, *Pulse of Asia*, 53; Pratt, *Imperial Eyes*, 38–56; Putnam et al., "Little Ice Age."

92. Huntington, *Pulse of Asia*, 359; Davis, "Coming Desert."
93. Ellsworth Huntington to Ruth Fletcher, 29 March 1910, Folder 260, Box 25, Ellsworth Huntington Papers, Manuscripts & Archives, Sterling Library, Yale University.
94. Huntington, *Secret of the Big Trees*, 3.
95. Huntington, *Secret of the Big Trees*, 14.
96. Huntington, *Secret of the Big Trees*, 20.
97. Ellsworth Huntington to Daniel T. MacDougal, 11 October 1911, Folder 429, Box 27, Ellsworth Huntington Papers, Manuscripts & Archives, Sterling Library, Yale University.
98. Huntington, "Fluctuating Climate of North America."
99. Huntington, *Climatic Factor as Illustrated in Arid America*, 173.
100. McGraw, *A. E. Douglass and the Role of the Giant Sequoia*, 75.
101. A. E. Douglass, "Survey of Sequoia Studies," *Tree-Ring Bulletin*, April 1945, 26–27, Folder 1, Box 123, Douglass Papers; A. E. Douglass, "Survey of Sequoia Studies, II," *Tree-Ring Bulletin*, October 1945, 11, Folder 1, Box 123, Douglass Papers.
102. Frederic E. Clements to A. E. Douglass, 18 February 1915, Folder 1, Box 80, Douglass Papers; Frederic E. Clements to A. E. Douglass, 14 August 1918, Folder 1, Box 80, Douglass Papers.
103. Douglass, "Survey of Sequoia Studies"; Douglass, "Survey of Sequoia Studies, II."
104. Douglass, *Climatic Cycles and Tree Growth*, 54–64.
105. Kohler, *Landscapes and Labscapes*, 98–124.
106. A. E. Douglass to Ellsworth Huntington, 17 June 1916, Box 35, Folder 816, Ellsworth Huntington Papers, Manuscripts & Archives, Sterling Library, Yale University; A. E. Douglass to Joseph Barrell, 19 February 1916, Folder 2, Box 64, Douglass Papers.
107. Douglass, *Climatic Cycles and Tree Growth*, 47.
108. A. E. Douglass to Dr. Gambio, 27 February 1923, Folder 6, Box 64, Douglass Papers.
109. Douglass, "Evidence of Climatic Effects in the Annual Rings of Trees," 25.
110. Douglass, "Evidence of Climatic Effects in the Annual Rings of Trees."
111. E. Walter Maunder to A. E. Douglass, 18 February 1922, Folder 5, Box 64, Douglass Papers. For the scientific paper, see Maunder, "Prolonged Sunspot Minimum 1645–1715."
112. A. E. Douglass to E. Walter Maunder, 23 March 1922, Folder 5, Box 64, Douglass Papers.
113. Eddy, "Maunder Minimum."
114. John R. White to A. E. Douglass, 22 May 1929, Folder 4, Box 65, Douglass Papers.
115. Fry and White, *Big Trees*, 38.
116. "Scientists Say Sherman Sequoia 3,500 Years Old," *Fresno Bee*, 25 August 1931; Stephenson, "Estimated Ages of Some Large Giant Sequoias."
117. Donahue, "Historian with a Chainsaw."
118. Christopher Baisan, interview by author, 12 May 2017; Parker, *Global Crisis*, 13–16.
119. A. E. Douglass to John C. Merriam, 17 September 1929, Folder 4, Box 78, Douglass Papers.
120. A. E. Douglass to John C. Merriam, 1 May 1930, Folder 4, Box 78, Douglass Papers.
121. A. E. Douglass, "Some Aspects of the Use of the Annual Rings of Trees in Climatic Study," *Scientific Monthly*, July 1922, 5–6; Andrew Ellicott Douglass, "The Secret of the Southwest Solved by Talkative Tree Rings," *National Geographic*, December 1929, 736–70.

Chapter 6

1. Tony Caprio, "Burned Areas—Redwood Mountain, Kings Canyon National Park" [map], 8 July 2013, in author's possession.
2. Rose, *Kings Canyon*, 109–18; Schrepfer, *Nature's Altars*, 168–74; Furmansky, *Rosalie Edge, Hawk of Mercy*, 221–34; "An Act to Establish the Kings Canyon National Park, California, to Transfer Thereto the Lands Now Included in the General Grant National Park, and for Other Purposes," Pub. L. 76-424, 54 Stat. 43 (1940).
3. Rumore, "Preservation for Science," 613–50; Patrick Kupper, *Creating Wilderness*.
4. Bruce Kilgore, "Restoring Fire to the Sequoias," *National Parks & Conservation Magazine*, October 1970, 16–22, Folder 6, Box 253, SEKI Park Files; Hartesveldt et al., *Giant Sequoias*, 72–74.
5. Nathan Stephenson, interview by author, 15 June 2017; Ambrose, "Hydraulic Constraints."
6. Stephenson interview; Stephenson et al., "Patterns and Correlates of Giant Sequoia"; Nydick et al., "Leaf to Landscape Responses."
7. Parsons, "Objects or Ecosystems?"; Stephenson, "Making the Transition."
8. "Fire in Sequoia Forest Declared to Be Checked," *Los Angeles Times*, 7 August 1928; "Fire Perils Sequoias," *Los Angeles Times*, 8 August 1928; "Brush Fire on Kaweah Stubborn," *Los Angeles Times*, 10 August 1928.
9. John R. White to Hon. Stephen T. Mather, 24 August 1928, Folder 2, Box 1, SEKI Fire Records; John D. Coffman, "Subject: South Fork of Kaweah Fire," 16 October 1928, Folder 3, Box 1, SEKI Fire Records; Horace M. Albright to John R. White, 10 November 1928, SEKI Fire Records; Pyne, *Fire in America*, 100–112; Rothman, *Blazing Heritage*, 33–43.
10. Swain, *Wilderness Defender*; Sutter, *Driven Wild*, 104–40; Horace M. Albright to Sir, 15 September 1924, Folder 2, Box 1, SEKI Fire Records.
11. Hydrick, "Genesis of National Park Management," 68–81; Tweed and Dilsaver, *Challenge of the Big Trees*, 111–12.
12. Tweed and Dilsaver, *Challenge of the Big Trees*, 111–12; Emilio Meinecke, "Memorandum on the Effects of Tourist Traffic on Plant Life, Particularly Big Trees, Sequoia National Park, California," unpublished report, May 1926, SEKI Park Files; White, "Atmosphere in the National Parks."
13. Rothman, *Blazing Heritage*, 40–42; White, *Superintendent's Annual Report, Sequoia National Park, 1926*, 10.
14. John R. White to Ansel F. Hall, 5 January 1929, Folder 2, Box 275, SEKI Park Files.
15. Fry and White, *Big Trees*, 79, 85.
16. Unknown typescript, "Hale Tharp, hunter and early settler in this region . . .," 14 December 1928, Folder 14, Box 179, SEKI Park Files.
17. Walter Fry, "History of Hospital Rock Country Ably Recounted by First White Man to Visit Indian Tenanted Regions," *Visalia Times Delta*, 13 November 1930, Folder 14, Box 179, SEKI Park Files.
18. M. Anderson, *Indian Fire-Based Management*, 245–46.
19. M. Anderson, *Indian Fire-Based*, 152–89; Anderson, *Tending the Wild*.
20. Moratto, *California Archaeology*, 329, 336–37.
21. M. Anderson, *Indian Fire-Based Management*, 298–99; Ron W. Goode, interview by author, 28 January 2019. For non-Native names, Goode's mother was Lena Adeline Kinsman (1911–79) and his great-grandmother was Julia Riley (1820/38–1941).

22. M. Anderson, *Indian Fire-Based Management*, 298–99.
23. M. Anderson, *Indian Fire-Based Management*, 295–96.
24. Kilgore and Taylor, "Fire History of a Sequoia–Mixed Conifer Forest," 129–42; Christensen et al., *Final Report*; Anderson, *Tending the Wild*, 183.
25. O'Neal, "Two Blades of Grass"; M. Anderson, *Indian Fire-Based Management*, 299.
26. George L. Hoxie, "How Fire Helps Forestry," *Sunset*, July–December 1910, 145.
27. John R. White to National Park Service director, 16 August 1924, Folder 2, Box 275, SEKI Park Files.
28. White, *Bullets and Bolos*, 43; Kida, "Colonel John Roberts White."
29. Albright and Albright Schenck, *Creating the National Park Service*, 279.
30. Cronon, "Trouble with Wilderness," 7–28; Pyne, *Fire in America*, 81–83.
31. Pyne, *Fire in America*, 243–49; DuBois, *Systematic Fire Protection in the California Forests*, 6.
32. Albright and Schenck, *Creating the National Park Service*, 279.
33. L. F. Cook, "Fire Control Plan—Sequoia National Park—Season of 1928," Folder 1, Box 1, SEKI Fire Records.
34. Price, "Fire Prevention Plan for the National Parks"; Schuft, "Prescribed Burning Program."
35. Maher, *Nature's New Deal*, 3–16, 55–56; Paige, *Civilian Conservation Corps*.
36. Lawrence Cook to John R. White, memorandum, 23 May 1934, Folder 17, Box 2, SEKI Fire Records.
37. "Cumulative Fire Statistics" and "Annual Fire Report," 1933–1942, Folder 13, Box 77, SEKI Park Files; Guy Hopping to Western Region director, memorandum, 30 July 1940, Folder 36, Box 4, SEKI Fire Records.
38. Gordon Kerr, "The Protection of Park's Resources Against Loss by Fire," Folder 11, Box 248, SEKI Park Files.
39. Kennedy, *Freedom from Fear*, 709–11, 776–82; "Sequoia Park Has First Woman Fire Watcher," *Fresno Bee*, 5 July 1943.
40. "Cumulative Fire Statistics" and "Annual Fire Report," 1929–1952, Folder 13, Box 77, SEKI Park Files.
41. Richard H. Boyer, "Supplemental Narrative Report on KC-8 McGee Ranch Fire 123.11—September 2 through November 13, 1955," Folder 70, Box 8, SEKI Fire Records; "Forest Fire Roars Up in Sierras," *Los Angeles Times*, 4 September 1955.
42. Eivind T. Scoyen, "Superintendent's Monthly Narrative Report—September 1955," Folder 53, Box 8, SEKI Supt. Records; Boyer, "Supplemental Narrative Report on McGee Ranch Fire."
43. R. H. Boyer, "Individual Forest Fire Report: McGee Ranch," 3 December 1955, Folder 70, Box 8, SEKI Fire Records; Boyer, "Supplemental Narrative Report on McGee Ranch Fire."
44. Henry H. Miwa to Whom It May Concern, 26 June 1960, Folder 75, Box 9, SEKI Fire Records.
45. Wayne R. Howe, "Narrative Report-Tunnel Rock-123.41," Folder 75, Box 9, SEKI Fire Records; "Fire Jumps East in Sequoia Park," *Los Angeles Times*, 28 June 1960.
46. John H. Davis, "Tunnel Rock Fire, Sequoia National Park, Threatens Giant Sequoia Groves of Giant Forest and Park Headquarters Development," *1960 Annual Report*, SEKI Park Files.
47. Rose Disbrow, "Tunnel Rock Fire Review," 20 July 1960, Folder 5, Box 248, SEKI Park Files.

48. Schuft, "Prescribed Burning Program."
49. Harold Biswell, "The Big Trees and Fire," *National Parks*, 11–14 April 1961, 14.
50. Sellars, *Preserving Nature in the National Parks*, 204–66.
51. Leopold et al., *Wildlife Management in the National Parks*, 6; emphasis added.
52. Haraway, "Situated Knowledges," 575–99; Latour, *Science in Action*, 215–57; Pyne, *Between Two Fires*, 47–48; Smith, *Engineering Eden*, 56–62, 101–17; Carle, *Burning Questions*, 32, 57–58, 98–130; Rothman, *Blazing Heritage*, 93–95.
53. Carle, *Burning Questions*, 121.
54. Biswell, Gibbens, and Buchanan, "Fuel Conditions and Fire Hazard Reduction," 2–4; Biswell and Weaver, "Redwood Mountain," 20–23.
55. Ron Taylor, "Underbrush Removal Pushes Sequoia Study Ahead," *Fresno Bee*, 7 June 1965, Folder 6, Box 253, SEKI Park Files; Hartesveldt and Harvey, "Fire Ecology of Sequoia Regeneration."
56. Rothman, *Blazing Heritage*, 112.
57. Kilgore, "Origin and History of Wildland Fire Use"; Kilgore and Briggs, "Restoring Fire to High Elevation Forests in California."
58. Thomas F. Ela to Western Region Director, 23 October 1969, Folder 2, Box 230, SEKI Park Files; Sequoia–Kings Canyon Superintendent to Western Region Director, memorandum, 1 February 1970, Folder 1, Box 230, SEKI Park Files; Kilgore, "Impact of Prescribed Burning on a Sequoia–Mixed Conifer Forest"; Carle, *Burning Questions*, 144.
59. Kilgore, "Fire Management in the National Parks," 54.
60. McLaughlin, "Restoring Fire to the Environment in Sequoia and Kings Canyon National Parks," 394.
61. Carle, *Burning Questions*, 140.
62. Rothman, *Blazing Heritage*, 107; Kilgore and Biswell, "Seedling Germination Following a Fire in a Giant Sequoia Forest."
63. Briggs, *Resource Management Objectives*, 5; H. Thomas Nichols, interview by author, 12 July 2017.
64. Nichols interview.
65. Kaufman, *National Parks and the Woman's Voice*, 121–26, 169–72; Sandra L. Graban, interview by author, 27 February 2019; "Sequoia–Kings Canyon National Parks: Redwood Mountain Prescribed Burn Report," Fall 1975, Folder 25, Box 2, SEKI Fire Effects.
66. Bancroft et al., "Evolution of the Natural Fire Management Program at Sequoia and Kings Canyon National Parks," 175.
67. William C. Tweed, interview by author, 23 August 2017.
68. Donald J. McGraw, interview by author, 11 May 2017.
69. Rothman, *Blazing Heritage*, 157–90; Carle, *Burning Questions*, 191–208.
70. Malinee Crapsey, interview by author, 14 January 2019.
71. Pyne, *Between Two Fires*, 198–99.
72. Stephen J. Pyne, "Passing the Torch," *American Scholar*, 1 March 2008, https://theamericanscholar.org/passing-the-torch/#.XC-cylxKjIU.
73. Leopold, "Speech to the National Park Service."
74. Bancroft et al., "Evolution of the Natural Fire Management Program," 176.
75. Kilgore and Taylor, "Fire History of a Sequoia–Mixed Conifer Forest," 139.

76. David J. Parsons, interview by author, 18 July 2017.
77. Smith, *Engineering Eden*, 297.
78. Bonnicksen and Stone, "Reconstruction of a Presettlement Giant Sequoia–Mixed Conifer Forest"; Bonnicksen and Stone, "Managing Vegetation"; Smith, *Engineering Eden*, 298.
79. David M. Graber to A. Starker Leopold, 5 April 1983, Carton 2, Series 1, A. Starker Leopold Papers, Bancroft Library, University of California, Berkeley.
80. Smith, *Engineering Eden*, 302.
81. Parsons interview.
82. A. Starker Leopold, interview by Carol Holleuffer, 14 June 1983, *Sierra Club Nationwide II* (San Francisco: Sierra Club Oral History Project, 1984), 15–17.
83. A. Starker Leopold to Boyd Evison, 9 June 1983, Carton 2, Series 1, A. Starker Leopold Papers, Bancroft Library, University of California, Berkeley.
84. Smith, *Engineering Eden*, 303.
85. Thomas W. Swetnam, interview by author, 14 July 2017.
86. John Todd, "Controlled Fires Under Sequoias Spark Concern," *San Francisco Examiner*, 17 November 1985, Folder 359, Box 9, SEKI Fire Effects; Eric Barnes, "Sequoia in Flames," 7 November 1985, Folder 359, Box 9, SEKI Fire Effects; Erik K. Barnes to John E. Davis, 27 November 1985, Folder 359, Box 9, SEKI Fire Effects.
87. David R. Brower to John H. Davis, 26 June 1986, Folder 377, Box 10, SEKI Fire Effects.
88. John B. Dewitt to Members of the Honorable Review Panel, 18 July 1986, Folder 377, Box 10, SEKI Fire Effects.
89. Dale Champion, "Growing Criticism Over Controlled Sequoia Burns," *San Francisco Chronicle*, 25 June 1986, Folder 359, Box 9, SEKI Fire Effects; Ronald B. Taylor, "Controlled Park Fires: A Burning Question," *Los Angeles Times*, 6 July 1986, Folder 359, Box 9, SEKI Fire Effects.
90. Christensen et al., "Review of Fire Management Program."
91. Christensen et al., "Review of Fire Management Program."
92. Christensen et al., "Review of Fire Management Program," 32; emphasis added.
93. Swetnam et al., *Giant Sequoia Fire History*.
94. Swetnam interview.
95. Swetnam interview.
96. Stephenson, Parsons, and Swetnam, "Restoring Natural Fire to the Sequoia–Mixed Conifer Forest"; Mutch and Swetnam, "Effects of Fire Severity and Climate."
97. Swetnam interview.
98. Swetnam, "Fire History and Climate Change in Giant Sequoia Groves," 885–89.
99. Howe, *Behind the Curve*, 147–48, 161; Houghton et al., *Climate Change 1995*, 4.
100. Swetnam interview; Swetnam et al., "Multi-Millennial Fire History of the Giant Forest," 120–50.
101. Adrian J. Das and Nathan L. Stephenson, "A Natural Resource Condition Assessment for Sequoia and Kings Canyon National Parks: Appendix 22—Climatic Change," Natural Resource Report NPS/SEKI/NRR—2013/665.22, June 2013; Safford et al., "Climate Change and the Relevance of Historical Forest Conditions."
102. Parsons, "Planning for Climate Change in National Parks," 255–69.
103. "Summary of Proceedings for Fire Effects/Prescribed Fire Workshop," 20–22 January 1993, Folder 57, Box 3, SEKI Fire Effects.

104. Anthony C. Caprio, interview by author, 14 June 2017; Caprio and Graber, "Returning Fire to the Mountains."
105. Tom Nichols, "The Fire Management Program of the National Park Service: Stall and Descent," *National Parks Traveler*, 7 November 2016, www.nationalparkstraveler.org/2016/11/fire-management-program-national-park-service-stall-and-descent; Harvey, *Brief History of Neoliberalism*.
106. Caprio interview; Sequoia and Kings Canyon National Parks, *Vignettes of the 2015 Rough Fire*.
107. Chin and Sillet, "Phenotypic Plasticity of Leaves"; Thayer Walker, "Are Giant Sequoia Trees Succumbing to Drought?" *Scientific American*, 29 December 2016, www.scientificamerican.com/article/are-giant-sequoia-trees-succumbing-to-drought/.
108. Walker, "Are Giant Sequoia Trees Succumbing to Drought?"; US Geological Survey, "Hotter Droughts, Forests and the Leaf to Landscape Project," 22 March 2016, www.usgs.gov/news/hotter-droughts-forests-and-leaf-landscape-project.
109. Martin et al., "Remote Measurement of Canopy Water Content"; Paz-Kagan et al., "Landscape-Scale Variation."
110. Nydick, *Climate-Smart Resource Stewardship*, 84.
111. Stephenson interview; Nathan Stephenson, personal communication with author, 1 February 2019.
112. Nydick, *Climate-Smart Resource Stewardship Strategy*, 8.
113. Goode interview.
114. Jane C. Allen, interview by author, 4 December 2018.
115. Ron W. Goode, "Cultural Burn" paper, March 2014, copy borrowed by author.
116. Adlam et al., "Keepers of the Flame"; Chris Clarke, "Burning for Acorns in Sequoia National Park: Native Peoples and the Park Service Working Together," KCET, 7 June 2017, www.kcet.org/shows/tending-the-wild/burning-for-acorns-in-sequoia-national-park-native-peoples-and-the-park.
117. Ron W. Goode, "Tribal-Traditional Ecological Knowledge," *Nuck-A-Hee: Newsletter of the Sierra Mono Museum*, Summer 2015.

Conclusion

1. Kaczynski, Cooper, and Jacobi, "Interactions of Sapsuckers and Cytospora Canker."
2. Wu et al., "Projected Avifaunal Responses to Climate Change"; Langham et al., "Conservation Status of North American Birds"; "How Climate Change Will Reshape the Range of the Wilson's Warbler," Audubon Society, accessed March 18, 2021, www.audubon.org/field-guide/bird/wilsons-warbler.
3. "Common Yellowthroat—Sightings Map," Cornell Lab's All About Birds, accessed 23 March 2021, www.allaboutbirds.org/guide/Common_Yellowthroat/maps-sightings; "Bison Bellows: Wrangell-St. Elias National Park and Preserve," National Park Service, March 2016, www.nps.gov/articles/bison-bellows-3-3-16.htm; Alaska Department of Fish and Game, "Wood Bison News," Spring 2018, www.adfg.alaska.gov/static/research/wildlife/species/woodbisonrestoration/pdfs/woodbison_news_10_spring_2018.pdf; Claire Stremple, "Mule Deer, White-Tail Deer Expand Range into Alaska," KHNS FM, 2 August 2019, https://khns.org/mule-deer-white-tail-deer-expand-range-into-alaska; Ben R. Bobowski, interview by author, 12 March 2021.

4. "IPCC, 2021: Summary for Policymakers," 3–32; Tollefson, "IPCC Climate Report"; Chen et al., "Rapid Range Shifts of Species," 1024–26; Craig Welch, "Half of All Species Are on the Move—And We're Feeling It," *National Geographic*, 26 April 2017, www.nationalgeographic.com/science/article/climate-change-species-migration-disease; Shah, *Next Great Migration*, 1–31; Miller, *Empire of Borders*, 248–57.
5. Fields, *Enclosure*, 1–22; Vallet, *Borders, Walls, Fences*, 2.
6. Ruiz Benedicto, Akkerman, and Brunet, *Walled World*, 6; Johnson, *Companions in Conflict*; Garet Bleir, "Endangered Species Are Casualties of Trump's Border Wall," *Sierra*, 18 February 2020, www.sierraclub.org/sierra/endangered-species-are-casualties-trump-s-border-wall; Andy Coghlan and Mićo Tatalović, "Fences Put Up to Stop Refugees in Europe Are Killing Animals," *New Scientist*, 17 December 2015, www.newscientist.com/article/dn28685-fences-put-up-to-stop-refugees-in-europe-are-killing-animals/; Kim Hjelmgaard, "Border Walls Aim to Keep Out Migrants, but Also Threaten Bears, Deer, Other Wildlife," *USA Today*, 24 May 2018, www.usatoday.com/story/news/world/2018/05/24/donald-trump-europe-border-walls-wildlife/567756002/.
7. Rigaud, et al., *Groundswell*.
8. Georgina Gustin, "Ravaged by Drought, a Honduran Village Faces a Choice: Pray for Rain or Migrate," Inside Climate News, 8 July 2019, https://insideclimatenews.org/news/08072019/climate-change-migration-honduras-drought-crop-failure-farming-deforestation-guatemala-trump/; Miranda Cady Hallett, "How Climate Change Is Driving Emigration from Central America," *The Conversation*, 6 September 2019, https://theconversation.com/how-climate-change-is-driving-emigration-from-central-america-121525.
9. "Lower Rio Grande Valley—Creating a Wildlife Corridor," U.S. Fish and Wildlife Service, accessed March 27, 2025, www.fws.gov/refuge/Lower_Rio_Grande_Valley/resource_management/wildlife_corridor.html; "Marianna Trevino Wright—Executive Director for the Rio Grande Valley Butterfly Center," YouTube, posted 27 February 2021, by The Ef'n Sonny Show, www.youtube.com/watch?v=H6c6moO-67s.
10. Marianna Treviño-Wright, "National Butterfly Center Hits a Wall—And Keeps Going," *American Butterflies*, Summer 2019.
11. Treviño-Wright, "National Butterfly Center Hits a Wall."
12. "Between the River and the Wall: Wild Creatures and Features on the Chopping Block," *Wild Without End* (blog), Defenders of Wildlife, 24 January 2019, https://medium.com/wild-without-end/between-the-river-the-wall-wild-creatures-and-features-on-the-chopping-block-397fc5495c45.
13. Gus Bova, "'We Build the Wall' Lands in South Texas, Vilifies Priest and Butterfly Refuge," *Texas Observer*, 21 November 2019, www.texasobserver.org/we-build-the-wall-south-texas-vilifies-priest-butterfly-refuge/; "Marianna Trevino Wright—Executive Director for the Rio Grande Valley Butterfly Center," YouTube.
14. "Marianna Trevino Wright—Executive Director for the Rio Grande Valley Butterfly Center," YouTube.; Ben Schreckinger, "Trump Acknowledges Climate Change—At His Golf Course," *Politico*, 23 May 2016, www.politico.com/story/2016/05/donald-trump-climate-change-golf-course-223436; Klein, *This Changes Everything*, 154; Chakrabarty, *Climate of History in a Planetary Age*, 45, 137; Carter and Poast, "Why Do States Build Walls?"

15. Walia, *Border and Rule*; "Marianna Trevino Wright—Executive Director for the Rio Grande Valley Butterfly Center," YouTube.
16. "Leaders of 'We Build the Wall' Online Fundraising Campaign Charged with Defrauding Hundreds of Thousands of Donors," US Department of Justice, 20 August 2020, www.justice.gov/usao-sdny/pr/leaders-we-build-wall-online-fundraising-campaign-charged-defrauding-hundreds-thousands; Jeremy Schwartz and Perla Trevizo, "Eroding Private Border Wall to Get an Engineering Inspection Just Months After Completion," *ProPublica*, 9 July 2020, www.propublica.org/article/eroding-private-border-wall-to-get-an-engineering-inspection-just-months-after-completion; Perla Trevizo and Jeremy Schwartz, "A Privately Funded Border Wall Was Already at Risk of Collapsing If Not Fixed. Hurricane Hanna Made It Worse," *ProPublica*, 29 July 2020, www.propublica.org/article/a-privately-funded-border-wall-was-already-at-risk-of-collapsing-if-not-fixed-hurricane-hanna-made-it-worse.
17. Peters et al., "Nature Divided, Scientists United."
18. Belle et al., *Protected Planet Report 2018*, 6; Stephen Leahy, "Half of All Land Must Be Kept in Natural State to Protect Earth," *National Geographic*, 19 April 2019.
19. John Wendle, "Syria's Climate Refugees," *Scientific American*, March 2016; Hussein A. Amery, "Climate, Not Conflict, Drove Many Syrian Refugees to Lebanon," *The Conversation*, 3 December 2019, https://theconversation.com/climate-not-conflict-drove-many-syrian-refugees-to-lebanon-127681; Kelley et al., "Climate Change in the Fertile Crescent"; Cook et al., "Spatiotemporal Drought Variability."
20. Linnell et al., "Border Security Fencing and Wildlife," 1–13.

BIBLIOGRAPHY

Primary Sources

Archives

ARIZONA
University of Arizona Libraries, Tucson
   Special Collections
      Papers of Andrew Ellicott Douglass

ARKANSAS
William J. Clinton Presidential Library, Little Rock
   Agricultural Biotechnology Files

CALIFORNIA
California State University, Fresno
   Madden Library
      Harold G. Schutt Collection
      June English Forestry Collection
Sequoia and Kings Canyon National Parks Museum Collection, Three Rivers
   Fire Monitoring and Fire Effects Records
   Fire Records
   Park Files
Tulare County Public Library, Visalia
   George W. Stewart Papers
University of California, Berkeley
   Bancroft Library
      A. Starker Leopold Papers

CANADA
University of Calgary, Calgary, Alberta
   Glenbow Western Research Centre
Library and Archives Canada, Ottawa, Ontario
   Microfilm Collection
   Record Group 10: Records of Indian Affairs
   Record Group 18: Records of Royal Canadian Mounted Police.
   Record Group 25: Records of Department of External Affairs
   Record Group 85: Records of North Affairs Program
Whyte Museum of the Canadian Rockies, Banff, Alberta
   Hubert Green Fonds
   Parks Canada Fonds
   Yellowstone to Yukon Collection

University of Toronto
    Special Collections Library
        Frederick Urquhart Fonds

CONNECTICUT
Yale University, New Haven
    Sterling Library Manuscripts & Archives
        Ellsworth Huntington Papers

KANSAS
University of Kansas, Lawrence
    Monarch Watch Library
        Insect Migration Studies Newsletters
        Monarch Watch Season Summaries

MARYLAND
National Archives and Records Administration, College Park
    Record Group 22: Records of U.S. Fish and Wildlife Service

MEXICO
Archivo General de la Nación, Mexico City
    Administración Presidencial de Ernesto Zedillo Ponce de León
    Administración Presidencial de Felipe Calderón Hinojosa
    Administración Presidencial de Lázaro Cárdenas
    Administración Presidencial de Miquel de la Madrid Hurtado
    Administración Presidencial de Vicente Fox Quesada

MICHIGAN
Michigan State University, East Lansing
    Archives and Historical Collections
        Charles H. Hackley and Thomas Hume Papers

MISSOURI
Washington University Archives, St. Louis
    Monsanto Company Records

MONTANA
Montana Historical Society Research Center, Helena
    Montana Stockgrowers Association Records
University of Montana, Missoula
    Mansfield Library Archives and Special Collections
        Glacier National Park Collection
        Robert Ream Papers

NORTH CAROLINA
Forest History Society Library and Archives, Durham
    Oral History Collection

TEXAS
Texas Tech University, Lubbock
    Special Collections Library
        Robert Michael Pyle Papers

WASHINGTON, DC
Smithsonian Institution Archives
  Record Unit 7176: Field Reports of the US Fish and Wildlife Service

Oral Histories

Allen, Jane C. Interview by author, 4 December 2018.
Anderson-Ramirez, Malou. Interview by author, 25 February 2020.
Baisan, Christopher. Interview by author, 12 May 2017.
Bobowski, Ben R. Interviews by author, 20 November 2020, 12 March 2021.
Boyd, Diane K. Interview by author, 4 September 2019.
Brower, Lincoln P. Interview by Christopher Koehler, 14 March 1994. University of Florida Libraries.
Brower, Lincoln P. Interview by Mark Madison, 21 May 2015. https://digitalmedia.fws.gov/digital/collection/videos/id/142/rec/1.
Burgess, Darlene A. Interview by author, 1 March 2021.
Caprio, Anthony C. Interview by author, 14 June 2017.
Crapsey, Malinee. Interview by author, 14 January 2019.
Davis, Donald A. Interview by author, 20 November 2018.
Esquivel Maya, Osvaldo. Interview by Ellen Sharp, 9 May 2020. Oral History Collection, Forest History Society Library and Archives.
Goode, Ron W. Interview by author, 28 January 2019.
Gottfried Joy, Carlos Federico. Interview by author, 25 November 2020.
Graban, Sandra L. Interview by author, 27 February 2019.
Hilty, Jodi A. Interview by author, 11 September 2018.
Leopold, A. Starker. Interview by Carol Holleuffer, 14 June 1983. In *Sierra Club Nationwide II*, 15–17. San Francisco: Sierra Club Oral History Project, 1984.
McGraw, Donald J. Interview by author, 11 May 2017.
Moreno de Jesús, Elidió. Interview by Patricio Moreno Rojas and Ellen Sharp, 27 April 2020. Oral History Collection, Forest History Society Library and Archives.
Moreno de Jesús, Melquiades. Interview by Patricio Moreno Rojas, 1 July 2020. Oral History Collection, Forest History Society Library and Archives.
Moreno Espinoza, José Leonel. Interview by Patricio Moreno Rojas and Ellen Sharp, 22 May 2020. Oral History Collection, Forest History Society Library and Archives.
Moreno Hernández, Francisco. Interview by Ellen Sharp, 9 May 2020. Oral History Collection, Forest History Society Library and Archives.
Moreno Rojas, Patricio. Interview by Ellen Sharp, 7 April 2020. Oral History Collection, Forest History Society Library and Archives.
Nichols, H. Thomas. Interview by author, 12 July 2017.
Oberhauser, Karen S. Interview by author, 16 June 2020.
Paquet, Paul C. Interview by author, 30 August 2019.
Pyle, Robert Michael. Interview by author, 4 September 2020.
Smith, Douglas W. Interview by author, 18 July 2018.
Stephenson, Nathan. Interview by author, 15 June 2017.
Swetnam, Thomas W. Interview by author, 14 July 2017.
Taylor, Orley R. Interview by author, 23 July 2020.
Tweed, William C. Interview by author, 23 August 2017.
Upson, Lisa. Interview by author, 4 February 2020.

Urquhart, Frederick A. Interviews by Paul A. Bator, 8–12 January 1979. Special Collections, University of Toronto.
Velázquez Moreno, Emilio. Interview by Ellen Sharp, 9 May 2020. Oral History Collection, Forest History Society Library and Archives.

Published Materials

Abbott, E. C., and Helen Huntington Smith. *We Pointed Them North: Recollections of a Cowpuncher*. 1939. Reprint, University of Oklahoma Press, 1955.
Albright, Horace M., and Marian Albright Schenck. *Creating the National Park Service: The Missing Years*. University of Oklahoma Press, 1999.
Albright, Horace Marden, Marian Albright Schenck, and William C. Tweed. *The Mather Mountain Party of 1915 and the Founding of the National Park Service*. Sequoia Natural History Association, 1990.
Anderson, J. B., and L. P. Brower. "Freeze-Protection of Overwintering Monarch Butterflies in Mexico: Critical Role of the Forest as a Blanket and an Umbrella." *Ecological Entomology* 21 (May 1996): 107–16.
Annual Report of the Department of Agriculture of Alberta 1916. J. W. Jeffery, 1917.
Annual Report of the Department of Agriculture of the North-West Territories 1898. John Alexander Reid, 1899.
Annual Report of the Department of Agriculture of the North-West Territories 1899. John Alexander Reid, 1900.
Annual Report of the Department of Agriculture of the Province of Alberta from the First of September, 1905, to the Thirty-First of December, 1906. Jas E. Richards, 1907.
Annual Report of the Department of Agriculture of the Province of Alberta 1907. Jas E. Richards, 1907.
Annual Report of the Department of Agriculture of the Province of Alberta 1908. Jas E. Richards, 1910.
Annual Report of the Department of the Interior. C. H. Parmelee, 1911.
Annual Report of the Entomological Society of Ontario, 1917. A. T. Wilgress, 1918.
Aridjis, Homero. "Conspiración contra la monarca." *Reforma*, 14 January 1996. In Aridjis and Ferber, *Noticias de la tierra*, 95–98.
———. "Declaración de 100 intelectuales y artistas contra la contaminación en la Ciudad de México." *Novedades*, 1 March 1985. In Aridjis and Ferber, *Noticias de la Tierra*, 27–29.
———. *El poeta niño*. 1971. Reprint, Fondo de Cultura Económica, 1997.
———. "El tamaño de la tala." *Reforma*, April 18, 1999. In Aridjis and Ferber, *Noticias de la Tierra*, 101–5.
———. "Grandeza y Miseria de la Mariposa Monarca," *La jornada*, 17 February 1993. In Aridjis and Ferber, *Noticias de la Tierra*, 89–93.
———. "La mariposa monarca: memoria y poesía." Earth of the Year 2000: PEN International-UNESCO Symposium, January 2000. In Aridjis and Ferber, *Noticias de la Tierra*, 79–86.
———. *La montaña de las mariposas*. Alfaguara, 2001.
———. "La nueva reserva de la mariposa monarca." *Reforma*, September 17, 2000. In Aridjis and Ferber, *Noticias de la Tierra*, 108–13.
———. "Los pájaros." In Aridjis and Ferber, *Noticias de la Tierra*, 19–25.
Aridjis, Homero, and Betty Ferber, eds. *Noticias de la Tierra*. Debate, 2012.
Aridjis, Homero, et al. "Letter to President Barack Obama, President Enrique Peña Nieto,

and Prime Minister Stephen Harper, 14 February 2014." In *News of the Earth*, edited by Homero Aridjis and Betty Ferber, 87–90, 382–83. Mandel Vilar Press, 2017.

Bailey, Vernon. *Animal Life of Yellowstone National Park*. Charles C. Thomas, 1930.

——. *Wolves in Relation to Stock, Game, and the National Forest Reserves*. Government Printing Office, 1907.

Bailey, Vernon, and Florence Merriam Bailey. *Wild Animals of Glacier National Park*. Glacier National Park, 1918.

Bancroft, Larry, Thomas Nichols, David Parsons, David Graber, Boyd Evison, and Jan van Wagtendonk. "Evolution of the Natural Fire Management Program at Sequoia and Kings Canyon National Parks." *Proceedings of the Symposium and Workshop on Wilderness Fire* (1985): 174–80.

Batty, Joseph H. *How to Hunt and Trap*. Orange Judd Company, 1884.

Belle, Elise, Naomi Kingston, Neil Burgess, Trevor Sandwith, Natasha Ali, and Kathy MacKinnon, eds. *Protected Planet Report 2018: Tracking Progress toward Global Targets for Protected Areas*. United Nations Environment Programme-World Conservation Monitoring Centre, 2018.

Biswell, Harold, and Harold Weaver. "Redwood Mountain." *American Forests* 74 (August 1968): 20–23.

Biswell, H. H., R. P. Gibbens, and Hayle Buchanan. "Fuel Conditions and Fire Hazard Reduction in a Giant Sequoia Forest." *California Agriculture* 22 (February 1968): 2–4.

Bonnicksen, Thomas M., and Edward C. Stone. "Managing Vegetation within U.S. National Parks: A Policy Analysis." *Environmental Management* 6 (March 1982): 109–22.

——. "Reconstruction of a Presettlement Giant Sequoia-Mixed Conifer Forest Community Using the Aggregation Approach." *Ecology* 63 (August 1982): 1134–48.

Boyd, Diane. "1995: The Return of the Wolf to Montana." In McIntyre, *War Against the Wolf*, 357–65.

——. "Food Habits and Spatial Relations of Coyotes and a Lone Wolf in the Rocky Mountains." Master's thesis, University of Montana, 1982.

Boyd, Diane K., Paul C. Paquet, Steve Donelson, Robert R. Ream, Daniel H. Pletscher, and Cliff C. White. "Transboundary Movements of a Recolonizing Wolf Population in the Rocky Mountains." In *Ecology and Conservation of Wolves in a Changing World*, edited by Ludwig. N. Carbyn, Dale R. Siep, and Steven H. Fritts, 135–40. Canadian Circumpolar Institute, 1995.

Boyd-Heger, Diane K. "Dispersal Genetic Relationships and Landscape Use by Colonizing Wolves in the Central Rocky Mountains." PhD diss., University of Montana, 1997.

Brett, Lloyd M. *Report of the Acting Superintendent of Yellowstone National Park to the Secretary of the Interior 1914*. Government Printing Office, 1914.

Brewster, Edwin Tenney. *Life and Letters of Josiah Dwight Whitney*. Houghton Mifflin Company, 1909.

Briggs, George. *Resource Management Objectives, Prescriptions, and Special Conditions Relating to Fall, 1976 Prescribed Burning*. Resources Management Office, 1976.

Brisbin, James S. *The Beef Bonanza; or, How to Get Rich on the Plains*. J. B. Lippincott & Company, 1881.

Brower, Lincoln P. "Biological Necessities for Monarch Butterfly Overwintering in Relation to the Oyamel Forest Ecosystem in Mexico." In Hoth et al., *1997 Reunión de América*, 5–7, 11–28.

——. "Monarch Migration." *Natural History* 86 (June–July 1977): 40–53.

Brower, Lincoln P., Jane Van Zandt Brower, and Joseph M. Corvino. "Plant Poisons in a Terrestrial Food Chain." *Proceedings of the National Academy of Sciences* 57 (April 1967): 893–98.

Brower, Lincoln P., William H. Calvert, Lee E. Hendrick, and John Christian. "Biological Observations of an Overwintering Colony of Monarch Butterflies (Danaus plexippus, Danaidae) in Mexico." *Journal of the Lepidopterists' Society* 31 (1977): 232–42.

Brown, John George. "Report of the Superintendent of Waterton Lakes Park." In *Annual Report of the Department of the Interior*. C. H. Parmelee, 1911.

Caprio, Anthony C., and David M. Graber. "Returning Fire to the Mountains: Can We Successfully Restore the Ecological Role of Pre-Euroamerican Fire Regimes to the Sierra Nevada?" In *Proceedings of the Wilderness Science in a Time of Change Conference*, edited by David N. Cole and Stephen F. McCool, 1–12. US Forest Service, 2000.

Carson, Rachel. *Silent Spring*. 1962. Reprint, Mariner Books, 2002.

Center for Biological Diversity, Center for Food Safety, Xerces Society, and Lincoln Brower. "Petition to Protect the Monarch Butterfly (*Danaus plexippus plexippus*) Under the Endangered Species Act." 26 August 2014. www.biologicaldiversity.org/species/invertebrates/pdfs/Monarch_ESA_Petition.pdf.

Christensen, Norman L., Lin Cotton, Thomas Harvey, et al. *Final Report: Review of Fire Management Program for Sequoia–Mixed Conifer Forests of Yosemite, Sequoia and Kings Canyon National Parks*. 22 February 1987. https://npshistory.com/publications/seki/fire-mgt-1987.pdf.

Comstock, Anna Botsford. *The Comstocks of Cornell: The Definitive Autobiography*. Edited by Karen Penders St. Clair. Cornell University Press, 2020.

———. *Handbook of Nature Study*. 1911. Reprint, Comstock Publishing Associates, 1967.

———. *Ways of the Six-Footed*. Ginn and Company, 1903.

Comstock, John Henry, and Anna Botsford Comstock. *How to Know Butterflies*. D. Appleton & Company, 1904.

Craig, John R. *Ranching with Lords and Commons; or, Twenty Years on the Range*. William Briggs, 1903.

Crolla, Jeffrey P., and J. Donald Lafontaine. *Status Report on the Monarch Butterfly (Danaus plexippus) in Canada*. Canadian Wildlife Service, 1996. https://monarchwatch.org/read/articles/canmon1.htm.

Diamond, Jared M. "The Island Dilemma: Lessons of Modern Biogeographic Studies for the Design of Natural Reserves." *Biological Conservation* 7 (February 1975): 129–46.

Dilsaver, Lary, ed. *America's National Park System: The Critical Documents*. Rev. ed. Rowman & Littlefield, 2016.

Dorst, J. H. *Report of the Acting Superintendent of Sequoia National Park*. Government Printing Office, 1891.

Douglass, A. E. "Evidence of Climatic Effects in the Annual Rings of Trees." *Ecology* 1 (January 1920): 22–32.

———. *Climatic Cycles and Tree Growth: A Study of the Annual Rings of Trees in Relation to Climate and Solar Activity*. Carnegie Institution, 1919.

———. "Weather Cycles in the Growth of Big Trees." *Monthly Weather Review* 37 (June 1909): 225–37.

DuBois, Coert. *Systematic Fire Protection in the California Forests*. Government Printing Office, 1914.

Eddy, John A. "The Maunder Minimum." *Science* 192 (June 1976): 1189–202.
Executive Office of the President, Office of Science and Technology Policy. "Coordinated Framework for Regulation of Technology," 26 June 1986. www.aphis.usda.gov/brs/fedregister/coordinated_framework.pdf.
Fancy, Steve G. *Satellite Telemetry: A New Tool for Wildlife Research and Management*, Resource Publication 172. US Fish and Wildlife Service, 1988.
Fifth Annual Report of the Bureau of Agriculture, Labor, and Industry for the Year Ending November 30, 1897. State Publishing Company, 1898.
Fink, Linda S., and Lincoln P. Brower. "Birds Can Overcome Cardenolide Defence of Monarch Butterflies in Mexico." *Nature* 291 (7 May 1981): 67–70.
Fischer, Hank. "1995: Supply-Side Environmentalism and Wolf Recovery in the Northern Rockies." In McIntyre, *War Against the Wolf*, 410–15.
———. *Wolf Wars: The Remarkable Inside Story of the Restoration of Wolves to Yellowstone*. Falcon Press, 1995.
Fletcher, Robert, and Bram Büscher. "Conservation Basic Income: A Non-market Mechanism to Support Convivial Conservation." *Biological Conservation* 244 (April 2020): 1–7.
Franklin, Jerry F. "The Biosphere Reserve Program in the United States." *Science* 195 (21 January 1977): 262–67.
Fry, Walter. *Report to the Secretary of the Interior by the Superintendent of Sequoia and General Grant National Parks 1915*. Government Printing Office, 1915.
Fry, Walter, and John R. White, *Big Trees*. Stanford University Press, 1930.
Gallegos, Carlos Melo, and José López García. "Contribucion Geográfica al Programa de Desarrollo Mariposa Monarca." *Investigaciones Geográficas* 19 (1989): 9–26.
Gottfried, Carlos F. *Monarcas*. Grupo Condumex, 1984.
Green, Herbert U. *Wolves of Banff National Park*. Ministry of Resources and Development, 1951.
Hansen Jesse, Laura C., and John J. Obrycki. "Field Deposition of Bt Transgenic Pollen: Lethal Effects on the Monarch Butterfly." *Oecologia* 125 (October 2000): 241–48.
Harding, A. R. *Steel Traps*. A. R. Harding Publishing Co., 1907.
Hartesveldt, R. J., and H. T. Harvey. "The Fire Ecology of Sequoia Regeneration." *Proceedings of the Tall Timbers Fire Ecology Conference* 7 (1967): 65–78.
Haskell, Burnette G. *A Pen Picture of the Kaweah Co-Operative Colony Co., Limited, a Joint Stock Company, Located in Kaweah Canyon, and the Giant Forest of Tulare Co., Cal*. Self-published, 1889.
Hodgson, John. "Monarch Bt-corn paper questioned." *Nature* 17 (July 1999): 627.
Hoth, Jürgen. "The Monarch: A Regal Opportunity for Working Together for Nature." In Hoth et al., *1997 Reunión de América*, 5–7.
Hoth, Jürgen, Leticia Merino, Karen Oberhauser, Irene Pisanty, Steven Price, and Tara Wilkinson, eds. *1997 Reunión de América del Norte sobre la Mariposa Monarca / 1997 North American Conference on the Monarch Butterfly*. Commission for Environmental Cooperation, 1999. www.fs.usda.gov/wildflowers/pollinators/Monarch_Butterfly/documents/1997_conference.pdf.
Houghton, J. T., L. G. Meira Filho, B. A. Callander, N. Harris, A. Kattenberg, and K. Maskell, eds. *Climate Change 1995: The Science of Climate Change*. Intergovernmental Panel on Climate Change. Cambridge University Press, 1996. www.ipcc.ch/site/assets/uploads/2018/02/ipcc_sar_wg_I_full_report.pdf.

Huntington, Ellsworth. *The Climatic Factor as Illustrated in Arid America*. Carnegie Institution, 1914.
———. "The Fluctuating Climate of North America." *Geographical Journal* 40 (October 1912): 410.
———. "A Geological and Physiographic Reconnaissance in Central Turkestan." In *Explorations in Turkestan with an Account of the Basin of Eastern Persia and Sistan*, edited by Raphael Pumpelly, 159–216. Carnegie Institution, 1905.
———. *The Pulse of Asia: A Journey in Central Asia Illustrating the Geographic Basis of History*. Houghton, Mifflin, and Company, 1907.
———. *The Secret of the Big Trees of Yosemite, Sequoia, and General Grant National Parks*. Government Printing Office, 1913.
Jepson, Willis Linn. *The Trees of California*. Cunningham, Curtis & Welch, 1909.
Kaweah Colony. *The By-Laws of the Joint-Stock Company, Limited, Kaweah*. J. J. Martin, 1888.
Kelly, L. V. *The Range Men: The Story of the Ranchers and Indians of Alberta*. William Briggs, 1913.
Kilgore, Bruce M. "Fire Management in the National Parks: An Overview." *Proceedings of the Tall Timbers Fire Ecology Conference* 14 (1976): 45–57.
———. "Impact of Prescribed Burning on a Sequoia-Mixed Conifer Forest." *Proceedings of the Tall Timbers Fire Ecology Conference* 12 (1973): 345–76.
Kilgore, Bruce M., and George S. Briggs. "Restoring Fire to High Elevation Forests in California." *Journal of Forestry* 70 (May 1972): 266–71.
Kilgore, Bruce M., and H. H. Biswell. "Seedling Germination Following a Fire in a Giant Sequoia Forest." *California Agriculture* 25 (February 1971): 8–10.
King, Calvin L. *Reasons for the Decline of Game in the Bighorn Basin of Wyoming*. Vantage Press, 1965.
Klein, T. M., E. D. Wolf, R. Wu, and J. C. Sanford. "High-Velocity Microprojectiles for Delivering Nucleic Acids into Living Cells." *Nature* 327 (7 May 1987): 70–73.
Koehler, Arthur. *Causes of Brashness in Wood*. Government Printing Office, 1933.
"Leonila Vázquez García: In Memoriam." *Anales de Instituto de Biología de Universidad Naciónal Autónomia de México* 66 (1995): 137–45.
Leopold, A. Starker. "A Speech to the National Park Service Western Region Superintendents' Resource Management Seminar." April 28, 1975. Reprinted in *George Wright Forum* 30 (2013): 203.
Leopold, A. Starker, S. A. Cain, C. M. Cottam, I. N. Gabrielson, and T. L. Kimball. *Wildlife Management in the National Parks: The Leopold Report*. Department of Interior, 1963.
Leopold, Aldo. *Game Management*. Charles Scribner's Sons, 1933.
———. *A Sand County Almanac*. 1949. Reprint. Ballantine Book, 1986.
Losey, John E., Linda S. Rayor, and Maureen E. Carter. "Transgenic Pollen Harms Monarch Larvae." *Nature* 399 (20 May 1999): 214.
Luxford. R. F., and L. J. Markwardt. *The Strength and Related Properties of Redwood*. Government Printing Office, 1932.
MacArthur, Robert H., and Edward O. Wilson, *The Theory of Island Biogeography*. Princeton University Press, 1967.
Maunder, E. Walter. "The Prolonged Sunspot Minimum 1645–1715." *Journal of the British Astronomical Association* 32 (1922): 140–45.

McDougall, John. *Pathfinding on Plain and Prairie: Stirring Scenes of Life in the Canadian North-West*. William Briggs, 1898.

McIntyre, Rick. "1992: Public Hearing on Wolf Environmental Impact Statement, Helena, Montana." In McIntyre, *War Against the Wolf*, 379–96.

———, ed. *War Against the Wolf: America's Campaign to Exterminate the Wolf*. Voyageur Press, 1995.

McLaughlin, John S. "Restoring Fire to the Environment in Sequoia and Kings Canyon National Parks." *Proceedings of the Tall Timbers Fire Ecology Conference* 12 (1972): 391–95.

McTaggert Cowan, Ian. "The Timber Wolf in the Rocky Mountain National Parks of Canada." *Canadian Journal of Research* 25 (October 1947): 139–74.

Merino, Leticia. "Reserva especial de la biosfera mariposa monarca: Problemática general de la region." In Hoth et al., *1997 Reunión de América*, 239–48.

Millard, Bailey. *History of the San Francisco Bay Region*. Volume 2. American Historical Society, 1924.

Mills, Enos. *The National Parks for All the People and Perhaps Our Greatest National Park (The Greater Sequoia)*. Government Printing Office, 1917.

Mollé, Nick, host. *A Walk in the Park with Nick Mollé*. "Birds Without Borders." Aired April 3, 2014, on PBS.

Mowat, Farley. *Never Cry Wolf*. 1963. Reprint, Back Bay Books, 2001.

Muir, John. "On the Post-glacial History of Sequoia Gigantea." *Proceedings of the American Association of the Advancement of Science* (August 1876): 3–15.

———. *Our National Parks*. Houghton, Mifflin, and Company, 1901.

Mutch, Linda S., and Thomas W. Swetnam. "Effects of Fire Severity and Climate on Ring-Width Growth of Giant Sequoia After Fire." *Proceedings of the Symposium on Fire in Wilderness and Park Management* (1995): 1–6.

Neubrech, W. LeRoy. *California Redwood and Its Uses*. Government Printing Office, 1937.

Newark, William D. "A Land-Bridge Island Perspective on Mammalian Extinctions in Western North American Parks." *Nature* 325 (29 January 1987): 430–32.

Nimmo, Joseph, Jr. *Report in Regard to the Range and Ranch Cattle Business of the United States*. Government Printing Office, 1885.

Nydick, Koren, editor. *A Climate-Smart Resource Stewardship Strategy for Sequoia and Kings Canyon National Parks*. National Park Service, 2017.

Oberhauser, Karen, and Dennis Frey. "Coercive Mating by Overwintering Male Monarchs." In Hoth et al., *1997 Reunión de América*, 67–78.

Oberhauser, Karen Suzanne. "Material Investment in Mating by Male Monarch Butterflies (*Danaus plexippus*)." PhD diss., University of Minnesota, 1989.

Oberhauser, Karen S., Michelle D. Prysby, Heather R. Mattila, Diane E. Stanley-Horn, Mark K. Sears, Galen Dively, Eric Olson, John M. Pleasants, Wai-Ki F. Lam, and Richard L. Hellmich. "Temporal and Spatial Overlap Between Monarch Larvae and Corn Pollen." *Proceedings of the National Academy of Sciences* 98 (9 October 2001): 11913–18.

O'Neal, John. "Two Blades of Grass Where Thousands Grew Before." *Western Livestock Journal* 31 (March 1953).

Otter, Floyd L. *The Men of Mammoth Forest: A Hundred-Year History of a Sequoia Forest and Its People in Tulare County, California*. Edwards Brothers, 1963.

Parsons, David J. "Objects or Ecosystems? Giant Sequoia Management in National Parks." *Proceedings of the Symposium on Giant Sequoias* (1994): 109–15.

———. "Planning for Climate Change in National Parks and Other Natural Areas." *Northwest Environmental Journal* 7 (1991): 255–69.

Parsons, J. A. "A Digitalis-Like Toxin in the Monarch Butterfly, *Danaus plexippus* L." *Journal of Physiology* 178 (May 1965): 290–304.

Peters, Robert, William J. Ripple, Christopher Wolf, Matthew Moskwik, Gerardo Carreón-Arroyo, Gerardo Ceballos, Ana Córdova, Rodolfo Dirzo, Paul R. Ehrlich, Aaron D. Flesch, Rurik List, Thomas E. Lovejoy, Reed F. Noss, Jesús Pacheco, José K. Sarukhán, Michael E. Soulé, Edward O. Wilson, and Jennifer R. B. Miller. "Nature Divided, Scientists United: US–Mexico Border Wall Threatens Biodiversity and Binational Conservation." *BioScience* 68 (October 2018): 740–43.

Pimlott, Douglas H. *Wolves: Proceedings of the First Working Meeting of Wolf Specialists and of the First International Conference on the Conservation of the Wolf.* International Union for the Conservation of Nature and Natural Resources, 1975.

Pinchot, Gifford. *Breaking New Ground*. 1947. Reprint, Island Press, 1998.

Pleasants, John M., and Karen S. Oberhauser. "Milkweed Loss in Agricultural Fields Because of Herbicide Use: Effect on the Monarch Butterfly Population." *Insect Conservation and Diversity* 6 (March 2013): 134–44.

Price, Jay H. "Fire Prevention Plan for the National Parks." In Dilsaver, *America's National Park System*, 65–70.

Province of British Columbia. *Preliminary Wolf Management Plan*. Ministry of Environment, 1979.

Pumpelly, Raphael, ed. *Explorations in Turkestan with an Account of the Basin of Eastern Persia and Sistan*. Carnegie Institution, 1905.

Pyle, Robert Michael. "The Eco-Geographic Basis for Lepidoptera Conservation." PhD diss., Yale University, 1976.

———. "International Efforts for Monarch Conservation." *Atala* 9 (1981–84): 21–22.

———. "International Problems in Insect Conservation." *Atala* 6 (1978): 56–58.

———. *Watching Washington Butterflies*. Seattle Audubon Society, 1974.

Reichstein, T., J. von Euw, J. A. Parsons, and Miriam Rothschild. "Heart Poisons in the Monarch Butterfly." *Science* 161 (30 August 1968): 861–66.

Report of Select Committee on Forest Reservations in California. Senate Report No. 1248. Government Printing Office, 1893.

Report on the Statistics of Agriculture in the United States at the Eleventh Census: 1890. Government Printing Office, 1895.

Rzedowski, Jerzy. "Nota sobre un vuelo migratorio de la mariposa Danaus Plexxippus L. observado el la region de Ciudad del Maíz, S.L.P." *Acta Zoológica Mexicana* 2 (March 1957): 1–4.

Safford, H. D., M. North, and M. D. Meyer. "Climate Change and the Relevance of Historical Forest Conditions." In *Managing Sierra Nevada Forests*, edited by Malcolm North, 23–45. US Forest Service, 2012.

Schuft, Peter H. "A Prescribed Burning Program for Sequoia and Kings Canyon National Parks." *Proceedings of the Tall Timbers Fire Ecology Conference* 12 (1973): 380.

Sears, Mark K., Richard L. Hellmich, Diane E. Stanley-Horn, Karen S. Oberhauser, John M. Pleasants, Heather R. Mattila, Blair D. Siegfried, and Galen P. Dively. "Impact of Bt Corn Pollen on Monarch Butterfly Populations: A Risk Assessment." *Proceedings of the National Academy of Sciences* 98 (October 9, 2001): 11937–42.

Second Annual Report of the Bureau of Agriculture, Labor, and Industry for the Year Ending November 30, 1894. State Publishing Company, 1895.
Seitz, Adalbert. *The Macrolepidoptera of the World*. Vol. 5. Alfred Kernan Verlag, 1924.
Sequoia and Kings Canyon National Parks. *Vignettes of the 2015 Rough Fire*. N.d. www.nps.gov/seki/learn/nature/upload/Rough_Fire_Vignettes-accessible-508.pdf.
Sigala Páez, Pascual. "La conservación de la monarca, reto para la organización campesina." In Hoth et al., *1997 Reunión de América*, 273–76.
Soulé, Michael E. "What Is Conservation Biology?" *BioScience* 35 (December 1985): 727–34.
Soule, Michael E., and John Terborgh, eds. *Continental Conservation: Scientific Foundations of Regional Reserve Networks*. Island Press, 1999.
Stephenson, Nathan L. "Estimated Ages of Some Large Giant Sequoias: General Sherman Keeps Getting Younger." *Madroño* 47 (January–March 2000): 61–67.
———. "Making the Transition to the Third Era of Natural Resources Management." *George Wright Forum* 31 (2014): 227–35.
Stephenson, Nathan L., David J. Parsons, and Thomas W. Swetnam. "Restoring Natural Fire to the Sequoia-Mixed Conifer Forest: Should Intense Fire Play a Role?" *Proceedings of the Tall Timbers Fire Ecology Conference* 17 (1991): 329–31.
Sudworth, George B. *Forest Trees of the Pacific Coast*. Government Printing Office, 1908.
———. "Present Condition of the California Bigtrees." *American Museum Journal* 12 (November 1912): 227–36.
Swetnam, Thomas W. "Fire History and Climate Change in Giant Sequoia Groves." *Science* 262 (5 November 1993): 885–89.
Swetnam, Thomas W., Christopher H. Baisan, Anthony C. Caprio, Peter M. Brown, Ramzi Touchan, R. Scott Anderson, and Douglas J. Hallett. "Multi-millennial Fire History of the Giant Forest, Sequoia National Park, California, USA." *Fire Ecology* 5 (2009): 120–50.
Swetnam, Thomas W., Christopher H. Baisan, Peter M. Brown, Anthony C. Caprio, and Thomas P. Harlan. *Giant Sequoia Fire History: A Feasibility Study*. Sequoia and Kings Canyon National Park, National Park Service, June 30, 1988. https://repository.arizona.edu/handle/10150/303521.
Tapia Torres, Silverio. "Retos para la conservación de los sitios de la mariposa monarca." In Hoth et al., *1997 Reunión de América*, 335–40.
Taylor, Joseph Henry. *Twenty Years on the Trap Line: Being a Collection of Revised Camp Notes Written at Intervals During a Twenty Years Experience in Trapping, Wolfing and Hunting, on the Great Northwestern Plains*. Published by the author, 1891.
Thew, Susan. *The Proposed Roosevelt-Sequoia National Park*. H. S. Crocker Company Press, 1926.
Toledo Manzur, Carlos. "Estrategia integral para el desarrollo sustentable de la región de la mariposa monarca." In Hoth et al., *1997 Reunión de América*, 29–45.
Urquhart, Frederick A. "Conservation Areas for the Eastern Population of the Monarch Butterfly, *Danaus plexippus plexippus*." *Proceedings of the Entomological Society of Ontario* 110 (1979): 109.
———. "An Ecological Study of Saltatoria at Point Pelee, Ontario." PhD diss., University of Toronto, 1940.
———. "Marked Monarchs." *Natural History* 61 (May 1952): 226–29.
———. *The Monarch Butterfly*. University of Toronto Press, 1960.

———. "Monarch Butterfly (*Danaus Plexippus*) Migration Studies: Autumnal Movement." *Proceedings of the Entomological Society of Ontario* 96 (1965): 23–33.
———. *The Monarch Butterfly: International Traveler.* Nelson-Hall, 1987.
———. "A Proposed Method for Marking Monarch Butterflies." *Canadian Entomologist* 73 (February 1941): 21–22.
———. *Report on the Studies of the Movements of the Monarch Butterfly in North America.* Royal Ontario Museum, 1955.
Urquhart, F. A., and N. R. Urquhart. "The Overwintering Site of the Eastern Population of the Monarch Butterfly (*Danauas p. plexippus; Danaidae*) in Southern Mexico." *Journal of the Lepidopterists' Society* 30 (September 1976): 153–58.
US Department of Labor. *History of Wages in the United States from Colonial Times to 1928.* Government Printing Office, 1934.
US Division of Forestry. *Report on the Big Trees of California.* Government Printing Office, 1900.
US Fish and Wildlife Service. *1997 North American Conference on the Monarch Butterfly: Roundtable Discussions and Priority Actions.* US Fish and Wildlife Service, 1998.
———. *Northern Rocky Mountain Wolf Recovery Plan.* US Fish and Wildlife Service, 1980.
———. *Northern Rocky Mountain Wolf Recovery Plan.* US Fish and Wildlife Service, 1987.
Vaeck, Mark, Arlette Reynaerts, Herman Höfte, Stefan Jansens, Marc De Beuckeleer, Caroline Dean, Marc Zabeau, Marc Van Montagu, and Jan Leemans. "Transgenic Plants Protected from Insect Attack." *Nature* 328 (July 2, 1987): 33–37.
Van Zandt Brower, Jane. "Experimental Studies of Mimicry in Some North American Butterflies: Part I—The Monarch, *Danaus plexippus*, and Viceroy, *Limentis archippus archippus*." *Evolution* 12 (March 1958): 32–47.
Villalobos, Walter Bello, Yaxine María Arias Núñez, and Vanessa Medina Padrón. "Cuidades Hermanas Estes Park—Monteverde." Reserva Bosque Nuboso Santa Elena, November 2016.
Warren, Conrad Kohrs, ed. *Conrad Kohrs: An Autobiography.* Platen Press, 1977.
Wells, Susan W., Robert M. Pyle, and N. Mark Collins. *IUCN Invertebrate Red Data Book.* Gland, Switzerland, International Union for the Conservation of Nature and Natural Resources, 1983.
White, John R. "Atmosphere in the National Parks." In Dilsaver, *America's National Park System*, 123–30.
———. *Bullets and Bolos: Fifteen Years in the Philippine Islands.* Century Company, 1928.
———. *Superintendent's Annual Report, Sequoia National Park, 1926.* Government Printing Office, 1926.
White, John R., and Samuel J. Pusateri. *Sequoia and Kings Canyon National Parks.* Stanford University Press, 1949.
Whitney, J. D. *Geology: Report of Progress and Synopsis of Field-Work from 1860 to 1864.* Sherman & Company, 1865.
Williams, C. B. *The Migration of Butterflies.* Oliver and Boyd, 1930.
World Wildlife Fund Mexico. *La tala ilegal y su impacto en la reserva de la biosfera mariposa monarca.* Mexico City: World Wildlife Fund Mexico, 2004.
Yellowstone to Yukon Conservation Initiative. *The Yellowstone to Yukon Vision: Progress and Possibility.* Yellowstone to Yukon Conservation Initiative, 2014.
Young, Stanley P. *The Last of the Loners.* Macmillan, 1970.

Secondary Sources

Adelman, Jeremy, and Stephen Aron. "From Borderlands to Borders: Empires, Nation-States, and the People in between in North American History." *American Historical Review* 104 (June 1999): 614–41.

Adlam, Christopher, Diana Almendariz, Ron W. Goode, Deniss J. Martinez, and Beth Rose Middleton. "Keepers of the Flame: Supporting the Revitalization of Indigenous Cultural Burning." *Society and Natural Resources* 32 (2021): 575–90.

Agrawal, Anurag. *Monarchs and Milkweed: A Migrating Butterfly, a Poisonous Plant, and Their Remarkable Story of Coevolution*. Princeton University Press, 2017.

Albright, Thomas P., Anna M. Pidgeon, Chadwick D. Rittenhouse, Murray K. Clyton, Curtis H. Flather, Patrick D. Culbert, Brian D. Wardlow, and Volker C. Radeloff. "Effects of Drought on Avian Community Structure." *Global Change Biology* 16 (August 2010): 2158–70.

Aldern, Jared Dahl. "Native Sustainment: The North Fork Mono Tribe's Stories, History, and Teaching of Its Land and Water Tenure in 1918 and 2009." PhD diss., Prescott College, 2010.

Ambrose, Anthony R. "Hydraulic Constraints Modify Optimal Photosynthetic Profiles in Giant Sequoia Trees." *Oecologia* 182 (2016): 713–30.

Anderson, Benedict. *Imagined Communities: Reflections on the Origin and Spread of Nationalism*. Verso, 1983.

Anderson, M. Kat. *Indian Fire-Based Management in the Sequoia-Mixed Conifer Forests of the Central and Southern Sierra Nevada*. Yosemite Research Center, 1993.

———. *Tending the Wild: Native American Knowledge and the Management of California's Natural Resources*. University of California Press, 2005.

Anderson, R. Scott. "Paleohistory of a Giant Sequoia Grove: The Record from Log Meadow, Sequoia National Park." *Proceedings of the Symposium on Giant Sequoias: Their Place in the Ecosystem and Society* (23–25 June 1992): 49–55.

Anderson, Virginia DeJohn. *Creatures of Empire: How Domestic Animals Transformed Early America*. Oxford University Press, 2004.

Andrews, Thomas G. *Coyote Valley: Deep History in the High Rockies*. Harvard University Press, 2015.

———. *Killing for Coal: America's Deadliest Labor War*. Harvard University Press, 2008.

Armitage, Kevin C. *The Nature Study Movement: The Forgotten Popularizer of America's Conservation Ethic*. University Press of Kansas, 2009.

Aycrigg, Jocelyn L., Craig Groves, Jodi A. Hilty, J. Michael Scott, Paul Beier, D. A. Boyce Jr., Dennis Figg, Healy Hamilton, Gary Machlis, Kit Muller, K. V. Rosenberg, Raymond M. Sauvajot, Mark Shaffer, and Rand Wentworth. "Completing the System: Opportunities and Challenges for a National Habitat Conservation System." *Bioscience* 66 (September 2016): 774–84.

Baillie, Jonathan, and Ya-Ping Zhang. "Space for Nature." *Science* 361 (14 September 2018): 1051.

Bangs, Ed. "Operation Wolfstock: Reports of the Wolf Tracker." In Schullery, *Yellowstone Wolf*, 283–334.

———. "Return of a Predator: Wolf Recovery in Montana." In Schullery, *Yellowstone Wolf*, 272–82.

Barnosky, Anthony D. Nicholas Matzke, Susumu Tomiya, Guinevere O. U. Wogan, Brian

Swartz, Tiago B. Quental, Charles Marshall, Jenny L. McGuire, Emily L. Lindsey, Kaitlin C. Maguire, Ben Mersey, and Elizabeth A. Ferrer. "Has the Earth's Sixth Mass Extinction Already Arrived?" *Nature* 471 (2 March 2011): 51–57.
Baud, Michiel, and Willem Van Schendel. "Toward a Comparative History of Borderlands." *Journal of World History* 8 (Fall 1997): 211–42.
Beattie, James. *Empire and Environmental Anxiety: Health, Science, Art, and Conservation in South Asia and Australasia, 1800–1920*. Palgrave Macmillan, 2011.
Beever, Erik A., Brady J. Mattsson, Matthew J. Germino, Max Post Van Der Burg, John B. Bradford, and Mark W. Brunson. "Successes and Challenges from Formation to Implementation of Eleven Broad-Extent Conservation Programs." *Conservation Biology* 28 (April 2014): 302–14.
Beinart, William. "The Night of the Jackal: Sheep, Pastures, and Predators in the Cape." *Past & Present* 158 (February 1998): 172–206.
Belich, James. *Replenishing the Earth: The Settler Revolution and the Rise of the Anglo-World, 1783–1939*. Oxford University Press, 2009.
Benjamin I. Cook, Kevin J. Anchukaitis, Ramzi Touchan, David M. Meko, and Edward R. Cook. "Spatiotemporal Drought Variability in the Mediterranean over the Last 900 Years." *JGR Atmospheres* 121 (16 March 2016): 2060–74.
Benson, Etienne. *Wired Wilderness: Technologies of Tracking and the Making of Modern Wildlife*. Johns Hopkins University Press, 2010.
Benton-Cohen, Katherine. *Borderline Americans: Racial Division and Labor War in the Arizona Borderlands*. Harvard University Press, 2009.
Berland, Oscar. "Giant Forest's Reservation: The Legend and the Mystery." *Sierra Club Bulletin* 47 (December 1962): 68–82.
Beutelspacher, Carlos R. *Las mariposas entre los antiguos Mexicanos*. Fondo de Cultura Económica, 1999.
Bingham, Heather C., Diego Juffe Bignoli, Edward Lewis, Brian MacSharry, Neil D. Burgess, Piero Visconti, Marine Deguignet, Murielle Misrachi, Matt Walpole, Jessica L. Stewart, Thomas M. Brooks, and Naomi Kingston. "Sixty Years of Tracking Conservation Progress Using the World Database on Protected Areas." *Nature Ecology and Evolution* 3 (May 2019): 737–43.
Black, Megan. *The Global Interior: Mineral Frontiers and American Power*. Harvard University Press, 2018.
Boardman, Robert. *International Organization and the Conservation of Nature*. Indiana University Press, 1981.
Bolster, W. Jeffrey. *The Mortal Sea: Fishing the Atlantic in the Age of Sail*. Harvard University Press, 2012.
Boxell, Mark. "Rapid Ascent: Rocky Mountain National Park in the Great Acceleration, 1945–Present." Master's thesis, Colorado State University, 2016.
Boxell, Mark, and Will Wright. "Postwar Play and Petroleum: Tourism and Energy Abundance in Rocky Mountain National Park." *Journal of Tourism History* 9 (2017): 119–38.
Boyer, Christopher R. "Contested Terrain: Forestry Regimes and Community Reponses in Northeastern Michoacán, 1940–2000." In *The Community-Managed Forests of Mexico: The Struggle for Equity and Sustainability*, edited by David Barton Bray, Leticia Merino-Pérez, and Deborah Barry, 27–47. University of Texas Press, 2004.

———. *Political Landscapes: Forests, Conservation, and Community in Mexico*. Duke University Press, 2015.
Brado, Edward. *Cattle Kingdom: Early Ranching in Alberta*. Heritage House, 2004.
Brenner, Ludger. "Aceptación de políticas de conservación ambiental: El caso de la Reserva de la Biosfera Mariposa Monarca." *Economía, Sociedad y Territorio* 9 (2009): 259–95.
———. "Áreas naturales protegidas y ecoturismo: El caso de la Reserva de la Biosfera Mariposa Monarca, México." *Relaciones: Estudios de Historia y Sociedad* 27 (2006): 237–65.
Brockington, Dan. *Fortress Conservation: The Preservation of the Mkomazi Game Reserve, Tanzania*. Indiana University Press, 2002.
Brower, Lincoln P. "Understanding and Misunderstanding the Migration of the Monarch Butterfly (Nymphalidae) in North America, 1857–1995." *Journal of the Lepidopterists' Society* 49 (1995): 304–85.
Brower, Lincoln P., Guillermo Castilleja, Armando Peralta, Jose Lopez-Garcia, Luis Bojorquez-Tapia, Salomon Diaz, Daniela Melgarejo, and Monica Missrie. "Quantitative Changes in Forest Quality in a Principal Overwintering Area of the Monarch Butterfly in Mexico, 1971–1999." *Conservation Biology* 16 (April 2002): 346–59.
Brown, Kate. "Gridded Lives: Why Kazakhstan and Montana Are Nearly the Same Place." *American Historical Review* 106 (February 2001): 17–48.
———. "A Place in Biography for Oneself." *American Historical Review* 114 (June 2009): 596–605.
Brüggemeier, Franz-Josef. "The Ecological Movement and Internationalism." *Moving the Social* 55 (October 2016): 75–92.
Bryan, Heather M., Judit E. G. Smits, Lee Koren, Paul C. Paquet, Katherine E. Wynne-Edwards, and Marco Musiani. "Heavily Hunted Wolves Have Higher Stress and Reproductive Steroids Than Wolves with Lower Hunting Pressure." *Functional Ecology* 29 (March 2015): 347–56.
Calvert, William H., Willow Zuchowski, and Lincoln P. Brower. "Monarch Butterfly Conservation: Interactions of Cold Weather, Forest Thinning and Storms on the Survival of Overwintering Monarch Butterflies (*Danaus plexippus* L.) in Mexico." *Atala* 9 (1981–84): 2–6.
Camas de Castro, María Elena. "Operative Programs in the Monarca A.C. Project." In Malcolm and Zalucki, *Biology and Conservation of the Monarch Butterfly*, 385–87.
Cameron, Maxwell A., and Brian W. Tomlin. *The Making of NAFTA: How the Deal was Done*. Cornell University Press, 2000.
Carle, David. *Burning Questions: America's Fight with Nature's Fire*. Prager, 2002.
Carruthers, Jane. *The Kruger National Park: A Social and Political History*. University of Natal Press, 1995.
Carter, David B., and Paul Poast. "Why Do States Build Walls? Political Economy, Security, and Border Stability." *Journal of Conflict Resolution* 61 (February 2017): 239–70.
Cassidy, Kira A., Douglas W. Smith, L. David Mech, Daniel R. MacNulty, Daniel R. Stahler, and Matthew C. Metz. "Territoriality and Inter-Pack Aggression in Gray Wolves: Shaping a Social Carnivore's Life History." *Yellowstone Science* 24 (June 2016): 37–42.
Ceballos, Gerardo, Paul R. Ehrlich, Anthony D. Barnosky, Andrés García, Robert M. Pringle, and Todd M. Palmer. "Accelerated Modern Human-Induced Species Losses: Entering the Sixth Mass Extinction." *Science Advances* 1 (19 June 2015): 1–5.

Chakrabarty, Dipesh. "The Climate of History: Four Theses." *Critical Inquiry* 35 (Winter 2009): 197–222.
———. *The Climate of History in a Planetary Age*. University of Chicago Press, 2021.
Chaney, Ralph W. "A Revision of Fossil Sequoia and Taxodium in Western North America Based on the Recent Discovery of Metasequoia." *Transactions of the American Philosophical Society* 40 (February 1951): 171–263.
Charles, Daniel. *Lords of the Harvest: Biotech, Big Money, and the Future of Food*. Perseus Publishing, 2001.
Chen, I-Ching, Jane K. Hill, Ralf Ohlemüller, David B. Roy, and Chris D. Thomas. "Rapid Range Shifts of Species Associated with High Levels of Climate Warming." *Science* 333 (19 August 2011): 1024–26.
Chester, Charles C. "Responding to the Idea of Transboundary Conservation: An Overview of Public Reaction to the Yellowstone to Yukon (Y2Y) Conservation Initiative." *Journal of Sustainable Forestry* 17 (June 2003): 103–25.
———. *Conservation across Borders: Biodiversity in an Interdependent World*. Island Press, 2006.
Chin, Alana O., and Stephen C. Sillet. "Phenotypic Plasticity of Leaves Enhances Water-Stress Tolerance and Promotes Hydraulic Conductivity in a Tall Conifer." *American Journal of Botany* 103 (May 2016): 796–807.
Clavin, Patricia. "Men and Markets: Global Capital and the International Economy." In *Internationalisms: A Twentieth-Century History*, edited by Glenda Sluga and Patricia Clavin, 85–109. Cambridge University Press, 2017.
Coleman, Jon T. *Vicious: Wolves and Men in America*. Yale University Press, 2004.
Colpitts, George. "Howl: The 1952–56 Rabie Crisis and the Creation of the Urban Wild at Banff." In *Animal Metropolis: Histories of Human-Animal Relations in Urban Canada*, edited by Joanna Dean, Darcy Ingram and Christabelle Sethna, 219–53. University of Calgary Press, 2017.
Cox, Thomas R. *The Lumberman's Frontier: Three Centuries of Land Use, Society, and Change in America's Forests*. Oregon State University Press, 2010.
Cronon, William. *Nature's Metropolis, Chicago and the Great West*. W. W. Norton, 1991.
———. "The Trouble with Wilderness; or, Getting Back to the Wrong Nature." *Environmental History* 1 (January 1996): 7–28.
Crosby, Alfred W. *Ecological Imperialism: The Biological Expansion of Europe, 900–1900*. Second Edition. Cambridge University Press, 2004.
Cullather, Nick. *The Hungry World: America's Cold War Battle Against Poverty in Asia*. Harvard University Press, 2010.
Curnow, Edward E. "The History of the Eradication of the Wolf in Montana." Master's thesis, University of Montana–Missoula, 1969.
Cushman, Gregory T. *Guano and the Opening of the Pacific World*. Cambridge University Press, 2013.
Davis, Heather, and Zoe Todd. "On the Importance of a Date, or Decolonizing the Anthropocene." *International Journal for Critical Geographies* 16 (December 2017): 761–80.
Davis, Mike. "The Coming Desert: Kropotkin, Mars, and the Pulse of Asia." *New Left Review* 97 (January–February 2016): 23–43.
De Bont, Raf. *Nature's Diplomats: Science, Internationalism, and Preservation, 1920–1960*. University of Pittsburgh Press, 2021.

De Bont, Raf, and Jens Lachmund, eds. *Spatializing the History of Ecology: Sites, Journeys, Maps.* Routledge, 2017.
De la Maza E., Javier, and William H. Calvert. "Investigations of Possible Monarch Butterfly Overwintering Areas in Central and Southeastern Mexico." In Malcolm and Zalucki, *Biology and Conservation of the Monarch Butterfly*, 295–300.
Debarbieux, Bernard, and Gilles Rudaz. *The Mountain: A Political History from the Enlightenment to the Present.* University of Chicago Press, 2015.
DeCesare, Nicholas J., Seth M. Wilson, Elizabeth H. Bradley, Justin A. Gude, Robert M. Inman, Nathan J. Lance, Kent Laudon, Abigail A. Nelson, Michael S. Ross, and Ty D. Smucker. "Wolf-Livestock Conflict and the Effects of Wolf Management." *Journal of Wildlife Management* 82 (May 2018): 711–22.
Dempsey, Hugh A. *Firewater: The Impact of the Whisky Trade on the Blackfoot Nation.* Fifth House Publishers, 2002.
Demuth, Bathsheba. *Floating Coast: An Environmental History of the Bering Strait.* W. W. Norton, 2019.
Dirzo, Rodolfo, Hillary S. Young, Mauro Galetti, Gerardo Ceballos, Nick J. B. Isaac, and Ben Collen. "Defaunation in the Anthropocene." *Science* 345 (25 July 2014): 401–6.
Dogliani, Patrizia. "The Fate of Socialist Internationalism." In *Internationalisms: A Twentieth-Century History*, edited by Glenda Sluga and Patricia Clavin, 38–60. Cambridge University Press, 2017.
Donahue, Brian. "Historian with a Chainsaw: Teaching Environmental History." *OAH Magazine* 25 (1 October 2011): 33–35.
Dorsey, Kurkpatrick. *The Dawn of Conservation Diplomacy: U.S.-Canadian Wildlife Protection Treaties in the Progressive Era.* University of Washington Press, 1998.
———. *Whales and Nations: Environmental Diplomacy on the High Seas.* University of Washington Press, 2013.
Dunbar-Ortiz, Roxanne. *An Indigenous Peoples' History of the United States.* Beacon Press, 2014.
Dunlap, Thomas R. "Beyond the Parks, Beyond the Borders: Some of the Places to Take Tyrrell's Perspective." *Journal of American Studies* 46, no. 1 (2012): 31–36.
———. *Saving America's Wildlife: Ecology and the American Mind, 1850–1990.* Princeton University Press.
Dunlop, Catherine Tatiana. *Cartophilia: Maps and the Search for Identity in the French-German Borderland.* University of Chicago Press, 2015.
Edwards, Paul N. *A Vast Machine: Computer Models, Climate Data, and the Politics of Global Warming.* MIT Press, 2010.
Ekbladh, David. *The Great American Mission: Modernization and the Construction of an American World Order.* Princeton University Press, 2010.
Ellis, Erle C., Kees Klein Goldewijk, Stefan Siebert, Deborah Lightman, and Navin Ramankutty. "Anthropogenic Transformation of the Biomes, 1700–2000." *Global Ecology and Biogeography* 19 (September 2010): 589–606.
Elmore, Bartow J. "The Commercial Ecology of Scavenger Capitalism: Monsanto, Fossil Fuels, and the Making of a Chemical Giant." *Enterprise and Society* 19 (March 2018): 153–78.
———. "Roundup from the Ground Up: A Supply-Side Story of the World's Most Widely Used Herbicide." *Agricultural History* 93 (Winter 2019): 102–38.
Elofson, Warren M. *Cowboys, Gentlemen, and Cattle Thieves: Ranching on the Western Frontier.* McGill-Queen's University Press, 2000.

———. *Frontier Cattle Ranching in the Land and Times of Charlie Russell*. University of Washington Press, 2004.
Estes, Nick. *Our History Is the Future: Standing Rock versus the Dakota Access Pipeline, and the Long Tradition of Indigenous Resistance*. Verso, 2019.
Evans, Sterling. *The Green Republic: A Conservation History of Costa Rica*. University of Texas Press, 1999.
Farmer, Jared. *Trees in Paradise: A California History*. W. W. Norton, 2013.
Farnham, Timothy J. *Saving Nature's Legacy: Origins of the Idea of Biological Diversity*. Yale University Press, 2007.
Farquhar, Francis P. *History of the Sierra Nevada*. University of California Press, 1966.
Fiege, Mark. *The Republic of Nature: An Environmental History of the United States*. University of Washington Press, 2012.
———. "Weedy West: Mobile Nature, Boundaries, and Common Space in the Montana Landscape." *Western Historical Quarterly* 36 (Spring 2005): 22–47.
Fields, Gary. *Enclosure: Palestinian Landscapes in a Historical Mirror*. University of California Press, 2017.
Finnegan, Diarmid A. "The Spatial Turn: Geographic Approaches in the History of Science." *Journal of the History of Biology* 41 (2008): 369–88.
Fish and Wildlife Historical Society. *Fish, Fur, and Feathers: Fish and Wildlife Conservation in Alberta, 1905–2005*. Federation of Alberta Naturalists, 2005.
Fleming, James Rodger. *Historical Perspectives on Climate Change*. Oxford University Press, 1998.
Frank, Jerry J. *Making Rocky Mountain National Park: The Environmental History of an American Treasure*. University Press of Kansas, 2013.
Franz, John E., Michael K. Mao, and James A. Sikorski. *Glyphosate: A Unique Global Herbicide*. American Chemical Society, 1997.
Fraser, Caroline. *Rewilding the World: Dispatches from the Conservation Revolution*. Picador, 2009.
Freemuth, John C. *Islands Under Siege: National Parks and the Politics of External Threats*. University of Kansas, 1991.
Furmansky, Dyana Z. *Rosalie Edge, Hawk of Mercy: The Activist Who Saved Nature from the Conservationists*. University of Georgia Press, 2009.
Gallegos, Carlos Melo. *Áreas naturales protegidas de México en el siglo XX*. Instituto de Geografía, Universidad Nacional Autónoma de México, 2002.
Geiger, Andrea. *Subverting Exclusion: Transpacific Encounters with Race, Caste, and Borders, 1885–1928*. Yale University Press, 2015.
Gerstell, Richard. *The Steel Trap in North America*. Stackpole Books, 1985.
Gissibl, Bernard, Sabine Höhler, and Patrick Kupper, editors. *Civilizing Nature: National Parks in Global Historical Perspective*. Berghahn Books, 2012.
González de Castilla, Susana Rojas. "The Importance of Alternative Sources of Income to 'Ejidatarios' (Local Residents) for Conservation of Overwintering Areas of the Monarch Butterfly." In Malcolm and Zalucki, *Biology and Conservation of the Monarch Butterfly*, 389–94.
Gonzalez-Duarte, Columba. "Butterflies, Organized Crime, and 'Sad Trees': A Critique of the Monarch Butterfly Biosphere Reserve Program in a Context of Rural Violence." *World Development* 142 (June 2021): 1–11.

Gottfried, Carlos F. *Mariposa Monarca*. Monarca, A.C., 1987.
———. "Monarch Conservation in Mexico: The Challenge of Membership and Fundraising for Monarca, A.C." In Malcolm and Zalucki, *Biology and Conservation of the Monarch Butterfly*, 379–85.
———. "One of Nature's Most Incredible Phenomenon: The Monarch Butterflies." *Royal Institution Proceedings* (1989): 33–34.
Grandin, Greg. *The End of the Myth: From the Frontier to the Border Wall in the Mind of America*. Metropolitan Books, 2019.
Grant, James Wilson. "Ranches with Wolves: How Straight Talk Is the Salvation of Open Range in the Northern Rockies." MA thesis, University of Montana, 2010.
Graybill, Andrew R. *Policing the Great Plains: Rangers, Mounties, and the North American Frontier, 1875–1910*. University of Nebraska Press, 2007.
Gressley, Gene M. *Bankers and Cattlemen: The Stock-And-Bonds, Havana-Cigar, Mahogany-And-Leather Side of the Cowboy Era*. Alfred A. Knopf, 1966.
Grove, Richard H. *Green Imperialism: Colonial Expansion, Tropical Island Edens and the Origins of Environmentalism*. Cambridge University Press, 1995.
Guha, Ramachandra. *Environmentalism: A Global History*. Penguin Books, 2000.
Gustafsson, Karin M., Anurag A. Agrawal, Bruce V. Lewenstein, and Steven A. Wolf. "The Monarch Butterfly Through Time and Space: The Social Construction of an Icon." *BioScience* 65 (June 2015): 612–22.
Haddad, Nick. *The Last Butterflies: A Scientist's Quest to Save a Rare and Vanishing Creature*. Princeton University Press, 2019.
Hahn, Steven. *A Nation Without Borders: The United States and Its World in the Age of Civil Wars, 1830–1910*. Viking, 2016.
Halpern, Sue. *Four Wings and a Prayer: Caught in the Mystery of the Monarch Butterfly*. Pantheon Books, 2001.
Hämäläinen, Pekka, and Samuel Truett. "On Borderlands." *Journal of American History* 98 (September 2011): 338–61.
Hampton, Bruce. *The Great American Wolf*. Henry Holt, 1997.
Hannibal, Mary Ellen. *The Spine of the Continent: The Most Ambitious Wildlife Conservation Project Ever Undertaken*. Lyons Press, 2012.
Haraway, Donna. "Anthropocene, Capitalocene, Plantationocene, Chthulucene: Making Kin." *Environmental Humanities* 6 (May 2015): 159–65.
———. "Situated Knowledges: The Science Question in Feminism and the Privilege of Partial Perspective." *Feminist Studies* 14 (1988): 575–99.
Hartesveldt, R. J., H. T. Harvey, H. S. Shellhammer, R. E. Stecker, and N. Stephenson. *Giant Sequoias*. Rev. ed. Sequoia Natural History Association, 2010.
Harvey, David. *A Brief History of Neoliberalism*. Oxford University Press, 2005.
———. *Social Justice and the City*. Rev. ed. University of Georgia, 2009.
Haugland, Scott, Ken Marcantonio, Hal Shaw, and the Sanger Historical Society. *Images of America: Sanger*. Arcadia Publishing, 2013.
Hays, Samuel P. *Conservation and the Gospel of Efficiency: The Progressive Conservation Movement, 1890–1920*. University of Pittsburgh Press, 1959.
Herren, Madeleine. "Fascist Internationalism." In *Internationalisms: A Twentieth-Century History*, edited by Glenda Sluga and Patricia Clavin, 191–212. Cambridge University Press, 2017.

Hennessey, Elizabeth. *On the Backs of Tortoises: Darwin, the Galápagos, and the Fate of an Evolutionary Eden.* Yale University Press, 2019.

Herberman, Ethan. *The Great Butterfly Hunt: The Mystery of the Migrating Monarch.* Simon and Schuster, 1990.

Hernández, Kelly Lytle. *Migra!: A History of the U.S. Border Patrol.* University of California Press, 2010.

Hilty, Jodi A. "Use of Riparian Corridors by Wildlife in the Oak Woodland Vineyard Landscape." PhD diss., University of California, Berkeley, 2001.

Hilty, Jodi A., Annika T. H. Keeley, Adina M. Merenlender, and William Z. Lidicker Jr. *Corridor Ecology: Linking Landscapes for Biodiversity Conservation and Climate Adaptation.* 2nd ed. Island Press, 2019.

Hogue, Michel. *Metis and the Medicine Line: Creating a Border and Dividing a People.* University of North Carolina Press, 2015.

Honey-Rosés, Jordi, José López-García, Eduardo Rendón-Salinas, Armando Peralta-Hiquera, and Carlos Galindo-Leal. "To Pay or Not to Pay? Monitoring Performance and Enforcing Conditionality When Paying for Forest Conservation in Mexico." *Environmental Conservation* 36 (June 2009): 120–28.

Honey-Rosés, Jordi, Kathy Baylis, and M. Isabel Ramírez. "A Spatially Explicit Estimate of Avoided Forest Loss." *Conservation Biology* 25 (October 2011): 1032–43.

Hoth, Jürgen. "Mariposa monarca, mitos y otras realidades aladas." *Ciencias* 37 (January–March 1995): 19–28.

Howe, Joshua P. *Behind the Curve: Science and the Politics of Global Warming.* University of Washington Press, 2014.

Howkins, Adrian, Jared Orsi, and Mark Fiege, eds. *National Parks Beyond the Nation: Global Perspectives on "America's Best Idea."* University of Oklahoma Press, 2016.

Hunt, Lynn. *Writing History in the Global Era.* W. W. Norton, 2014.

Hurt, R. Douglas. *American Agriculture: A Brief History.* Rev. ed. Purdue University Press, 2002.

Hydrick, Rick. "The Genesis of National Park Management: John Roberts White and Sequoia National Park, 1920–1947." *Journal of Forest History* 28 (April 1984): 68–81.

"IPCC, 2021: Summary for Policymakers." In *Climate Change 2021: The Physical Science Basis,* edited by V. Masson-Delmotte, P. Zhai, A. Pirani, et al., 1–32. Cambridge University Press, 2021.

Isenberg, Andrew C. *The Destruction of the Bison: An Environmental History, 1750–1920.* Cambridge University Press, 2000.

———. *Mining California: An Ecological History.* Hill & Wang, 2005.

Jacobs, Nancy J. *Environment, Power, and Injustice: A South African History.* Cambridge University Press, 2003.

Jacoby, Karl. *Crimes Against Nature: Squatters, Poachers, Thieves, and the Hidden History of American Conservation.* University of California Press, 2001.

Jenkins, E. W. "Science, Sentimentalism, or Social Control? The Nature Study Movement in England and Wales, 1899–1914." *History of Education* 10 (1981): 33–43.

Jiminez, Michael D., Edward E. Bangs, Diane K. Boyd, Douglas W. Smith, Scott A. Becker, David E. Ausband, Susannah P. Woodruff, Elizabeth H. Bradley, Jim Holyan, and Kent Laudon. "Wolf Dispersal in the Rocky Mountains, Western United States: 1993–2008." *Journal of Wildlife Management* 81 (May 2017): 581–92.

Johnson, Penny. *Companions in Conflict: Animals in Occupied Palestine*. Melville House Publishing, 2019.
Johnston, Hank. *They Felled the Redwoods: A Saga of Flumes and Fails in the High Sierra*. Stauffer Publishing, 1996.
Jones, Karen R. "From Big Bad Wolf to Ecological Hero: Canis Lupus and the Culture(s) of Nature in the American-Canadian West." *American Review of Canadian Studies* 40 (September 2010): 338–50.
———. "Never Cry Wolf: Science, Sentiment, and the Literary Rehabilitation of Canis Lupus." *Canadian Historical Review* 84 (March 2003): 65–93.
———. *Wolf Mountains: A History of Wolves Along the Great Divide*. University of Calgary Press, 2002.
Jones, Ryan Tucker. *Empire of Extinction: Russians and the North Pacific's Strange Beasts of the Sea*. Oxford University Press, 2014.
Jørgensen, Dolly, Finn Arne Jørgensen, and Sara B. Pritchard, editors. *New Natures: Joining Environmental History with Science and Technology Studies*. University of Pittsburgh Press, 2013.
Kaczynski, Kristen M., David J. Cooper, and William R. Jacobi. "Interactions of sapsuckers and Cytospora Canker Can Facilitate Decline of Riparian Willows." *Botany* 92 (July 2014): 485–93.
Kass, Dorothy. *Educational Reform and Environmental Concern: A History of School Nature Study in Australia*. Routledge, 2018.
Kasten, Kyle, Carl Stenoien, Wendy Caldwell, and Karen S. Oberhauser. "Can Roadside Habitat Lead Monarch on a Route to Recovery?" *Journal of Insect Conservation* 20 (2016): 1047–57.
Kati, Vassiliki, Tasos Hovardas Martin Dieterich Pierre L. Ibisch Barbara Mihok and Nuria Selva. "The Challenge of Implementing the European Network of Protected Areas Natura 2000." *Conservation Biology* 29 (February 2015): 260–70.
Kaufman, Polly Welts. *National Parks and the Woman's Voice: A History*. Updated ed. University of New Mexico Press, 2006.
Kelley, Colin P., Shahrzad Mohtadi, Mark A. Cane, Richard Seager, and Yochanan Kushnir. "Climate Change in the Fertile Crescent and Implications of the Recent Syrian Drought." *Proceedings of the National Academy of Sciences* 112 (17 March 2015): 3241–46.
Kennedy, David M. *Freedom from Fear: The American People in Depression and War, 1929–1945*. Oxford University Press, 1999.
Kida, Geoffrey P. "Colonel John Roberts White and Sustainable Management from the Philippines to Sequoia National Park." Master's thesis, California State University, Northridge, 2011.
Kilgore, Bruce. "The Origin and History of Wildland Fire Use in the U.S. National Park System." *George Wright Forum* 24 (2007): 101–7.
Kilgore, Bruce M., and Dan Taylor. "Fire History of a Sequoia-Mixed Conifer Forest." *Ecology* 60 (February 1979): 129–42.
Klein, Naomi. *This Changes Everything: Capitalism vs. The Climate*. Simon & Schuster, 2014.
Knox, John H., and David L Markell. "The Innovative North American Commission for Environmental Cooperation." In *Greening NAFTA: The North American Commission for Environmental Cooperation*, edited by David L. Markell and John H. Knox, 1–21. Stanford University Press, 2003.

Kohler, Robert E. *Landscapes and Labscapes: Exploring the Lab-Field Border in Biology.* University of Chicago Press, 2002.

Kohlstedt, Sally Gregory. *Teaching Children Science: Hands-On Nature Study in North America, 1890–1930.* University of Chicago Press, 2010.

Kolbert, Elizabeth. *The Sixth Extinction: An Unnatural History.* Henry Holt & Company, 2014.

Kupper, Patrick. *Creating Wilderness: A Transnational History of the Swiss National Park.* Berghahn Books, 2014.

Langham, Gary M., Justin G. Schuetz, Trisha Distler, Candan U. Soykan, and Chad Wilsey. "Conservation Status of North American Birds in the Face of Future Climate Change." *PLOS One* 10 (September 2015): 1–16.

Latour, Bruno. *Science in Action: How to Follow Scientists and Engineers through Society.* Harvard University Press, 1987.

Lear, Linda. *Rachel Carson: Witness for Nature.* Henry Holt and Company, 1997.

LeCain, Timothy J. "Against the Anthropocene: A Neo-Materialist Perspective." *International Journal for History, Culture, and Modernity* 3 (April 2015): 1–24.

———. *The Matter of History: How Things Create the Past.* Cambridge University Press, 2017.

"Leonila Vázquez García: In Memoriam." *Anales de Instituto de Biología de Universidad Naciónal Autónomia de México* 66 (1995): 137–45.

Lewis, Michael L. *Inventing Global Ecology: Tracking the Biodiversity Ideal in India, 1947–1997.* Ohio University Press, 2004.

Limerick, Patricia Nelson. *The Legacy of Conquest: The Unbroken Past of the American West.* W. W. Norton, 1987.

Lindenmayer, Dave, Simon Thorn, and Reed Noss. "Countering Resistance to Protected-Area Extension." *Conservation Biology* 32 (April 2018): 315–21.

Linnell, John D. C., Arie Trouwborst, Luigi Boitani, et al. "Border Security Fencing and Wildlife: The End of the Transboundary Paradigm in Eurasia?" *PLOS Biology* 14 (22 June 2016): 1–13.

Locke, Harvey. "The International Movement to Protect Half the World: Origins, Scientific Foundations, and Policy Implications." Reference Module in *Earth Systems and Environmental Sciences* (2018): 1–10.

———. "Nature Needs Half: A Necessary and Hopeful New Agenda for Protected Areas in North American and around the World." *George Wright Forum* 31 (2014): 359–71.

Loo, Tina. *States of Nature: Conserving Canada's Wildlife in the Twentieth Century.* UBC Press, 2006.

Lothian, W. F. *A Brief History of Canada's National Parks.* Minister of Environment Canada, 1987.

Lovejoy, Thomas E. "Extinction Tsunami Can Be Avoided." *Proceedings of the National Academy of Sciences* 114 (8 August 2017): 8440–41.

MacDonald, Graham A. *Where the Mountains Meet the Prairies: A History of Waterton Lakes National Park.* Parks Canada, 1992.

MacDowell, Laurel Sefton. *An Environmental History of Canada.* UBC Press, 2012.

MacEachern, Alan. "The Sentimentalist: Science and Nature in the Writing of H. U. Green, a.k.a Tony Lascelles." *Journal of Canadian Studies* 47 (Fall 2013): 16–41.

MacKinnon, Doris Jeanne. *The Identities of Marie Rose Delorme Smith: Portrait of a Métis Woman.* Canadian Plains Research Council, 2012.

Madley, Benjamin. *An American Genocide: The United States and the California Indian Catastrophe, 1846–1873*. Yale University Press, 2016.

Maher, Neil M. *Nature's New Deal: The Civilian Conservation Corps and the Roots of the American Environmental Movement*. Oxford University Press, 2008.

Maier, Charles S. "Consigning the Twentieth Century to History: Alternative Narratives for the Modern Era." *American Historical Review* 105 (June 2000): 807–31.

Malcolm, Stephen B., Barbara J. Cockerell, and Lincoln P. Brower. "Spring Recolonization of Eastern North America by the Monarch Butterfly: Successive Brood or Single Sweep?" In Malcolm and Zalucki, *Biology and Conservation of the Monarch Butterfly*, 253–55.

Malcolm, Stephen B., and Myron P. Zaluki, eds. *Biology and Conservation of the Monarch Butterfly*. Natural History Museum of Los Angeles County, 1993.

Malm, Andreas, and Alf Hornborg. "A Geology of Mankind?: A Critique of the Anthropocene Narrative." *Anthropocene Review* 1 (April 2014): 62–69.

Malone, Michael P., Richard B. Roeder, and William L. Lang. *Montana: A History of Two Centuries*. Rev. ed. University of Washington Press, 1991.

Martin, Geoffrey J. *Ellsworth Huntington: His Life and Thought*. Archon Books, 1973.

———. "Robert Lemoyne Barrett, 1871–1969; Last of the Founding Members of the Association of American Geographers." *Professional Geographer* 24 (1972): 29–31.

Martin, Roberta E., Gregory P. Asner, Emily Francis, Anthony Ambrose, Wendy Baxter, Adrian J. Das, Nicolas R. Vaughn, Tarin Paz-Kagan, Todd Dawson, Koren Nydick, and Nathan L. Stephenson. "Remote Measurement of Canopy Water Content in Giant Sequoias (*Sequoiadendron giganteum*) During Drought." *Forest Ecology and Management* 419–20 (2018): 279–90.

Mazerolle, Daniel F., Kevin W. Dufour, Keith A. Hobson, and Heidi E. den Haan. "Effects of Large-Scale Climatic Fluctuations on Survival and Production of Young in a Neotropical Migrant Songbird, the Yellow Warbler *Dendroica petechia*." *Journal of Avian Biology* 36 (February 2005): 155–63.

McAlester, Virginia Savage. *A Field Guide to American Houses*. Alfred A. Knopf, 2013.

McGraw, Donald J. *A. E. Douglass and the Role of the Giant Sequoia in the Development of Dendrochronology*. Edwin Mellen Press, 2001.

McManus, Sheila. *The Line Which Separates: Race, Gender, and the Making of the Alberta-Montana Borderlands*. University of Nebraska Press, 2005.

Mech, L. David, and Luigi Boitani, eds. *Wolves: Behavior, Ecology, and Conservation*. University of Chicago Press, 2003.

———. "Wolf Social Ecology." In Mech and Boitani, *Wolves: Behavior, Ecology, and Conservation*, 11–18.

Merino Pérez, Leticia, and Mariana Hernández Apolinar. "Destrucción de instituciones comunitarias y deterioro de los bosques en la Reserva de la Biosfera Mariposa Monarca, Michoacán, México." *Revista Mexicana de Sociología* 66 (2004): 261–308.

Miller, Daegan. *This Radical Land: A Natural History of American Dissent*. University of Chicago Press, 2018.

Miller, Todd. *Empire of Borders: The Expansion of the U.S. Border Around the World*. Verso, 2019.

Moore, Jason W. *Capitalism in the Web of Life: Ecology and the Accumulation of Capital*. Verso, 2015.

Moratto, Michael J. *California Archaeology*. Academic Press, 1984.

Morton, Desmond. *A Short History of Canada*. 7th ed. McClelland & Stewart, 2017.
Netz, Reviel. *Barbed Wire: An Ecology of Modernity*. Wesleyan University Press, 2004.
Nydick, Koren, Nathan L. Stephenson, Anthony R. Ambrose, et al. "Leaf to Landscape Responses of Giant Sequoia to Hotter Drought: An Introduction and Synthesis for the Special Section." *Forest Ecology and Management* 419–20 (2018): 249–56.
O'Connell, Jay. *Co-Operative Dreams: A History of the Kaweah Colony*. Raven River Press, 1999.
Ogarrio, Rodolfo. "Conservation Actions Taken by Monarca, A.C., to Protect the Overwintering Sites of the Monarch Butterfly in Mexico." In Malcolm and Zalucki, *Biology and Conservation of the Monarch Butterfly*, 77–90.
———. "Development of the Civic Group, Pro Monarca, A.C., for the Protection of the Monarch Butterfly Wintering Grounds in the Republic of Mexico." *Atala* 9 (1981–84): 11–13.
Orsi, Jared. "Construction and Contestation: Toward a Unifying Methodology for Borderlands History." *History Compass* 12 (May 2014): 433–43.
Orsi, Richard J. *Sunset Limited: The Southern Pacific Railroad and the Development of the American West, 1850–1930*. University of California Press, 2005.
Ortíz Monasterio, Fernando, and Valentina Ortíz Monasterio Garza. *Mariposa monarca: Vuelo de papel*. Centro de Información y Desarrollo de la Comunicación y la Literatura Infantiles, 1987.
Ortíz Monasterio, Fernando, V. Sanchez, H. G. Liquidano, and M. Venegas. "Magnetism as a Complementary Factor to Explain Orientation Systems Used by Monarch Butterflies to Locate Their Overwintering Areas." *Atala* 9 (1981–84): 14–16.
Paddle, Robert. *The Last Tasmanian Tiger: The History and Extinction of the Thylacine*. Cambridge University Press, 2002.
Paige, John C. *The Civilian Conservation Corps and the National Park Service, 1933–1942: An Administrative History*. National Park Service, 1985. www.nps.gov/parkhistory/online_books/ccc/index.htm.
Parker, Geoffrey. *Global Crisis: War, Climate Change & Catastrophe in the Seventeenth Century*. Yale University Press, 2013.
Pawson, Eric, and Tom Brooking, eds. *Making a New Land: Environmental Histories of New Zealand*. Otago University Press, 2013.
Paz-Kagan, Tarin, Nicholas R. Vaughn, Roberta E. Martin, et al. "Landscape-Scale Variation in Canopy Water Content of Giant Sequoias during Drought." *Forest Ecology and Management* 419–20 (2018): 291–304.
Pérez Talavera, Víctor Manuel. *La expotación de los bosques en Michoacán, 1881–1917*. Gobierno del Estado de Michoacán de Ocampo, 2016.
Philip, Justine. "The Institutionisation of Poison: A Historical Review of Vertebrate Pest Control in Australia, 1814 to 2018." *Australian Zoologist* 40 (2019): 129–39.
Piirto, Douglas Donald. "Factors Associated with Tree Failure of Giant Sequoia." PhD diss., University of California-Berkeley, 1977.
———. "Wood of Giant Sequoia: Properties and Unique Characteristics." *Proceedings of the Workshop on Management of Giant Sequoias* (December 1986): 19–23.
Pleasants, John. "Monarch Butterflies and Agriculture." In *Monarchs in a Changing World: Biology and Conservation of an Iconic Butterfly*, edited by Karen S. Oberhauser, Kelly R. Nail, and Sonia Altizer, 169–78. Cornell University Press, 2015.

Post, Robert C. *Who Own's America's Past?: The Smithsonian and the Problem of History.* Johns Hopkins University Press, 2013.
Pratt, Mary Louise. *Imperial Eyes: Travel Writing and Transculturation.* 2nd ed. Routledge, 2008.
Price, Jennifer. *Flight Maps: Adventures with Nature in Modern America.* Basic Books, 1999.
Prysby, Michelle D., and Karen S. Oberhauser. "Temporal and Geographic Variation in Monarch Densities: Citizen Scientists Document Monarch Population Patterns." In *The Monarch Butterfly: Biology and Conservation,* edited by Karen S. Oberhauser and Michelle J. Solensky, 9–20. Cornell University Press, 2004.
Putnam, Aaron E., David E. Putnam, Laia Andreu-Hayles, et al. "Little Ice Age Wetting of Interior Asian Deserts and the Rise of the Mongol Empire." *Quaternary Science Reviews* 131 (1 January 2016): 33–50.
Pyle, Robert Michael. "Conclusion." *Atala* 9 (1981–84): 21–22.
———. "A History of Lepidoptera Conservation, with a Special Reference to Its Remingtonian Debt." *Journal of the Lepidopterists' Society* 49 (1995): 399–401.
———. "International Red Data Book of Invertebrates." *Atala* 7 (1979): 60.
———. "The Lepidoptera Specialist Group Holds Its Inaugural Meeting." *Atala* 4 (1976): 29.
———. "Recent IUCN Activity in Insect Conservation." *Atala* 7 (1979): 26.
———. "Symposium on the Biology and Conservation of Monarch Butterflies." *Atala* 9 (1981–84): 1.
Pyne, Stephen J. *Between Two Fires: A Fire History of Contemporary America.* University of Arizona Press, 2015.
———. *Fire in America: A Cultural History of Wildland and Rural Fire.* Princeton University Press, 1982.
Quammen, David. *The Song of the Dodo: Island Biogeography in an Age of Extinctions.* Scribner, 1996.
Raby, Megan. *American Tropics: The Caribbean Roots of Biodiversity Science.* University of North Carolina Press, 2017.
Radkau, Joachim. *Nature and Power: A Global History of the Environment.* Cambridge University Press, 2008.
Ramírez, M. Isabel, Cuauhtémoc Sáenz-Romero, Gerald Rehfeldt, and Lidia Salas-Canela. "Threats to the Availability of Overwintering Habitat in the Monarch Butterfly Biosphere Reserve." In *Monarchs in a Changing World: Biology and Conservation of an Iconic Butterfly,* edited by Karen S. Oberhauser, Kelly R. Nail, and Sonia Altizer, 157–68. Cornell University Press, 2015.
Ramler, Joseph P., Mark Hebblewhite, Derek Kellenberg, and Carolyn Sime. "Crying Wolf? A Spatial Analysis of Wolf Location and Depredations on Calf Weight." *American Journal of Agricultural Economics* 96 (April 2014): 631–56.
Reidy, Michael S. *Tides of History: Ocean Science and Her Majesty's Navy.* University of Chicago Press, 2008.
Reinalda, Bob. *Routledge History of International Organizations: From 1815 to the Present Day.* Routledge, 2009.
Rensink, Brenden W. *Native But Foreign: Indigenous Immigrants and Refugees in the North American Borderlands.* Texas A&M University Press, 2018.
Rice, Mark. *Making Machu Picchu: The Politics of Tourism in Twentieth-Century Peru.* University of North Carolina Press, 2018.

Richards, John F. *The Unending Frontier: An Environmental History of the Early Modern World*. University of California Press, 2003.
Rigaud, Kanta Kumari, Alex de Sherbinin, Bryan Jones, et al. *Groundswell: Preparing for Internal Climate Migration*. World Bank, 2018.
Ripple, William J., and Robert L. Beschta. "Wolves and the Ecology of Fear: Can Predation Risk Structure Ecosystems?" *BioScience* 54 (August 2004): 755–66.
Ritvo, Harriet. "On the Animal Turn." *Daedalus* 36 (Fall 2007): 118–22.
Robbins, Jim. "Wolves Across the Border." *Natural History* 5 (1986): 6–16.
Robbins, William G. *Colony and Empire: The Capitalist Transformation of the American West*. University Press of Kansas, 1994.
Robin, Libby, Sverker Sörlin, and Paul Warde, eds. *The Future of Nature: Documents of Global Change*. Yale University Press, 2013.
Robin, Marie-Monique. *The World According to Monsanto: Pollution, Corruption, and the Control of Our Food Supply*. New Press, 2010.
Robinson, Michael J. *Predatory Bureaucracy: The Extermination of Wolves and the Transformation of the West*. University Press of Colorado, 2005.
Rodney, William. *Kootenai Brown: The Unknown Frontiersman*. Heritage House Publishing, 2002.
Rodriquez, Juan Jose Reyes. "Mariposa Monarca." *Atala* 9 (1981–84): 9–10.
Rome, Adam. *The Bulldozer in the Countryside: Suburban Sprawl and the Rise of American Environmentalism*. Cambridge University Press, 2001.
———. *The Genius of Earth Day: How a 1970 Teach-In Unexpectedly Made the First Green Generation*. Hill and Wang, 2013.
Rose, Debra A. "The Politics of Mexican Wildlife: Conservation, Development, and the International System." PhD diss., University of Florida, 1993.
Rose, Gene. *Kings Canyon: America's Premier Wilderness National Park, A History*. Sequoia Natural History Association, 2011.
Rothman, Hal K. *Blazing Heritage: A History of Wildland Fire in the National Parks*. Oxford University Press, 2007.
Ruiz Benedicto, Ainhoa, Mark Akkerman, and Pere Brunet. *A Walled World: Towards a Global Apartheid*. Transnational Institute, 2020.
Rumore, Gina. "Preservation for Science: The Ecological Society of America and the Campaign for Glacier Bay National Monument." *Journal of the History of Biology* 45 (November 2012): 613–50.
Runte, Alfred. *National Parks: The American Experience*. 4th ed. Taylor Trade Publishing, 2010.
Russell, Philip L. *The History of Mexico: From Pre-Conquest to Present*. Routledge, 2010.
Sánchez-Bayo, Francisco, and Kris A. G. Wyckhuys. "Worldwide Decline of the Entofauna: A Review of Its Drivers." *Biological Conservation* 232 (April 2019): 8–27.
Sayre, Nathan F. *The Politics of Scale: A History of Rangeland Science*. University of Chicago Press, 2017.
Schrepfer, Susan R. *Nature's Altars: Mountains, Gender, and American Environmentalism*. University Press of Kansas, 2005.
Schullery, Paul, ed. *The Yellowstone Wolf*. University of Oklahoma Press, 2003.
Scott, James C. *Seeing Like a State: How Certain Schemes to Improve the Human Condition Have Failed*. Yale University Press, 1998.

Sellars, Richard West. *Preserving Nature in the National Parks: A History*. Yale University Press, 1997.
Shah, Sonia. *The Next Great Migration: The Beauty and Terror of Life on the Move*. Bloomsbury Publishing, 2020.
Sharp, Ellen, and Will Wright. "'We Were in Love with the Forest': Protecting Mexico's Monarch Butterfly Biosphere Reserve." *Forest History Today* 26 (Spring/Fall 2020): 4–16.
Sharp, Ellen. "Mexico's Monarch Migration: Dealing with Deforestation in the Butterfly Forest." *Saving Earth Magazine*, Fall 2020.
Sharp, Paul F. *Whoop-Up Country: The Canadian-American West, 1865–1885*. Historical Society of Montana, 1960.
Shetler, Jan Bender. *Imagining Serengeti: A History of Landscape Memory in Tanzania from Earliest Times to the Present*. Ohio University Press, 2007.
Simonian, Lane. *Defending the Land of the Jaguar: A History of Conservation in Mexico*. University of Texas Press, 1995.
Smalley, Andrea L. *Wild by Nature: North American Animals Confront Colonization*. Johns Hopkins University Press, 2017.
Smith, Diane. *Yellowstone and the Smithsonian: Centers of Wildlife Conservation*. University Press of Kansas, 2017.
Smith, Douglas W., and Gary Ferguson. *Decade of the Wolf: Returning the Wild to Yellowstone*, Updated Edition. Lyons Press, 2012.
Smith, Douglas, Daniel R. Stahler, Erin Albers, et al. *Yellowstone Wolf Project: Annual Report 2008*. Yellowstone Center for Resources, 2009.
Smith, Douglas, Daniel Stahler, Erin Albers, et al. *Yellowstone Wolf Project: Annual Report 2009*. Yellowstone Center for Resources, 2010.
Smith, Douglas, Daniel Stahler, Erin Stahler, et al. *Yellowstone Wolf Project: Annual Report 2011*. Yellowstone Center for Resources, 2012.
Smith, Douglas, Daniel Stahler, K. Cassidy, et al. *Yellowstone National Park Wolf Project: Annual Report 2017*. Yellowstone Center for Resources, 2018.
Smith, Jordan Fisher. *Engineering Eden: The True Story of a Violent Death, a Trial, and the Fight over Controlling Nature*. Crown, 2016.
Snook, Laura C. "Conservation of the Monarch Butterfly Reserves in Mexico: Focus on the Forest." In Malcolm and Zalucki, *Biology and Conservation of the Monarch Butterfly*, 363–70.
Snow, Chief John. *These Mountains are our Sacred Places: The Story of the Stoney Indians*. Samuel Stevens, 1977.
Spence, Mark David. "Crown of the Continent, Backbone of the World: The American Wilderness Ideal and Blackfeet Exclusion from Glacier National Park." *Environmental History* 1 (July 1996): 29–49.
———. *Dispossessing the Wilderness: Indian Removal and the Making of National Parks*. Oxford University Press, 1999.
St. John, Rachel. *Line in the Sand: A History of the Western U.S.-Mexico Border*. Princeton University Press, 2011.
Stahler, Erin E., Douglas W. Smith, and Daniel R. Stahler. "Wolf Turf: A Glimpse at 20 Years of Wolf Spatial Ecology in Yellowstone." *Yellowstone Science* 24 (June 2016): 50–54.
Stein, Sarah Abravaya. *Plumes: Ostrich Feathers, Jews, and a Lost World of Global Commerce*. Yale University Press, 2010.

Stelfox, John G. "Wolves in Alberta: A History, 1800–1969." *Alberta Lands, Forests, Parks, Wildlife* 12 (1969): 18–27.

Stephenson, Nathan L., Adrian J. Das, Nicholas J. Ampersee, et al. "Patterns and Correlates of Giant Sequoia Foliage Dieback During California's 2012–2016 Hotter Drought." *Forest Ecology and Management* 419–20 (2018): 268–78.

Strode, Paul K. "Implications of Climate Change for North American Wood Warblers (Parulidae)." *Global Change Biology* 9 (August 2003): 1137–44.

Strong, Douglas Hillman. "A History of Sequoia National Park." PhD diss., Syracuse University, 1964.

———. *Trees—or Timber? The Story of Sequoia and Kings Canyon National Parks.* Sequoia Natural History Association, 1986.

Sutter, Paul S. *Driven Wild: How the Fight Against Automobiles Launched the Modern Wilderness Movement.* University of Washington Press, 2002.

———. "The Trouble with 'America's National Parks'; or Going Back to the Wrong Historiography: A Response to Ian Tyrrell." *Journal of American Studies* 46, no. 1 (2012): 23–29.

Swain, Donald C. *Wilderness Defender: Horace M. Albright and Conservation.* University of Chicago Press, 1970.

Swenson, Astrid. "Response to Ian Tyrrell, 'America's National Parks: The Transnational Creation of National Space in the Progressive Era.'" *Journal of American Studies* 46, no. 1 (2012): 37–43.

Taylor, Joseph E. "Boundary Terminology." *Environmental History* 13 (July 2008): 454–81.

Thogmartin, Wayne E., Laura López-Hoffman, Jason Rohweder, et al. "Restoring Monarch Butterfly Habitat in the Midwestern U.S.: 'All Hands on Deck.'" *Environmental Research* 12 (29 June 2017): 1–10.

Thogmartin, Wayne E., Ruscena Wiederholt, Karen Oberhauser, et al. "Monarch Butterfly Population Decline in North America: Identifying the Threatening Processes." *Royal Society Open Science* 4 (September 2017): 1–16.

Todd, Molly. "'We Were Part of the Revolutionary Movement There': Wisconsin Peace Progressives and Solidarity with El Salvador in the Reagan Era." *Journal of Human and Civil Rights* 3 (July 2017): 1–56.

Tollefson, Jeff. "IPCC Climate Report: Earth Is Warmer Than It's Been in 125,000 Years." *Nature* 596 (9 August 2021): 171–72.

Tompkins, Andrew S. "An 'Ecological Internationale'? Nuclear Energy Opponents in Western Europe, 1975–1980." In *Leftist Internationalisms: A Transnational Political History*, edited by Michele Di Donatio and Mathieu Fulla, 219–32. Bloomsbury Academic, 2024.

Tsing, Anna Lowenhaupt. *The Mushroom at the End of the World: On the Possibility of Life in Capitalist Ruins.* Princeton University Press, 2015.

Turner, James Morton. *The Promise of Wilderness: American Environmental Politics since 1964.* University of Washington Press, 2012.

Tweed, William C. *Kaweah Remembered: The Story of the Kaweah Colony and the Founding of Sequoia National Park.* Sequoia Natural History Association, 1986.

———. *King Sequoia: The Tree That Inspired a Nation, Created Our National Park System, and Changed the Way We Think about Nature.* Sierra College Press, 2016.

———. *Uncertain Path: A Search for the Future of National Parks.* University of California Press, 2010.

Tweed, William C., and Lary M. Dilsaver. *Challenge of the Big Trees: The History of Sequoia and Kings Canyon National Parks*. Rev. ed. George F. Thompson Publishing, 2016.
Tyrrell, Ian. "America's National Parks: The Transnational Creation of National Space in the Progressive Era." *Journal of American Studies* 46, no. 1 (2012): 1–21.
———. *Crisis of the Wasteful Nation: Empire and Conservation in Theodore Roosevelt's America*. University of Chicago Press, 2015.
———. "Ian Tyrrell Replies." *Journal of American Studies* 46, no. 1 (2012): 45–49.
———. *Reforming the World: The Creation of America's Moral Empire*. Princeton University Press, 2010.
———. *True Gardens of the Gods: Californian-Australian Environmental Reform, 1860–1930*. University of California Press, 1999.
Vallet, Elisabeth. *Borders, Walls, Fences: State of Insecurity?* Routledge, 2016.
Van der Meer, Sjoerd. "The Butterfly Effect: Butterflies, Forest Conservation and Conflicts in the Monarch Butterfly Reserve (Mexico)." Master's thesis, Wageningen University, Netherlands, 2007.
Van Nuys, Frank. *Varmints and Victims: Predator Control in the American West*. University Press of Kansas, 2015.
Vane-Wright, Richard I. "The Columbus Hypothesis: An Explanation for the Dramatic 19th Century Range Expansion of the Monarch Butterfly." In Malcolm and Zalucki, *Biology and Conservation of the Monarch Butterfly*, 179–85.
Vázquez G., Leonila, and Héctor Perez R. "The Monarch Butterfly as a Resource for Ecological Research in Mexico." *Atala* 9 (1981–84): 7–8.
Vermaas, Lori. *Sequoia: The Heralded Tree in American Art and Culture*. Smithsonian Books, 2003.
Vetter, Jeremy. *Field Life: Science in the American West during the Railroad Era*. University of Pittsburgh Press, 2016.
———, ed. *Knowing Global Environments: New Historical Perspectives on the Field Sciences*. Rutgers University Press, 2010.
Vik, Hanne Hagtvedt. "Indigenous Internationalism." In *Internationalisms: A Twentieth-Century History*, edited by Glenda Sluga and Patricia Clavin, 315–39. Cambridge University Press, 2017.
Villalobos, Walter Bello, Yaxine María Arias Núñez, and Vanessa Medina Padrón. *Proyecto de Sistematización: Reserva Bosque Nuboso Santa Elena*. Sistema Nacional de Áreas de Conservación, 2017.
Von Hardenberg, Wilko Graf, Matthew Kelly, Claudia Leal, and Emily Wakild, eds. *The Nature State: Rethinking the History of Conservation*. Routledge, 2017.
Wadewitz, Lissa. "The Scales of Salmon: Diplomacy and Conservation in the Western Canada-U.S. Borderlands." In *Bridging National Borders in North America: Transnational and Comparative Histories*, edited by Benjamin H. Johnson and Andrew R. Graybill, 141–65. Duke University Press, 2010.
Wakild, Emily. "Border Chasm: International Boundary Parks and Mexican Conservation, 1935–1945." *Environmental History* 14 (July 2009): 453–75.
———. *Revolutionary Parks: Conservation, Social Justice, and Mexico's National Parks, 1910–1940*. University of Arizona Press, 2011.
Walia, Harsha. *Border and Rule: Global Migration, Capitalism, and the Rise of Racist Nationalism*. Haymarket Books, 2021.

Walker, Brett L. "Animals and the Intimacy of History." *History and Theory* 52 (December 2013): 45–67.
———. *The Lost Wolves of Japan*. University of Washington Press, 2005.
Walter, Dave. *Montana Campfire Tales: Fourteen Historical Narratives*. TwoDot, 2011.
Warren, Louis S. *The Hunter's Game: Poachers and Conservationists in Twentieth-Century America*. Yale University Press, 1997.
Wassenaar, Leonard I., and Keith A. Hobson. "Natal Origins of Migratory Monarch Butterflies at Wintering Colonies in Mexico: New Isotopic Evidence." *Proceedings of the National Academy of Sciences* 95 (December 1998): 15436–39.
Wayne, Robert K., Niles Lehman, Marc W. Allard, and Rodney L. Honeycutt. "Mitochrondrial DNA Variability of the Gray Wolf: Genetic Consequences of Population Decline and Habitation Fragmentation." *Conservation Biology* 6 (December 1992): 559–69.
Weaver, John C. *The Great Land Rush and the Making of the Modern World, 1650–1900*. McGill-Queen's University Press, 2003.
Weaver, John L. *Big Animals and Small Parks: Implications of Wildlife Distribution and Movements for Expansion of Nahanni National Park Reserve*. Wildlife Conservation Society-Canada, 2006.
Webb, George Ernest. *Tree Rings and Telescopes: The Scientific Career of A. E. Douglass*. University of Arizona Press, 1983.
Weisiger, Marsha. *Dreaming of Sheep in Navajo Country*. University of Washington Press, 2011.
Wetherell, Donald G. *Wildlife, Land, and People: A Century of Change in Prairie Canada*. McGill-Queen's University Press, 2016.
White, Richard. "Animals and Enterprise." In *The Oxford History of the American West*, edited by Clyde A. Milner II, Carol A. O'Connor, and Martha A. Sandweiss, 237–73. Oxford University Press, 1994.
———. "'Are You an Environmentalist or Do You Work for a Living?': Work and Nature." In *Uncommon Ground: Rethinking the Human Place in Nature*, edited by William Cronon, 171–85. W. W. Norton, 1995.
———. *"It's Your Misfortune and None of My Own": A New History of the American West*. University of Oklahoma Press, 1991.
———. "The Nationalization of Nature." *Journal of American History* 86 (December 1999): 976–86.
Willard, Dwight. *Giant Sequoia Groves of the Sierra Nevada: A Reference Guide*. Self-published, 1994.
Williams, Michael. *Americans and Their Forests: A Historical Geography*. Cambridge University Press, 1989.
———. *Deforesting the Earth: From Prehistory to Global Crisis, An Abridgement*. University of Chicago Press, 2006.
Wilson, Edward O. *Half-Earth: Our Planet's Fight for Life*. Liveright, 2016.
Wilson, Robert M. "Mobile Bodies: Animal Migration in North American History." *Geoforum* 65 (October 2015): 465–72.
———. *Seeking Refuge: Birds and Landscapes of the Pacific Flyway*. University of Washington Press, 2010.
Wilson, Seth M., Elizabeth H. Bradley, and Gregory A. Neudecker. "Learning to Live with

Wolves: Community-Based Conservation in the Blackfoot Valley of Montana." *Human-Wildlife Interactions* 11 (Winter 2017): 245–57.

Wise, Michael D. "Killing Montana's Wolves: Stockgrowers, Bounty Bills, and the Uncertain Distinction between Predators and Producers." *Montana the Magazine of Western History* 63 (Winter 2013): 51–67.

———. *Producing Predators: Wolves, Work, and Conquest in the Northern Rockies.* University of Nebraska Press, 2016.

Wohlleben, Peter. *The Hidden Life of Trees: What They Feel, How They Communicate.* Greystone Books, 2016.

Wonderley, Anthony. "The Most Utopian Industry: Making Oneida's Animal Traps, 1852–1925." *New York History* 91 (Summer 2010): 175–95.

Worster, Donald. *A Passion for Nature: The Life of John Muir.* Oxford University Press, 2008.

———. "World Without Borders: The Internationalizing of Environmental History." *Environmental Review* 6 (Autumn 1982): 8–13.

Wright, Will. "Geophysical Agency in the Anthropocene: Engineering a Road and River to Rocky Mountain National Park." *Environmental History* 22 (October 2017): 668–95.

———. "Monarch Butterfly Conservation (Mexico)." In *Oxford Research Encyclopedia of Latin American History*, edited by Stephen Webre. Oxford University Press, 2022. https://doi.org/10.1093/acrefore/9780199366439.013.1084.

Wu, Joanna X., Chad B. Wilsey, Lotem Taylor, and Gregor W. Schuurman. "Projected Avifaunal Responses to Climate Change across the U.S. National Park System." *PLOS One* 13 (March 2018): 1–18.

Yeager, Brooks B. "Roundtable Address: Policies and Laws." In Hoth et al., *1997 Reunión de América*, 225.

INDEX

Adan (horse handler), 95
Agapito (guide), 113
Agricultural Biotechnology Stewardship Working Group, 145–46, 150
Aguado, Catalina, 112–15, 121
Aichi Target 11, 9
air pollution, 135
alar tag, 107–8
Alberta Fish and Game Association, 58
Albright, Horace, 192, 210–12, 215
Alianza de Ejidos y Comunidades de la Reserva Mariposa Monarca, 138–39, 155–56
Allen, Durward, 53
Allen, Jane, 239–40
Ambrose, Anthony, 238
Anderson, Hannibal, 89–90
Anderson, Julie, 90
Anderson-Ramirez, Malou, 89–91
Andrews, H. V., 99
Animal Division of Dominion Parks Branch, 46–47
anthropogenic biomes (anthromes), 10
anti-GMO activists, 132
Arenal Tempisque Conservation Area, 9
ARGOS (satellite observation tools), 74, 75
Arias Núñez, Yaxine María, 2, 11
Aridjis, Homero, 102–3, 135, 154, 160–61
Asner, Greg, 238

Babbitt, Bruce, 156
Bailey, Vernon, 41, 44, 49–51
Baisan, Chris, 195–96, 204–5, 232–36
Ballantyne, A. A., 60
Bancroft, Hubert H., 197
Banff National Park, 55, 58, 60, 75, 81–83
Banff Wildlife Crossings Project, 82–83
Bangs, Ed, 73
Bannon, Steve, 247–48
barbed wire, 36–37

Barbour, Henry, 193–94
Bárcena Ibarra, Alicia, 128
Barnes, Eric, 231
Barrett, Robert LeMoyne, 198–99
Barton, James L., 198
Bateman, Robert, 43, 49–51
Baxter, Wendy, 238
Bayer Corporation, 131
BC Forest Alliance, 84
beaver population, 3–4, 6
Beecher, Melba, 213
Behring, Ken, 32
Bennett, Ira, 179
Beringer, John, 144
Bevan, George, 46
Billy (wolf pup), 50
biological diversity, definition of, 5
Biological Survey. *See* Bureau of Biological Survey
bison, 24
Biswell, Harold, 220, 222–24
Blackfeet Nation, 33, 46, 72
Blackfoot Challenge, 88–89
Bobowski, Ben, 2–3, 6–10, 11, 243–44, 250
Bonnicksen, Tom, 228–29
Boone and Crockett Club, 193
Border Patrol, US, 246
border wall, US, 246–48
borders: borderlands, distinction, 7; as ecological divides, 10, 244–50; and ecological internationalism, 15; national borders, hardening of, 23, 247; national borders, softening of, 7, 63; national park boundaries as islands, 4–7, 8, 229–30, 236; as political, overview, 10–14
Border Wolf Technical Committee, 67
boundary-work, 115, 120, 121
bounty system, 32, 37–39, 43–44, 48, 51
Bowdler, John, 224

326 | Index

Boyd, Diane, 64–69, 75–76
Boyer, Richard, 218
Brannin, Barney, 51
Bridge, David, 110
Brisbin, James, 34
bristlecone pines, 235
Brockington, Dan, 4
Brooks, Jennie, 99
Brower, David, 231
Brower, Jane van Zandt, 104–6
Brower, Lincoln: and Lepidoptera Specialist Group, 122, 123; and milkweed and GMOs, 145–46; on milkweed corridor, need for, 160–61; milkweeds and toxicity, 103–6; and monarch biosphere, need for, 124, 125–29, 133, 139; monarch overwintering site, location of, 115–18; stepping-stone hypothesis, 148–49; on trinational threats to monarchs, 155
Brown, John George "Kootenai," 25–31
Brown, Peter, 232–36
Brugger, Kenneth, 112–15, 121
buffer zones for monarchs, 126, 135–36, 156–57
Bulyea, G. H. V., 38
Bureau of Biological Survey, 23, 39–40; desire for, 48, 50; and Glacier National Park, 44; and wolf eradication, 42–43, 47–51
Burgess, Darlene, 164–66
Bushway, Catherine, 72
butterflies. *See* monarch butterflies
Butterflies and Their People (organization), 159, 164–66
butterfly gardens, 164. *See also* Monarch Waystations

Calgary Fish and Game Association, 60
California Geological Survey, 174–75
California Gold Rush, 173
Calvert, William, 116–17, 126–27, 155
Camacho Solís, Manuel, 135, 136
Camas de Castro, María Elena, 135
Caprio, Tony, 207, 232–37
Cárdenas, Cuauhtémoc, 127, 140
Cárdenas, Lázaro, 101, 102, 127

Caro Gómez, Rosendo, 155–56
Carson, Rachel, 61
cattle, 33–35
cattle thieves, 35
CCC (Civilian Conservation Corps), 216–17
Ceballos, Gerardo, 248–49
CEC (Commission for Environmental Cooperation), 12–13, 132, 154–57
Central Pacific Railroad, 173
Central Rockies Wolf Project, 55, 75
CEPANAF (State Commission of National Parks and Wildlife), 118–20, 139–41, 158, 163–66
Cerro Pelón (sanctuary), 95, 100, 114, 118–20, 139, 158
Chaparro, Jamie, 101–2
Chappo (Mono chief), 212
Charlie, Clara, 214
chemical ecology, 14, 103–6. *See also* Brower, Lincoln; milkweeds; monarch butterflies
Cheney, Amby, 22
Christensen, Norman, 231–32
Christensen Report, 231–32
Cieslar, A., 197
citizen science, 12, 96, 106–7, 108–12, 149–50
Civilian Conservation Corps (CCC), 216–17
civil society, definition, 14
Clements, Frederic, 201
climate refugees, 245–50
Close, Al, 22
Cochrane, Matthew, 34, 35
Cochrane, William, 35
Coffman, John, 216
Collins, Mark, 128
Commission for Environmental Cooperation (CEC), 12–13, 132, 154–57
Competitive Enterprise Institute, 84
Comstock, Anna Botsford, 97–99, 149
Comstock, John Henry, 98–99, 149
Comstock, Smith, 171, 177, 188
Comstock Lumber Company, 171
conferences, about monarch butterfly. *See* MonCons

conservation biology, emergence of, 73–74
Convention on Biological Diversity, 9, 249
Cook, Lawrence, 210, 216–17
Coordinated Framework for the Regulation of Biotechnology, 143–44
core zones: for monarchs, 126–28, 132–33, 135–36; for wolves, 41–43, 70, 75, 85. *See also* corridors
Correa, Juan, 101–2
corridors: and corridor ecology, 13–14, 55; federal designation of, 78; need for, generally, 249; for wolves, need for, 47, 54, 57, 67–71, 74, 77. *See also* milkweed corridor; Yellowstone to Yukon (Y2Y)
Cowan, Ian McTaggert, 58–59
cows, 33–35
coyotes, 37, 43
Craighead, Lance, 84
Crapsey, Malinee, 226–27
Crees, 36
Criddle, Norman, 48
Crocker, Charles, 190–92
Crowfoot (Siksika chief), 29
Cueller, Henry, 246

Davis, Don, 106–7, 109
Davis, John, 220
Davis, William, 198
Defenders of Wildlife, 73, 88, 90
Defenders of Wildlife v. Hall, 87
De Greef, Willy, 144–45
Dehcho First Nations, 85
Dekalb Seeds, 148
dendrochronology. *See* Douglass, A. E.; tree rings
Department of Forestry, Game, and Fish (Mexico), 101–2
Dewitt, John, 231
Division of Economic Ornithology and Mammalogy. *See* Bureau of Biological Survey
D'Lonais, Olivia, 26–29
Doggett, Stuart, 71
Dolbeer, John, 180
Dominion Parks Branch, establishment of, 31

donkey engines, 180
Dorst, J. H., 192
Douglass, Andrew Ellicott (A. E.), 195, 196, 199, 200–204, 205, 231, 233
D-Plex listserv, 149
drug cartels, 159
DuBois, Coert, 215

ecological internationalism, definition of, 3, 15
ecological science, 220–24
ecoregion, 79. *See also* Yellowstone to Yukon (Y2Y)
ecotourism: and exclosures, 3; as insufficient to replace income from resource extraction on protected lands, 133; and monarchs, 127–28, 129, 133, 135, 138, 163–66; and sequoias, 185, 190, 210–12, 215–16, 231–32
Edge, Rosalie, 207
Ehrhardt, Mac, 147–48
Ejido El Capulín, 100, 102
ejidos: and ecotourism versus logging income, 127–28, 129, 133, 135, 138–40, 155–60; and forest rights, 101–2, 113, 125, 126; lease of forest, 139; and outmigration, 157; population growth and poverty, 133, 134–35
Elizabeth II, Queen, 128
Elk and Vegetation Management Plan, 6
Ell, Louis, 29
El Niño–Southern Oscillation and La Niña, 2, 204–5
Emery, Mrs. C., 114
Emiliano Zapata Forestry Cooperative, 101–2
Endangered Species Act (ESA), 63; and monarchs, 161–63; Section 10(j), 72; and wolves, 52, 67, 86–87
Esquivel Maya, Osvaldo, 165
Evison, Boyd, 228–30
exclosures, 1, 3
extinction, mass, 8

Fairchild, Mike, 66, 68
Federal Environmental Protection Agency (PROFEPA), 140, 158

fencing, 36–37, 89, 90–91
FFA (Future Farmers of America), 89, 91
fieldwork versus lab work, 116–17, 201–2
fires: prescribed burns, 13, 207, 220–30, 232–36, 240; prescribed natural fires, 223; total suppression, 13, 209–10, 212, 215–17; wildfires, 209–10, 217–20, 237
Fischer, Hank, 72–73
Fisher, Tommy, 248
Fisher, W. C., 58
fladry, for fencing, 89, 90–91
flumes, in logging, 181
Fondo Monarca, 156–57
Forestry Code, 101
fortress conservation, 4–12
foxtail pines, 235
Fraley, Robert, 147
Francis, Wendy, 78, 85
Franco, Hector, 214
Freese, Curtis, 136
Frémont, John C., 173
Fry, Walter, 188–90, 192–93, 194, 204, 212
Fuller, Kathryn, 141
Fumento, Michael, 145
Future Farmers of America (FFA), 89, 91

Gallegos, Carlos Melo, 127
Game Act, 39
General Agreement on Tariffs and Trade (GATT), 12, 132
General Grant National Park, 172, 189–92, 207–8. *See also* Kings Canyon National Park
General Sherman Tree (also called Karl Marx Tree), 177, 204
genetic engineering, 142–53
George, Lillie, 97
George, William, 97
Giant Forest, 176–77, 191, 192, 209, 211, 214, 218–20
giant sequoias. *See* sequoias
Gillett, Frederick, 192–93
Glacier National Park, 31, 33, 44, 57, 64, 68–69, 75
Gladstone, Jack, 72
Glickman, Dan, 143–44, 145

global warming, 2, 235–39, 243–50; and humanitarian crises, 245–50
glyphosate-tolerant crops, 12, 132, 141, 142–45, 146–53, 160–63
Godsal, F. W., 31
González de Castilla, Susana Rojas, 135
Goode, Ron, 214, 239–41
Gottfried Joy, Carlos Federico, 134, 135, 136–38, 139, 141
Graban, Sandy, 226
Graber, David, 228–29
Graham, Maxwell, 46–47
Grant Grove, 172, 217–18
Graves, Henry, 197
gray wolves. *See* wolves
Great Bear Wilderness, 67
Greater Sequoia, 192–94, 207–8
Greeley, William, 193–94
Green, Hubert (alias Tony Lascelles), 47, 59–60
Greer, Dan, 71
Grigg, William, 84
Grosvenor, Gilbert, 192–93
ground squirrels, 37
Grupo de los Cien (Group of 100), 133, 135
Guest, Robert, 62–63
Guzmán Reyes, Vicente, 157

habitat fragmentation, 8
Half Earth, 8–9, 86, 248–49
Hanify, Martha, 181
Hansen, James, 235
Hardwick, Thomas, 30
Harkin, James B., 45, 47, 48
Hartesveldt, Richard, 208, 223
Harvey, Thomas, 208, 223
Haskell, Burnette, 176–77
Helle, Joe, 71
Henry P. Kendall Foundation, 81
Henshaw, Henry, 47
herbicide-tolerant crops. *See* glyphosate-tolerant crops
Heuer, Karsten, 84
Hewitt, C. Gordon, 48
Hilty, Jodi, 55–57, 77
HIPPO (factors in species decline), 32

Hoth, Jürgen, 155
Howe, Wayne, 219
Hoxie, G. L., 215
Huckvale, Walter, 44
Hudson's Bay Company, 29
Hughes, Sally, 122, 123
Huidekoper, Wallis, 43
Hume, Thomas, 179
Hume-Bennett Lumber Company, 179, 183, 187, 200
Huntington, Ellsworth, 197–200, 201, 205
Hutchinson, J. A., 58

Ickes, Harold, 207
Immigration and Customs Enforcement (ICE), US, 159
Indigenous peoples: alcohol and trading, 29–30; and burning, use of, 13, 209, 212–15, 220, 227–29, 232, 239–41; displacement, land cessions, and reservations, 4–12, 13, 24, 33, 173–74; and NPS collaboration, 13, 14, 229, 239–41; on strychnine, use of, 30; and wolves, 36, 44–46; and Y2Y, 85
Insect Migration Association, 12, 14, 96, 106–7, 109–12, 114, 131–32
Intergovernmental Panel on Climate Change (IPCC), 235–36, 249
International Network of Monarch Butterfly Reserves, 154
International Union for the Conservation of Nature (IUCN), 12, 122, 124–25, 128, 160
International Wolf Center, 66
IPCC (Intergovernmental Panel on Climate Change), 235–36, 249
island biogeography, 5–6, 8, 79, 194
Isle Royale National Park, 53
IUCN (International Union for the Conservation of Nature), 12, 122, 124–25, 128, 160

Jackson, J. S., 47
Jackson, Maddy, 64
Jaworski, Ernest, 142
Jepson, Willis, 181–82

Jestrup, H. H., 197
JM Butterfly B&B, 164
John Birch Society, 84
Juárez, Benito, 121

Kainai Nation, 33
Kanawyer, Cilicy, 178
Karl Marx Tree (also called General Sherman Tree), 177, 204
Kaweah Co-operative Commonwealth, 176–77, 190–92
Kay (wolf), 69, 73
Keepers of the Flame, 13, 14, 229, 239–41
Kerr, Gordon, 62
Kerr, Irvin, 217
Kilgore, Bruce, 223–24, 227–29
Kings Canyon National Park, 172, 207–8, 217, 221, 223, 225. *See also* Sequoia National Park
Kings River Lumber Company, 176, 177–79
Kishinena (wolf), 64–65, 66, 67–68
Knowles, M. E., 47
Koch Brothers, 84
Kohrs, Conrad, 34, 35
Kolfage, Brian, 247–48
Kootenay Lakes Forest Reserve, 31

Laboratory of Tree-Ring Research, 195–96, 204–5, 230
landscape permeability, 57–58
large landscape conservation: cautions for, 14; necessity of, 3, 8–9. *See also* core zones; corridors; Half Earth; Insect Migration Association; Keepers of the Flame; Sequoia National Park; Yellowstone to Yukon (Y2Y)
Lascelles, Tony (alias of Hubert Green), 47, 59–60
Lawton, Benjamin, 38–39
Leaf to Landscape Project, 238
Leopold, Aldo, 59
Leopold, Starker, 220, 221–24, 227, 228–30
Leopold Report, 209, 221–22, 224, 227, 232
Lepidoptera Specialist Group, 123. *See also* IUCN
Lepidopterists' Society, 121–22, 125

330 | Index

Lindsey, Tipton, 189–90
Linnell, John, 249–50
Lipscomb, Charles, 110
Little Ice Age, 199
local extinctions, 8
Locke, Harvey, 74, 76–77, 79, 81, 86
logging: and monarch butterflies' overwintering site, 127–28, 129, 133, 135, 138–40, 155–60, 164–65; of sequoias, 13, 171–84, 185, 187–88
Long Point National Wildlife Area, 154
Losey, John, 144–46
Lovejoy, Thomas, 123, 125, 127

Macario (ejidatario), 134, 138
MacArthur, Robert, 5
MacDougal, Daniel, 199
MacLeod, James, 30, 33
Magic Pack, 68
Mahoney, Richard, 142
Man and the Biosphere program, 7, 126
mange (disease), 47
Manuel de Jesús, Jesús, 156
Marble, Charles, 43
marcas. *See* tagging of monarch butterflies
Marshall, Robert, 193–94
Martin, J. J., 190–92
Martínez, Artemio, 134
Martínez, Ronis, 245
mass extinction, 8
Mather, Stephen, 192–93, 194, 210–11
Mather Mountain Party, 192–93
Mattson, Ursula, 67, 68, 71
Maunder, Annie Russell, 203–4
Maunder, E. Walter, 203–4
Maunder Minimum, 203–4
Maussan, Jaime, 134, 138
Mayberry, Syble, 111
McClure, S. W., 42
McClusky, John, 114
McCormick, Aiden, 221
McCormick, Ernest, 192–93
McDevitt, James, 28
McDougall, John, 24
McGraw, Don, 226
McGregor, James, 36

McKay, W. H., 36
McLaughlin, John, 223, 224
Mech, L. David, 65–66
Medieval Warm Period, 235
Mejia, Nelson, 245
Merino Pérez, Leticia, 156
Merriam, C. Hart, 40
Merriam, John, 205
Métis people, 26–27
Migratory Bird Treaty, 7
milkweed corridor, 132–33, 151–55, 160–63, 164, 165–66
milkweeds: chemical makeup and toxicity, 104–6; and glyphosate-tolerant crops, 12, 141, 144–46, 148–53, 160–63
mimicry hypothesis, 104
Ministry of Urban Development and Ecology, Mexico, 128
Miwa, Henry, 218
Miwok Nation, 14, 173–74, 241
Molloy, Don, 87
monarch butterflies: and Bt pollen, 150; buffer zones for, 126, 135–36, 156–57; core zones for, 126–28, 132–33, 135–36; and ecotourism, 127–28, 129, 133, 135, 138, 163–66; foreign capital, 136–38, 141, 156–57; population decline, 104, 122, 126–27, 132, 160–63; reproduction, 149–50; tagging of, 12, 16, 96, 106–12, 114, 131–32; and vehicle collisions, 161; views on, 100, 102–3, 132. *See also* milkweeds; Monarch Butterfly Biosphere Reserve
—migration: as "endangered phenomenon," 129–30, 162; milkweed corridor, 132–33, 151–55, 160–63, 164, 165–66; stepping-stone hypothesis, 132, 148–50; theories of, 97–100
—overwintering locations: and deforestation, 133; discovery of, 12, 116–17, 118–20, 121; protection of, 115–16, 117–20, 123, 124–29; search for, 106–15; and winter storm, 154
Monarch Butterfly Biosphere Reserve (Reserva de la Biosfera Mariposa Monarca): and citizen science, 95–96; establishment and management of, 135–41; expansion

of, 155–57; logging and deforestation, 133, 138–40, 164–65; proposed, 126–28, 133. *See also* ejidos
Monarch Butterfly Protection Trust, 135
Monarch Highway, 161–63
Monarch Larva Monitoring Project, 149–50
Monarch Project, 129–30
Monarch Watch, 96, 131, 149, 151–53
Monarch Waystations, 131–32, 151–53, 164, 165–66
Monasterio, Valentina, 136
MonCons (symposia): MonCon I, 127–28; MonCon II, 135; MonCon III, 155–57
Mono Nation, 13, 14, 174, 212–16, 239–41
Monsanto Company, 12, 131–33, 141–53, 161–63
Montana Livestock Loss Board, 88
Montana Stock Growers' Association, 35
Montes de Oca, Jesús Ávila, 118–20, 139, 158
Moore, Austin, 176, 177–79
Moreno de Jesús, Elidió, 102, 118–20, 140
Moreno de Jesús, Melquiades, 96, 103, 139, 163
Moreno Espinoza, Leonel, 100
Moreno Hernández, Francisco, 165
Moreno Rojas, Anayeli, 121
Moreno Rojas, Joel, 163–66
Moreno Rojas, Patricio "Pato," 95–96, 159, 164–66
Mowat, Farley, 61–62
Muir, John, 184–85, 186–87, 192

NAFTA (North American Free Trade Agreement), 12, 81, 132, 141, 153
Nahanni National Park, 85
Nakoda people, 44–45
National Butterfly Center, 245–48
National Origins Act, 23
National Park Service, US. *See* US National Park Service
National Wool Growers Association, 42
natural density, 86
Naturalmente Juntos–Naturally Together, 2
natural regulation, 6
nature-study movement, 98
Neill, Earl, 22

Neilson, Len, 240
Nelson, Edward, 48–49
Newark, William, 8
Newhouse, Sewell, 52
Newhouse traps, 52
Nichols, Tom, 225, 228–29
North American Agreement on Environmental Cooperation, 153–54
North American Butterfly Association, 245
North American Free Trade Agreement (NAFTA), 12, 81, 132, 141, 153
Northern Rocky Mountain Wolf Recovery Team, 69–71
North Face, 74
Norwegian, Herb, 85
Noss, Reed, 74, 76, 77
Novartis, 145–46

Oberhauser, Karen, 149–50, 153, 155, 160
Ogarrio, Rodolfo, 124, 125, 127, 130, 134–35, 136, 155
oil boom, 30
Old Man Topna, 239
Oneida Community, 52
Operation Wolfstock, 73
Orr, Mrs. C. R., 110
Ortíz Monasterio, Fernando, 124, 125, 134
Oswood, John, 23
Otomí peoples, 102–3
Owl Woman, 214

Panic of 1893, 178
Paquet, Paul, 55, 74–76, 82
Parsons, Dave, 228–29, 232
paternalism, 178
Path of the Pronghorn (wildlife corridor), 78
Pemex, 134, 135
People and Carnivores (organization), 88–89, 90
Peplinski, John, 110
Pérez, Héctor, 125, 127
Pérez, Santos, 101–2
Pérez Artavia, Johnny, 2, 250
Philip, Prince (Duke of Edinburgh), 128, 138

Phyllis (wolf), 68
Pikani Nation, 33
Pimlott, Douglas, 63
Pinchot, Gifford, 40
Pioneer Hi-Bred, 145, 147–48
Plan of Ayala, 101
Plant Genetic Systems, 143
Pleasants, John, 151, 160
Pluie (wolf), 55–57, 75, 84
Point Pelee National Park, 154, 164
pollinator gardens. *See* Monarch Waystations
Popocatépetl-Iztaccíhuatl National Park, 102
Power, Thomas C., 35–36
prairie dogs, 37
predator-deterrence strategies, 88–91
predator eradication. *See* Bureau of Biological Survey; strychnine; wolves
Prince Edward Point National Wildlife Area, 154
PROFEPA (Federal Environmental Protection Agency), 140, 158
Pro-Monarca, 124, 126–27, 128, 133, 136–41; foreign capital, 134–35
Prysby, Michelle, 149–50
Pumpelly, Raphael, 198–99
Pyle, Robert Michael, 121–30

Queen Alexandra's birdwing butterfly, 123–24
Queeny, John Francis, 141

Rabe, Jack, 64
rabies, 60
radio collars, 11, 16, 64–69, 74, 75, 87
railroads, 30, 34; Central Pacific Railroad, 173; Southern Pacific Railroad, 173, 178, 181, 185, 188, 190–92
rainforest, 123–24
range riding, 88–89, 90–91
Ream, Robert, 65–72
Red Data Books, 123, 129
Red Earth Wildlife Overpass, 78
Redwood Mountain, 174, 185, 207–8, 222, 223, 224

Remington, Charles, 104, 122
Rendón-Salinas, Eduardo, 155
Reserva de la Biosfera Mariposa Monarca. *See* Monarch Butterfly Biosphere Reserve
resource extraction, 14. *See also* logging
Reyes Rodriguez, Juan José, 127
Ririe, J. B., 50
Riviere, Henry, 46
Robinson, Bart, 81
Roosevelt-Sequoia National Park, proposal for, 193–94
Roundup Ready crops. *See* glyphosate-tolerant crops
Royal Ontario Museum, 107
Ruiz Barranco, Hector Luis, 155
Rummel, Charles, 104
Rzedowski, Jerzy, 103

Sachs, Eric, 162
Sage (wolf), 68
Salazaar, Dimas, 138–39
Sanford, John, 146–47
Sanger Lumber Company, 179
Santa Elena Cloud Forest Reserve, 2, 9
satellite transmitting systems, 55–57
Sawchuck, Wayne, 76
Schell, Jozef, 143
Schotte, Gunnar, 197
Schultz, James Willard, 24–25
Scott, Peter, 122–23, 125, 128
Scoyen, Eivind, 218
Seifert, Tim, 148
Seitz, Adalbert, 99–100
Sequoia National Park: boundaries of, 185, 188–95; and CCC, 216–17; establishment of, 172; fire, full suppression, 216–17; and Indigenous peoples, collaboration with, 239–41; resource extraction, 193–94. *See also* sequoias; US National Park Service
—prescribed burns: 13, 207, 220–30, 232–36; public response to, 231–32; and climate change, 236–39
Sequoia Railroad, 178
sequoias: age of, 188, 195, 199–200, 201; biophysical traits, 185–88; and climate change, response to, 208–9, 237–39; and

climate change data, 195–205, 235–36; conservation, calls for, 176–77, 183–84; as cultural icon, 185; distribution of, 186; durability of fallen trees, 184–85, 187; and ecotourism, 185, 190, 210–12, 215–16, 231–32; and fire, 212, 221–26, 233–36, 238–39, 240–41; and hydrological cycle, 185; logging of, 13, 171–84, 185, 187–88; reproduction of, 223, 224; tree-ring data, 227–28, 230. *See also* Sequoia National Park
Shapiro, Robert, 141, 142
Sharp, Ellen, 163–66
Shearer, Chad, 72
Sibbald, Howard, 44
Sierra Chincua (butterfly colony), 114, 117, 120–21, 124, 125, 126–27, 129
Sierra Club, 61
Sigala Páez, Pascual, 156
Siksika Nation, 33
Sing, Tie, 192–93
Sister Cities International, 2–3, 11, 14
Slim, Carlos, 165
Smith, Doug, 51–54, 86
Smith, Hiram, 176, 177–79
Smith, Orson Kirk "O. K.," 174
Smithsonian Museum of Natural History, 31–32, 42
Snipes, Roy, 247
Snowden, Herb, 60
Snowslide (wolf), 51
Sociedad Mexicana de Lepidopterología, 125
Society for Conservation Biology, 74
Soulé, Michael, 73–74
Southern Pacific Railroad, 173, 178, 181, 185, 188, 190–92
species-area relationship, 5
Species Survival Commission (Survival Service Commission), 123, 124–25
Standish, Walter, 49
State Commission of National Parks and Wildlife (CEPANAF), 118–20, 139–41, 158, 163–66
Stephenson, Nathan, 208–9, 238, 239
Stewart, George, 176, 189–90, 194
Stone, Edward, 228

Stouder, Wayne, 148
"string of pearls" conservation strategy, 126
Strode, Thomas, 23
strychnine, 23, 25, 27–28, 30, 31, 63
Sudworth, George, 183–84
Sully, Alfred, 33
sunspot cycles, 196–97, 201, 203–4
Survival Service Commission (Species Survival Commission), 123, 124–25
Sutter, Johann, 173
Swetnam, Thomas, 204, 230–31, 232–36

Tabor, Gary, 78–79
tagging of monarch butterflies, 96, 106–12, 114, 131–32
Tapia Torres, Silverio, 138
Taylor, Joseph Henry, 25
Taylor, Michael, 143
Taylor, Orley "Chip," 121, 131–32, 151–53, 155, 161
Telmex-Telcel, 165
Tharp, Hale, 212
Thew, Susan, 194
Thomas, Gomer, 23–24
Three Cs of rewilding, 74, 77
thylacines, 32
Timber and Stone Act, 13, 175, 176, 189
Tomkiewicz, Stan, 74, 75
Tom Miner Basin Association, 90–91
Tompkins, Doug, 74
Touhy, John, 189–90
tracking technology, 11, 16, 40–41, 53, 64–65, 66, 74–77, 87; of migratory birds, 2–3; and tagging of monarch butterflies, 12, 96, 106–12, 114, 131–32
Trans-Canada Highway, 81–83
transnational conservation. *See* Insect Migration Association; Keepers of the Flame; Sister Cities International; Yellowstone to Yukon (Y2Y)
Treaty of Ghent, 27
Treaty Seven, 33
tree rings, 13, 14, 16, 195–97, 199–205, 227–28, 230, 232–36
Tres Amigos summit, 160–61
Treviño-Wright, Marianna, 246–48

"triangle of power relations," 10–11
Tulare Expedition, 174
Tule River Company, 177
Tweed, Bill, 226

Udall, Stewart, 221
Union of Concerned Scientists, 144
Upson, Lisa, 88
Urquhart, Fred, 96, 106–12, 123, 124, 127, 131–32
Urquhart, Norah Patterson, 96, 106–7, 108–12, 131–32
US Border Patrol, 246
US border wall, 246–48
US Immigration and Customs Enforcement (ICE), 159
US National Park Service: budget, 237; dual mandate of, 210–12, 215–16, 231–32; fire, full suppression, 13, 209–10, 212, 215–20; fire, prescribed burns, 13, 220–30; and Indigenous Nations, collaboration, 229, 239–41; women's employment, 217, 226. *See also* names of specific parks

Valdez, Jesse, 241
Van Deutzen, John, 49–50
Vandever, William, 189–90
van Montagu, Marc, 143
van Wagtendonk, Jan, 232
Vaughn, Rose, 217
Vázquez García, Leonila, 125–28, 130, 133
Velázquez, Miguel, 158
Velázquez, Valentín, 118–20, 158
Velázquez Moreno, Emilio, 158–60
Viacrucis del Refugiados (Refugees' Way of the Cross), 245
viceroy butterflies, 98–99, 104
Victoria Gold Rush, 173
Volcker, Paul, 134

Wagner, Theodore, 172
Walker, Frank, 189–90
Warren, E. B. "Burt," 49–50
Warren, Francis, 42
Waterton Lakes National Park, 31, 44–46

Weaver, John, 72
We Build the Wall, 247–48
Wells, Susan, 128
Western Stock Growers' Association, 36, 38–39
Wheatley, William, 179–80
whiskey, 29–30
White, John Roberts, 194–95, 204, 210, 211–12, 215
white supremacy, 247
White Wolf, 21–23
Whitney, Josiah, 174–75
Wilburforce Foundation, 81
Wilderness Act, 3, 5, 67
Wildlands Network, 74, 77
wildlife overpasses, 78, 81–83
Willcox, Louisa, 79
Williams, C. B., 99
Wilson, Audrey, 110
Wilson, E. O., 5, 8–9, 86, 248–49
Wilson, Seth, 89
Wilson's warblers, 243–44
Witt, Dave, 72
Wolf, Edward, 147
Wolf-Come-Into-View, 46
Wolf Ecology Project (WEP), 65–72
wolves: and core zones, 41–43, 70, 75, 85; corridors, need for, 13–14, 47, 54, 57, 67–71, 74, 77; ESA protections lifted, 86–87; livestock, killing of, 21, 23, 35–36, 42, 49, 71, 87–89; loner wolves, 6–7, 49; population numbers, 24–25, 48–49, 51, 73, 86; rabies crisis, 60; social structure of pack, 66; as territorial, 11, 41; tracking and dispersal, 11, 16, 40–41, 47, 49, 64–65, 66–71, 74–77, 87; views on, 52–53, 57, 61–62. *See also* Yellowstone to Yukon (Y2Y)
—eradication, 23–24, 42–43, 47–51; international collaboration attempt, 47–48; and wolfing and bounty system, 25–31, 35–39, 42–44, 48, 51
—protection and restoration: population recovery efforts, 53, 70–73; protection by grassroots environmentalism, 61–63; state and federal agencies, coordination,

70–73, 86–87; Yellowstone National Park wolf-restoration goals, 51
World Database on Protected Areas, 9
World Wildlife Fund, 122–23, 127, 128, 138, 141, 157, 165
World Wildlife Fund Canada, 61
Wurdeman, David, 151

Xerces blue, 122
Xerces Society, 122, 129–30

Yeager, Brooks, 156
Yeager, Kerry, 110
Yellowstone National Park: elk population, 6; and wolves, 40–41, 51–54, 86
Yellowstone to Yukon (Y2Y): acres under protection, 85; as civil society, overview, 14; and ecological internationalism, example of, 54, 55–57, 91; education and outreach for, 84–85; funding for, 81, 90; opposition from corporate influencers, 84; origins of, 11, 76–77, 78–81; and predator-deterrence measures, 88–89; and Trans-Canada Highway, 81–83
Yellowstone Wolf Project, 86
yellow warblers, 2–3, 7, 11, 241, 250
Yokut Nation, 14, 174, 213, 239, 241
Yoon, Carol, 146
Young, Stanley Paul, 23

Zapata, Emiliano, 100–101
Zaranek-Anderson, Hilary, 90–91
Zumbo, Jim, 71–72
Zumwalt, Daniel K., 190

www.ingramcontent.com/pod-product-compliance
Lightning Source LLC
Chambersburg PA
CBHW030520230426
43665CB00010B/699